HCS08 Unleashed

Designer's Guide to the HCS08 Microcontrollers

2nd Edition

ISBN: 1-4196-8592-9

ISBN-13: 978-1-4196-8592-7

Visit: www.booksurge.com to order additional copies.

Acknowledgments

This is my first book in English and the seventh overall (my six first books were written in Portuguese). I would like to acknowledge some people without whom this book would not to be possible:

- Paulo Knirsch, from the Brazilian Freescale office, for support and ideas for this book. This is a relationship that started some years ago, when I published a Portuguese book about the HC908 chip. Ever since, Paulo provided me with all the support, kits and samples I needed to research for my books. Thank you very much!

- John Anguiano, Shannon Reid and Huda Aldahhan, from Freescale USA, for believing and supporting the U.S. publishing. I know the book went out of schedule, but I believe it is worth the wait!

- Luiza Meneghim, from Target English, for the grammar review and support. She is a very special person, with a great knowledge of English grammar, without her this book would never exist. Thank you Luiza!

- Glaucia Milani, Jose Palazzi, Renato Frias and Bruno Castelucci from the Brazilian Freescale office, for local support and demonstration kits.

- Johnny Andrey Sierascky, for the ideas, suggestions and the great reviewing work.

- Daniel Friederich, from Freescale USA, for the technical corrections (especially on the compiler and build tools).

- Jim Donelson, for additional ideas and corrections.

- Luiz Eduardo "Dado" Sutter, for all support and ideas to improve the book.

- Roberto do Amaral, for the additional reviewing and ideas.

- Lynn Eang and Jenny Parnow, from BookSurge.

- And special thanks to the people that supported me during this task and through my life: my father Pedro, my mother Cristina, my grandmother Juçá, my brother Luis Fernando, my sister Cristiane and especially, my beloved wife Mariani.

Thank you for believing and making a dream come true!

About the Author

Fábio Pereira is the co-founder of ScTec, a design house located in Joinville, Santa Catarina, Brazil. He also taught microcontrollers and C programming language in some local schools and courses.

Along the last eighteen years Fábio has developed applications using several architectures such as Z-80, 8051, PIC, Z8-Encore, HC908, HCS08, MSP430, ARM, Cortex and Coldfire, using Assembly, BASIC, Pascal and C languages.

He is the author of six other Portuguese books:

- Microcontroladores PIC: Técnicas Avançadas (about PIC programming in assembly);

- Microcontroladores PIC: Programação em C (about PIC programming in C);

- Microcontroladores HC908Q: Teoria e Prática (about Freescale's HC908QT and QY);

- Microcontroladores MSP430: Teoria e Prática (about TI's MSP430);

- Microcontroladores HCS08: Teoria e Prática (about Freescale's HCS08);

- Tecnologia ARM: Microcontroladores de 32 bits (about ST's STR711 ARM7 MCU);

Fábio also enjoys programming computers using visual programming languages such as Visual Basic and Delphi. Currently, he has been playing with VHDL and LUA.

About ScTec

ScTec is a design house established in Joinville, SC, Brazil. Its main business is the designing of electronic devices and circuits using programmable devices such as 8, 16 and 32-bit microcontrollers and programmable logic devices.

ScTec is a Freescale Alliance member and has a great expertise in embedded applications designing with Freescale products.

Most of the design service we do is related to the automotive and industrial markets. Some of these projects are listed below:

- Cold heading monitoring system.

- Automotive instrumentation (tachometers, voltmeters, lambda meters, etc.).

- Proportional solid state relay (with an asynchronous serial communication protocol for easier interfacing with any industrial PLC).

- Motorcycle digital sparking ignitions.

- Professional motorcycle digital sparking ignition (fully customizable by using a computer or remote management unit).

- Programmable fuel injection system (mainly for mono-cylinder engines).

- Graphical and touch screen interfaces.

- RF communication devices.

- RFID and embedded security systems.

- Probes and automated testing machines.

- Digital servomechanisms for animatronics and robotics.

For more information visit: www.sctec.com.br.

Contents

1

Introduction

The idea behind this book is to show how to use and to program the Freescale's HCS08 microcontrollers. Before we start, it is important to review some important concepts on electronics and programming.

1.1. Computer's History

The first computing machines date from 1623 (Wilhelm Schickard's calculating clock) and 1642 (Blaise Pascal's Pascaline). They were mechanical machines capable only of add operation.

After them, in 1671, the German mathematician and philosopher Gottfried von Leibniz designed the Staffelwalze (also known as Stepped Reckoner), the first mechanical calculator capable of performing the four basic operations (addition, subtraction, multiplication and division).

All these machines were only mechanical calculators and could not be programmed in any way.

The first known programmable machines date from the 19th century. Those machines were very different from our current concept of a computer. The Jacquard loom (created by Joseph Marie Jacquard in 1801) is probably the first machine that could be programmed by the user. By using punched paper cards, it was possible to program complicated weaving patterns to be woven into the textiles, increasing the flexibility of those machines (the same machine could be used to make different looms).

Some years later, in 1837, Charges Babbage designed his "Analytical Engine" that would be probably the first fully programmable mechanical computer if it had ever been built.

The progress on computers advanced on the US census in 1890 with the tabulating machines of Herman Hollerith (manufactured by the Computing Tabulating Recording Corporation, which later became IBM).

Source: http://commons.wikimedia.org/wiki/Image:Jacquard.loom.full.view.jpg

Figure 1.1 – The Jacquard loom

The tabulating machines also used punched cards and allowed the 1890 census to be completed in eighteen months (the 1880 census was completed in seven years!). These machines were not programmable but were an important step into the direction of the programmable computers.

The Atanasoff-Berry Computer (ABC) designed and built in 1941 at the Iowa State College by John Vincent Atanasoff, was the first American electronic computer. It was designed specifically to help solve linear equations. Despite not being programmable nor a Turing complete computer, the ABC featured some important concepts such as:

- An organized system with separate computation and memory units (much like today's computers).

- A regenerative capacitor memory with about 3,200 capacitors organized into two drums composed of 32 "bands" of 50 capacitors each (two bands were always left as spare), giving a total capacity of 60 numbers of 50 bits (the regeneration system was the predecessor of the DRAM refresh circuits used today and worked rotating the drums into a common shaft).

- Fully electronic arithmetic and logic functions implemented by using about 280 vacuum tubes.

- The I/O system was composed by a primary system and a secondary system for intermediate results (used when the problems being solved were too large to be handled entirely by the internal electronic memory).

Source: http://irb.cs.tu-berlin.de/~zuse/Konrad_Zuse/en/Rechner_Z3.html

Figure 1.2 – Zuse's Z3

The world's first fully programmable digital (binary) computer was the Z3, which was designed and built by the German engineer Konrad Zuse in 1941. This early computer used about 2,000 relays to implement the switches and therefore was an electromechanical computer. The program was stored externally into a tape (a punched tape) making it possible to run different programs in a relatively easy way. The Z3 contained almost all the features of modern computers with the exception of the jump instructions (these were implemented into the microcode for floating point calculations).

In 1943, the researchers at the Post Office Research Station, at Dollis Hill, UK, presented a prototype of one of the first electronic computers: the Colossus Mark 1. This computer (Colossus Mark 2 actually) was used to help the decoding of German's encrypted messages during World War II.

Source: http://en.wikipedia.org/wiki/Image:Classic_shot_of_the_ENIAC.jpg

Figure 1.3 – The ENIAC

Another famous computer of the era was the ENIAC (Electronic Numerical Integrator And Computer), an electronic computer designed and built for the US Army's Ballistics Research Laboratory in 1946 to help calculate artillery firing tables and other complex mathematical problems (such as those related to the design of the first atomic bombs). The ENIAC was a massive computer with impressive numbers such as:

- It was composed of 17,468 vacuum tubes, 7,200 crystal diodes and about 4,100 magnetic elements (such as relays).

- The estimated power consumption was about 174 kW and the computer needed a special air-conditioning system to operate without damage. The total occupied area was about 1,800 sq ft.

Other interesting numbers:

- There were twenty 10-digit signed decimal accumulators (the ENIAC was not a binary computer).

- Running at a clock speed of about 100 kHz it could perform up to 5,000 simple additions or subtractions, 385 multiplications, 40 divisions or 3 square root operations per second. It was also possible to wire the accumulators to perform parallel operations and even expand the word size by wiring the carry of one accumulator into another.

One of the greatest caveats of ENIAC was the program storage: the program was wired inside the computer and therefore, could not be changed in an easy way. In 1948 some improvements were added to ENIAC; the most important was a primitive read-only programming mechanism using function tables (proposed by John von Neumann). This mechanism worked like a primitive program ROM (Read-Only Memory).

Another famous and important computer was the IBM Automatic Sequence Controlled Calculator ASCC (also known as Mark I), designed by Howard H. Aiken from the Harvard University and built by IBM in 1944. It was the first fully automatic computer that worked pretty much like the modern computers.

The ASCC was an electromechanical computer and had separate memories for storing instructions and data. This architecture was thereafter known as "Harvard architecture".

Another computer, the IBM Selective Sequence Electronic Calculator (SSEC), built in 1948, implemented another approach for its internal architecture: a unified memory space where both instruction and data were stored. This architecture was then known as "Von Neumann" or Princeton, after the mathematician John von Neumann from the Princeton University, even so John Presper Eckert (one of the men behind ENIAC) had proposed the same architecture years before him.

These two computer architectures (Von Neumann and Harvard) are still in use even in the most recent computers. Figure 1.4 shows the simplified schematic for the two architectures.

The Von Neumann architecture is characterized for using three buses; one for addressing (selection of the memory position to be read/written), another for controlling (selection of reading/writing operation among other things) and finally one for carrying data read or written into the memory.

The Harvard architecture differs from the Von Neumann one because it implements separate spaces for program and data storage (sometimes peripheral registers, too). Dedicated address/data/control buses access each space.

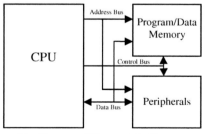

Von Neumann (Princeton) Architecture

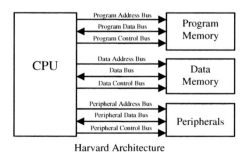

Harvard Architecture

Figure 1.4

Each architecture presents pros and cons: Von Neumann offers more flexibility because it makes no distinction between instructions and data. On the other hand, it imposes a unique width to the data bus, implying that an 8-bit CPU fetches instructions with 8 or multiple of 8-bit instructions. Another drawback of the Von Neumann architecture is the issue known as "Von Neumann bottleneck", caused by the limited memory data transfer rate when compared to the CPU necessity for accessing the memory for instruction fetch and also to read/write data. Despite the cons, the Von Neumann architecture was and still is vastly used by almost all computers and the majority of microprocessors and microcontrollers on the market.

On the Harvard architecture, the speed is the most relevant benefit of the multiple bus arrangement: while executing one instruction, the CPU can read or write data in the data memory and fetch the next instruction from the program memory. In this case, the pros and cons mix up because of the formal distinction between program and data areas: normally it is not possible to have data in the program memory or instructions in the data memory. Some Harvard machines include special instructions to allow accessing the program memory for reading/writing data.

1.2. Basic Concepts

Electronics is the science that studies and applies electricity (electrical charge movement) with components (resistors, capacitors, inductors, semiconductors, etc.) to construct complex circuits that can interact with human beings or with nature.

From TV sets to the Microwave oven, from light bulbs to cell phones, from desktop computers to wristwatches, we interact daily with a number of different electric and electronic devices, some of which are so small or simple that we do not even know they exist.

Whereas the universe of electronics is so wide, it is important to focus on specific areas to be able to understand the way a microcontroller works and how to use it to build circuits and equipments that can help us.

The base behind microcontrollers and microcomputers is the digital electronics. To better understand what it is, first we need to think about how our universe works.

In our world, we are used to dealing with analog quantities: the room temperature, the light intensity, the time, the physical dimensions of a box, the sound, the distance between our home and our jobs; all of them are analog quantities. They can be measured in some kind of unit and may vary considerably (just think that almost nothing in our world is exactly equal. Even the most precise machined pieces are slightly different when measured with the right precision).

On the other hand, modern computers are binary machines. For them there are only two types of information: 0 (false) or 1 (true). The pictures, movies and colors we see in a computer screen or on a DVD, the music we listen to from a CD player, all of them can be reduced to just 0s and 1s.

The reasons for using a binary (digital) system instead of an analog system are very simple:

1. Having only two voltage levels makes it easier to design electronic circuits for computers. By using transistors as voltage switches, it is possible to generate two distinct voltage levels (0V for false, +3 or +5V for true).

2. The math behind modern computers is based on the Boolean math created by George Boole (a British mathematician and philosopher, that lived in the 18th century and created the Boolean algebra, which is based on two states: true and false).

It is easy to notice the application of the digital circuits (based on 0s and 1s, or OFF state and ON state) to deal with the Boolean algebra. Mix them and we have digital (logic) circuits, which are the basics for computers and microcontrollers.

Digital circuits can be built with transistors (TTL circuits) or MOSFETs (Complementary Metal-Oxide-Semiconductor (CMOS) circuits). On TTL chips we have NPN and PNP transistors (figure 1.5 B) to control the logic levels inside the chip. On CMOS circuits, the building blocks are the N-channel and P-channel MOSFETs (figure 1.5 C). The purpose is always the same: to act like a switch to control the current flow.

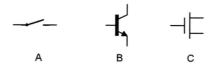

Figure 1.5 – Three switching elements

Based on the Boolean algebra there are four basic functions on digital electronics: NOT, AND, OR and Exclusive-OR (or simply EOR). The NOT function acts only inverting the input logic level: NOT 0 = 1 and NOT 1 = 0. The AND function returns true (1) when all inputs are true, the OR function returns true when any of the inputs are true while the EOR function returns true only the inputs are different. Table 1.1 shows the truth table for the three basic logic functions.

Input		Output		
A	B	AND	OR	EOR
0	0	0	0	0
0	1	0	1	1
1	0	0	1	1
1	1	1	1	0

Table 1.1 – Basic logical functions on digital electronics

On figures 1.6 A and 1.6 B we can see how to use some N-channel and P-channel MOSFETs to build a CMOS NOT gate and a CMOS NOT-AND (NAND) logic gate.

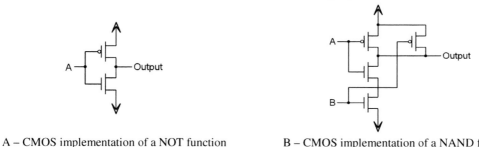

A – CMOS implementation of a NOT function   B – CMOS implementation of a NAND function

Figure 1.6

The standard symbols used to represent the basic logic functions (called logic gates) are drawn on figure 1.7.

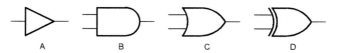

Figure 1.7 – Basic logic gates: NOT (A), AND (B), OR (C) and EOR (D)

By using standard logic functions, it is possible to create more complex digital circuits that can be classified into two categories: combinational circuits and sequential circuits.

On combinational circuits, the output is a function of the input state. Some examples of combinational logic circuits are adders, subtractors, digital comparators, etc. Figure 1.8 shows the implementation of a full adder (the basic block of any computer arithmetic circuit).

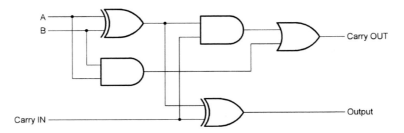

Figure 1.8 – Full adder

On sequential circuits, the output is a function not only of the present input state but also of the previous states before it. Some examples of sequential logic circuits are the latches, flip-flops, counters, registers, etc. Figure 1.9 shows a Data Latch, which is the basic memory element in modern computers.

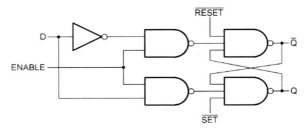

Figure 1.9 – Data latch

The sequential and combinational logic circuits are the building blocks of the CPUs (Central Processing Units) of the microprocessors and microcontrollers we use daily.

However, what is a microprocessor? How does it work? What is the difference between a microprocessor and a microcontroller?

To understand what a microprocessor is, first we need to understand some basic concepts about programming.

A program is the codification in computational language of a logical algorithm created to solve a specific problem.

For one to write a program, it is necessary to analyze carefully the problem to be solved, identifying all the steps to reach the solution.

Once the steps for the solution are determined, it is necessary to create an algorithm describing it. Using a graphic tool such as a flowchart can help in the designing of the algorithm.

As an example, imagine the simple problem of counting from 0 to 10. The steps to solve the problem could be:

```
1.  Start the counting at 0.
2.  The current counting is equal to 10?
3.  Yes: end.
4.  No: increase the counting by one.
5.  Return to step 2.
```
Listing 1.1

The flowchart describing the suggested algorithm can be seen on figure 1.10.

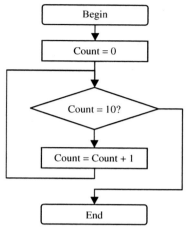

Figure 1.10

Now we can start codifying the program in a way a computer can understand. However, what way is this?

The answer is: by using the binary language of 0s and 1s! At first sight it may appear incredible that a computer which can do so many incredible things (surfing the internet, playing music, games, etc.) can only differentiate 0 from 1, but it is exactly what happens.

In fact, it is possible to group bits to form larger sequences; these sequences can represent numbers, instructions, colors or whatever the programmer wants.

Every computer, microcomputer, microprocessor, microcontroller, etc., recognizes some sequences of bits as instructions. These instructions tell it what to do: add numbers, store something into the memory, branch to another place within the program, etc. The **Central Processing Unit (CPU)** is responsible for reading these instructions from the memory, interpreting and executing them.

We call **Operation Codes** (or simply **opcodes**) the bit groups that have special meaning to the CPU.

Different computers frequently have different CPUs and different CPUs have different opcodes for the instructions they recognize (this explains why you can not run your PALM® application directly on your Windows® desktop computer).

As an example of an opcode, let us take the clear instruction. It clears (fill with zero) a register content or the content of a memory address. The basic opcode for this instruction is 00111111 binary. After reading the opcode, the CPU decodes it and finds that this instruction has an operand (the memory address that is going to be cleared). Therefore, the CPU reads the next memory address (the next following the opcode) and takes this value as the address.

After clearing the specified memory address, the execution of the instruction is completed and the CPU starts the process of fetching the memory for another instruction. Figure 1.11 shows a simplified diagram of the instruction execution cycles.

Inside the CPU we can find some important modules (this can vary from one CPU to another): the instruction decoder (responsible for the decoding of each opcode), the PC (Program Counter, for controlling the program sequence), the Arithmetic and Logic Unit (ALU, responsible for most calculations in the CPU) and the accumulator and other general purpose registers (for temporary storage of results).

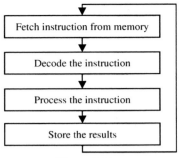

Figure 1.11

The instruction decoder is the piece of hardware that actually decodes the opcodes into smaller internal operations that perform the instruction task. Some CPUs implement a hardwired decoder while others implement a microcoded decoder. Hardwired decoders are faster because the opcode bits directly activate the internal hardware to execute the instruction. Microcoded decoders use state machines to perform instruction decoding. This makes them slower albeit more flexible than hardwired ones.

Most modern CPUs also include some type of internal memory. The basic internal memory cells are the registers: they are generally used for temporary storage of data and for controlling the operation of the CPU. The most important registers in almost all architectures are the program counter (PC) and the accumulator.

The program counter (PC) is a register responsible for controlling and sequencing the program flow. Being (in most cases) a sequential machine, the CPU must execute a sequence of instructions stored in some kind of memory. The function of the PC is to point the memory address where the next instruction to be executed is stored. The instruction pointed by the PC is fetched and feeds the decoder following the steps described on figure 1.11. Before completing the current instruction, the CPU automatically increments the PC so that it points to the memory address of the next instruction.

The accumulator is mostly used in arithmetic and logic operations. Most CPUs use the accumulator as one operand of every arithmetical or logical operation (the other operand can be another register or a value taken from the memory).

Another important component of the CPU is the arithmetic and logic unit (ALU). This is a combinational logic circuit utilized by all arithmetic and logic instructions. In most systems, the ALU has two inputs (the operands of the operations): one is generally connected to the accumulator and the other is connected to another register or read from memory. The ALU often outputs the result to the accumulator.

Before we get back to the programming topic, let us see the difference between a microprocessor and a microcontroller: put in plain words, a microprocessor is a chip with a CPU, ALU and basic support circuitry. A microprocessor relies on external memories and chips to be able to do useful things. On the other hand, a microcontroller is a chip which integrates in the same package the CPU, ALU, support circuitry, memories and peripherals. A microcontroller is also referred as a single chip computer.

The target application for a microcontroller and a microprocessor is also different: while a microprocessor is designed to be used as a general purpose computer, able to run complex softwares and operating systems, a microcontroller is designed to do specific tasks and to run simpler software (generally without an operating system).

Back to the programming and opcode topic, it is easy to figure out that using binary numbers and opcodes to write a program can be very difficult, especially because of poor program readability and the increasing probability of typing errors. Therefore, in 1950s another kind of computer language was developed: the assembly language.

In fact, the assembly is solely a symbolic representation of the opcodes. By using a special program (the assembler), statements in assembly source code are directly translated into binary codes. This translation is performed on a one-to-one basis; that is, each assembly instruction is translated into a single opcode (opcodes can be single or many bytes long).

Therefore, to be able to program a machine, first we need to know what instructions this machine understands and how we should use these instructions.

Focusing on the HCS08 CPU, we have six instructions (from a set of ninety-one) that can be used to implement listing 1.1:

CLR – Clear the content of the specified memory address or register

LDA – Load the accumulator with a specified value

CMP – Compare the content of the accumulator with a specified value

BEQ – Conditional branch to a specified address (only if last operation resulted equal)

INC – Increment the content of a register or memory address

BRA – Unconditional branch to a specified address

Most instructions are followed by one or more bytes with the instruction operands. For example: CLR 5, this instruction tells the CPU to clear the content of the address 5 of the memory.

The assembly listing for our program can be seen below. It makes the following assumptions:

1. The code is stored starting from address $F000 (the $ identify a hexadecimal number in the assembler). This is the first FLASH address on the 4 KiB HCS08 devices.

2. The Count variable is located on address $80 (128 decimal). This is a valid RAM address on all HCS08 devices.

```
Address   Opcode            Assembly       Comments
$F000     $3F80             CLR Count      Count = 0
$F002     $A60A             LDA #10        Load the accumulator with 10 (A=10)
$F004     $B180    Again:   CMP Count      Compare Count with A (A=10!)
$F006     $2704             BEQ End        If Count=A then branch to end
$F008     $3C80             INC Count      If not, increment count …
$F00A     $20F8             BRA Again      … And branch to again
$F00C              End:
```
Listing 1.2

Figure 1.12 shows how we can fit listing 1.2 inside the flowchart of figure 1.10:

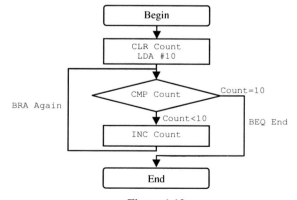

Figure 1.12

It is very easy to understand the flow of the program. It is also easy to see that the CMP instruction plays a major role in the control of the program.

In fact, the CMP instruction makes a comparison between two values: the "Count" variable and the content of the accumulator. Looking at the opcode column, we find $3F80. The instruction opcode is $3F and the operand address ("Count" variable) is $80.

Note that the comparison is actually made by subtracting the "Count" content from the accumulator content. According to the result, some bits are set/reset on a special register (called Condition Code Register or simply CCR on HCS08 devices). The bits inside this register are used by conditional branch instructions (such as BEQ) to decide whether to branch: if "Count" is equal to A, the BEQ instruction evaluates its condition as true and branches to "End", otherwise, the BEQ instruction does not branch and the next instruction (INC Count) is executed.

Another important thing to notice is the encoding of the BEQ instruction: $2704. The first byte is the instruction opcode ($27) and the second is the relative address of the target address ($04). On the HCS08, branch instructions use a relative addressing scheme where the target address is calculated ahead/aback the current instruction. On these instructions, the operand is a signed 8-bit value that is added to the current value of PC.

Considering the encoding of the instruction ($2704) we can figure out that the target address is located at the current PC + $04. So, to calculate the target address, first we need to know the current PC value. The first idea that comes to mind is that the current PC is equal to $F006, but that is not true because at the time the instruction BEQ is decoded, the PC is already pointing to the next instruction at address $F008. Therefore, the target address is equal to $F008 + $4 = $F00C.

In case the BEQ condition is evaluated false, the branch is not taken and the program executes the INC instruction. Looking at its encoding ($3C80) we can see that the first byte is the instruction opcode ($3C) and the second one is the operand address ($80). This address is the one of the Count variable.

After the INC instruction, the BRA instruction always branches to the CMP instruction ("Again" label). As we said before, this branch instruction uses a relative addressing scheme. To calculate the target address we need to add the current PC ($F00C) to the instruction offset ($F8): $F00C + $F8 = $F104.

Of course this address ($F104) is different from what we expected ($F004). But what is wrong?

The answer is: we forgot the offset is a signed value and so it is represented using the 2's complement format. The correct calculation is thus the following:

$F8 = -8

$F00C + (-8) = $F004 or $F00C + (NOT($F8)+1) = $F004

Well, this was just a review of some basic programming concepts and also a preview of some HCS08 architectural concepts. Later on, we will explore the many architectural aspects of the HCS08 CPU more deeply, but before continuing, let us take a tour through the history of the HCS08 microcontrollers.

1.3. HCS08's History

It all started in 1979, when Motorola first released the M6805 and the M146805. The first ones were a NMOS microcontroller (MCU) family and the others were a CMOS microprocessor (MPU) family. Both families had an 8-bit CPU and were developed over the MC6800, one of the first 8-bit microprocessors on the market.

The improvements on the silicon technology led to the release of the CMOS based M68HC05 family.

Some years later, another set of improvements in the M68HC05 resulted in the creation of the HC08 family. Those chips feature a better C code support by including an extra register (H), new addressing modes and instructions.

Another set of years and the introduction of the HCS08 family came to address three major fronts: the inclusion of a state-of-art on-board debugging hardware, increasing in the clock speed and reducing the general power consumption of the devices.

The on-chip debugger used into the HCS08 devices is based on the same debugging technology that debuted on the 16-bit HC12 devices: it comprises an internal background debug controller (BDC), an internal background debug module (BDM) and an external device, which interfaces the chip to the host computer, known as BDM pod.

The increase in the speed was achieved by multiplying the speed of the internal bus clock signal by 2.5 times, resulting in a great increase in the clock speed over the HC08 devices.

The low power profile of the HCS08 is due to some improvements on the semiconductor technology (all devices are manufactured on 0.25µm technology), the new low power modes, lower voltage operation (down to 1.8V) and a better clock management.

Let us see some of HCS08 family features:

- Advanced HCS08 CPU running up to 40 MHz (up to 50 MHz on the QE devices);

- Advanced programming and debugging through just one pin (BDM);

- Background Debug Controller (BDC) with advanced breakpoint features and 8-level-deep instruction trace buffer;

- up to 60 KiB of FLASH, and up to 4 KiB of RAM (newer Flexis devices offer even more Flash and RAM);

- Low power consumption and low-voltage operation (down to 1.8 Volts);

- Advanced peripherals like 16-bit timers with capture, compare and PWM channels, 8-bit timers, real-time interrupt, analog comparator, 10-bit analog-to-digital converter (some devices include a 12-bit ADC) with internal reference and internal temperature sensor , LCD controller (MC9S08MC9S08LCxx, LGxx and LLxx families), serial interfaces such as USB, CAN, UART, SPI and I^2C;

1.4. Controller Continuum

The HCS08 family is also a key element on the Freescale's "Controller Continuum" initiative, which provides the migration path from low-end 8-bit microcontrollers to high-end 32-bit microcontrollers.

Freescale's microcontroller portfolio ranges from the 8-bit low-end RS08 devices to the 32-bit high-performance devices such as the PowerPC, i.MX and ColdFire MCUs and MPUs.

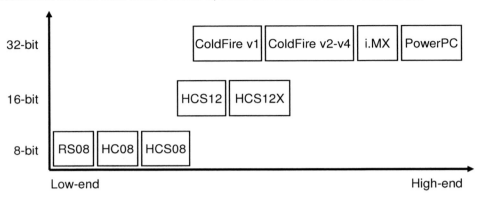

Figure 1.13 – Freescale's 8-bit to 32-bit MCU and MPU families

The controller continuum initiative comprises a comprehensive roadmap of compatible 8-bit and 32-bit devices that share the same hardware and software development tools, pin-out and peripheral registers compatibility, allowing easier migration from 8-bit devices to powerful 32-bit devices.

The roadmap portfolio starts with the ultra-low-cost 6/8-pin RS08 devices (such as the MC9RS08KA2) which are pin-compatible with the lower-end 8-pin HCS08 devices (as the MC9S08QD2). The roadmap extents to higher complexity devices: from the 64-pin/8-bit/50MHz HCS08 devices (such as the MC9S08QE128) to the 32-bit/50 MHz Coldfire-V1 devices (such as the MCF51QE128).

By using the same peripherals, the high-level interface of the application is almost the same, no matter using an 8-bit or a 32-bit device. Of course, the use of a high-level language such as C can help even more the migration to another device/platform.

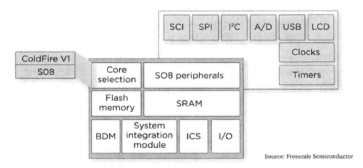

Figure 1.14 – Sharing the same base chip is the key for the controller continuum

1.5. Adoption of the C Language

In this book we will focus on the C language. Some of the reasons which led us to choose that language are:

- ◆ Easier to learn and to program (when compared to the assembly language), shortening the learning curve of a new microcontroller;

- ◆ Code portability, allowing the programmer to reuse code already written and tested on other platforms;

- ◆ High efficiency: the C language is very close to the assembly language allowing very efficient, short and fast code to be produced;

- ◆ It suits the controller continuum initiative;

The C language is a standard in the programming of microcontrollers in commercial code development and that cannot be ignored.

Of course we are not saying that the knowledge of the assembly language is unnecessary. In fact, knowing it is very important to better understand the way the compiler generates the code and the way to improve the C code efficiency.

1.6. Development Hardware

The examples in this book were developed to run on the following Freescale's demonstration kits:

Figure 1.15 - M68DEMO908GB60 (for the MC9S08GB examples)

The demonstration board for the MC9S08GB60 includes a prototype area, a central bus with all microcontroller signals, two EIA-232 ports, five leds and four push-buttons. This board is mainly intended to demonstrate the low power capabilities of the HCS08 MCUs.

Figure 1.16 - DEMO9S08AW60E (for the MC9S08AW examples)

The demonstration board for the MC9S08AW60 features an on-board BDM device (for in-circuit debugging and programming), on-board two-axis accelerometer (MMA6260Q), one EIA-232 port, ten leds, eight-port rocker switch, one potentiometer, one light sensor and four push-buttons.

Figure 1.17 - DEMO9S08QG8 (for the MC9S08QG4 and QG8 examples)

The demonstration board for the MC9S08QG4 and QG8 devices includes an on-board BDM device (for in-circuit debugging and programming), one EIA-232 port, a bus with all microcontroller signals, two user leds, two push-buttons, one potentiometer and one light sensor.

Figure 1.18 – DEMOQE128 board (for the MC9S08QE examples)

The DEMOQE128 demonstration board includes an on-board BDM device (for in-circuit debugging and programming), an on-board three-axis accelerometer (MMA7260Q), an EIA-232 port, nine user leds, five push-buttons, a potentiometer, a piezoelectric buzzer and a bus with all microcontroller signals.

Figure 1.19 – DEMO9S08LC60 board (for the MC9S08LC examples)

The DEMO9S08LC60 demonstration board includes an on-board BDM device (for in-circuit debugging and programming), a custom LCD display, an on-board three-axis accelerometer (MMA7260Q), an EIA-232 port, eight user leds, five push-buttons, a potentiometer, a piezoelectric buzzer, a light sensor (LDR), a temperature sensor (NTC), an on-board 32,768 Hz quartz crystal and a bus with all microcontroller signals.

For in-circuit programming and debugging, it is also possible to use the BDM USB Multilink, manufactured and sold by P&E Microcomputer Systems Inc. (www.pemicro.com) or the Softec's Spyder08 (only for the RS08KA, S08QD and S08QG devices).

Figure 1.20 – BDM USB Multilink   **Figure 1.21 – USB SPYDER08**

For professional development, production and standalone programming, there is the Cyclone Pro tool, a very powerful and easy-to-use programming and debugging tool (www.pemicro.com).

The Cyclone PRO can work connected to a host PC or in stand-alone mode. When connected to a host PC (through a serial, USB or Ethernet connection) the device can act like an in-circuit debugger/programmer. In the stand-alone mode, the device can work like a production programmer. It is possible to store multiple device images into its memory and select the one to be programmed on-the-fly.

Figure 1.22 – Cyclone PRO

1.7. Standard Conventions

Through this book we used some standardized conventions to represent the special symbols and names:

1. We are using the new information units defined by the International Electrotechnical Commission (IEC) in 2000. They were defined to help differentiate the standard decimal values from the binary ones:

Standard SI Prefix			Binary SI Prefix		
Prefix	Symbol	Weight	Prefix	Symbol	Weight
Kilo	k	10^3	kibi	Ki	2^{10}
Mega	M	10^6	Mebi	Mi	2^{20}
Giga	G	10^9	Gibi	Gi	2^{30}
Tera	T	10^{12}	Tebi	Ti	2^{40}
Peta	P	10^{15}	Pebi	Pi	2^{50}
Exa	E	10^{18}	Exbi	Ei	2^{60}

Table 1.1

2. Devices belonging to the same family are sometimes represented using a global descriptor such as: Ax (for AC and AW devices), Dx (for DN, DV and DZ devices), Ex (for EL and

EN devices), Gx (for GB and GT devices), Lx (for LC, LG and LL devices), Qx (for QA, QB, QD, QE and QG devices), Sx (for SG, SH and SL devices) and Rx (for RC, RD, RE and RG devices).

3. Bit and register names are always written in capital letters.

4. References to specific bits occur by their name or by the association of the register name and the bit name in the following format: REGISTER:BIT. I.e.: the bit I of the CCR register can be represented by the letter I or by CCR:I.

5. To generically represent a set of registers or bits available in a peripheral or register, we use the letter "x". I.e.: PTxD represents any of the following registers: PTAD, PTBD, PTCD, etc. The same applies to numbers: TPMxCxV can be TPM1C0V, TPM1C1V and so on.

6. In the C language, 16-bit registers such as the TPMCNT can be referenced in three different manners: TPMCNT for the whole 16-bit register (represented as an **int** type), or TPMCNTH and TPMCNTL (both **char** types), representing the higher byte (the most significant one) and the lower byte (the lowest significant one).

7. We created a special symbolic file called "HCS08.h". This file defines several symbols used through the text to specify the bit names, allowing an easier configuration of attribution statements.

8. Through the text, the C-reserved words are always written in boldface.

9. Important notes are presented using the following box:

 This is an important note!

10. Tricks and hints are presented using the following box:

 This is a hint!

11. The microcontroller registers are often represented by a table such as the following:

Example MTIMSC Register

Name	Model		BIT 7	BIT 6	BIT 5	BIT 4	BIT 3	BIT 2	BIT 1	BIT 0
MTIMSC	QG	Read	TOF	TOIE	0	TSTP	0	0	0	0
		Write	-		TRST		-	-	-	-
		Reset	0	0	0	1	0	0	0	0

Let us take a look into how to read the table:

The register named MTIMSC is located on the 0x003C memory address and it is available only in the QG series.

There are three lines labeled "Read", "Write" and "Reset". The first and the second indicate the function of each bit when they are read or written by the CPU. The "reset" line represents the bit state after a Power-On Reset (POR).

In the table above we can see there is a bit named TOF (bit 7) which is read-only. This means that the bit can only be read by the CPU, but not changed. Any writes on it have no result.

On the other hand, the 6th bit named TOIE has a read/write function as both lines (read and write) have the TOIE function. This way, when the bit is read by the CPU, the state of TOIE is returned and when the bit is written, the state of TOIE is changed accordingly.

Another case is the bit 5, TRST: this bit can only be changed by the CPU, but TRST state cannot be read and always returns "0". In this case, we say that TRST is a write-only bit.

Finally, we have the not-implemented bits (0 to 3 in the register example). These bits cannot be modified by the CPU and when read, they always return "0"

In the example, all bits, except the bit 4 (TSTP) are reset to "0" after a POR. The TSTP bit is set to "1" after a POR event. When the state is represented by "?", the bit is not modified by the operation.

We also provided some C symbols to make it easier to access bits in peripheral registers. These symbols are always presented in a specific column on the register description (most symbols start with a "b" character). Writing:

```
MTIMSC = bTRST;
```

Writes "1" into the TRST bit and "0" into all other bits of the MTIMSC register.

Besides, it is also possible to use bit field structures included automatically in the project. That allows direct access to all bits within a register.

By using these structures and related symbols, the bits in MTIMSC can be accessed by simply writing the register name followed by an underline and the bit name. Examples:

```
MTIMSC_TOIE = 0;          // clear bit TOIE
MTIMSC_TRST = 1;          // set bit TRST
```

 The HCS08.h file and all other examples presented in this book are available for download at www.sctec.com.br/hcs08

2

The HCS08 Architecture

The HCS08 series is an enhancement of the original HC08 family. Both use the same instruction set and peripherals. The HCS08 series also has several new features that make it even more powerful and efficient. Some of them are:

- Object-code compatible with the HC05 and HC08 families;

- Two-level-deep instruction queue, which allows greater speed in the opcode fetch and instruction execution;

- New addressing modes for the LDHX, STHX and CPHX instructions, allowing improvements in the compiler efficiency. In most cases the HCS08 code is smaller and faster than the HC08 code at the same clock speed;

- Advanced internal debugging hardware (BDC - Background Debug Controller), which communicates with the host through one pin and includes advanced features like: multiple breakpoints, advanced triggers, instruction tracing with CPU state capturing and real-time data visualization;

- The new BGND instruction, which allows the insertion of as many software breakpoints as necessary;

- Extended memory addressing: devices with a memory management unit (MMU), which can address up to 4 MiB (256 pages of 16 KiB each). Two new instructions allow manipulating subroutines using these pages: CALL and RTC. Until now, only the QE devices offer this feature;

- Higher clock speed: with a maximum frequency of 40 MHz (external crystal) and 20 MHz (in the internal bus system). The HCS08 series is about 3 times faster than the original HC08 (QE devices operate at a higher 50 MHz frequency for external oscillator and 25 MHz bus);

- Up to five operating modes, including three new low power stop modes, which allow a lower power consumption;

- Due to the low operating voltage (down to 1.8 V on some devices), the new power modes and the low power peripheral set, the HCS08 series can operate with as low as 500 nA or even 20 nA when totally stopped (stop1 mode);

 The HCS08 family can be divided into two supply-voltage categories: the 1.8V devices, which have supply voltages from 1.8 up to 3.6V and the 5V devices, with supply voltages from 2.7 up to 5.5V. The stop1 mode is only available on some of the 1.8V devices.

In the following pages we will see some of the greatest features of the CPU as well as the new improvements in the HCS08 series.

2.1. HCS08 Programmer's Model

The HCS08 CPU is based on the Von Neumann (also known as Princeton) architecture with a CISC (Complex Instruction Set Computer) instruction set. This architecture is characterized by the use of a common data bus shared between the memories and peripherals, one address bus (responsible for the selection of the memory address or the peripheral register to be accessed) and one control bus (for controlling the kind of operation to be executed). That means the CPU views only one address space shared between the application code and the user data. On the HCS08 CPU, the data bus is split into two separate buses: one for writing operations and the other for reading operations.

The address bus is 16-bit wide, allowing the addressing of up to 2^{16} bytes or 65536 (64 Kibi) bytes. The data buses are 8-bit wide, allowing instructions and data to be moved to/from the CPU in 8-bit blocks.

 Values larger than 8 bits are stored in memory using big endian format: the least significant data byte is stored into the highest memory address.

Using a CISC instruction set these days may appear to be in the wrong-way of today's architectures, but the use of a CISC set means that there are many complex operations coded into single instructions. On one hand this can make the assembly instruction set harder to be learned, but on the other hand, CISC processors are frequently very C-friendly, and this helps the C compilers produce a more efficient code.

The instructions also appear in a variable length fashion, which means that the no-operand instructions and those with shorter operands need less space and run faster than those with larger or more complex operands.

Another key concept of the HCS08 architecture is the memory mapped peripherals: some address spaces are used for accessing the peripheral registers. This way, the same instructions used to manipulate data can also read/write onto the peripheral registers.

The HCS08 CPU also uses five special registers (inside the CPU core):

- An 8-bit accumulator (A);
- A 16-bit program counter (PC);
- A 16-bit index register (H:X);
- A 16-bit stack pointer (SP);
- An 8-bit condition code register (CCR).

2.1.1. Accumulator "A"

This 8-bit register is used to temporarily store the results of the CPU operations and is located in the output of the arithmetic and logic unit (ALU).

The accumulator is usually present in arithmetic operations, as an operand or as the destination of the results (some operations can also use the X register).

There are special instructions to load the accumulator with data (LDA) as well as to store the content of the accumulator into the memory (STA), and to exchange data with X and CCR registers.

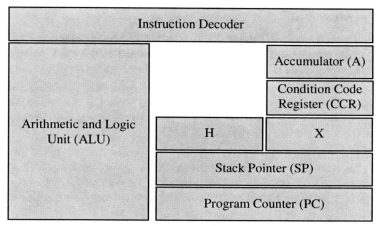

Figure 2.1 – HCS08 CPU Simplified Block Diagram

2.1.2. Program Counter "PC"

The program counter (PC) is a binary counter which points to the memory address of the next instruction or operand to be fetched. After each fetch, the PC content is added by one, so that it is always pointing to the next instruction or operand.

In the HCS08 architecture, as in the HC08 architecture, the 16-bit wide program counter allows addressing up to 64 KiB (65,536) memory addresses.

One important thing to note is that the PC content cannot be directly modified by the application code. The only way the application code can change it, is through the change-of-flow instructions like jump and branch.

The PC content can also be modified by the debugging hardware, allowing the debug software to freely change the program flow.

On the CPU reset, the PC register is initialized with the 16-bit value stored in the addresses 0xFFFE and 0xFFFF (the address of the first instruction of the application program). The lowest significant byte of the instruction address is stored at 0xFFFF and the most significant byte is stored at 0xFFFE (the big endian format)

2.1.3. Index Register "H:X"

The HCS08 architecture implements a 16-bit index register called "H:X". This register is actually formed by joining the two 8-bit registers H and X. The X register is the same index register used in the former HC05 architecture. By the use of the H:X register it is now possible to index any address in the 64 KiB available addresses.

To reduce any compatibility issues, the H register is initialized with zero and the X register with 0xFF after a CPU reset. This way, the H:X register pair behaves like the old single X index register.

There are some special instructions to load X (LDX), H:X (LDHX), and to store X (STX) and H:X (STHX). There are also instructions to transfer the content between H:X and the stack pointer (SP).

As we said earlier, the X register can also work as an auxiliary accumulator on some ALU-related operations (bit shift, bit rotation, increment, decrement and complement operations are some of them).

The H:X register and the new addressing modes related to it are especially useful to deal with data arrays and tables. We will learn about the addressing modes later in this chapter.

2.1.4. Stack Pointer "SP"

Another HCS08 core register is the Stack Pointer (SP). Its main function is to point to the top of the software stack, being automatically incremented/decremented by the stack-manipulation instructions and stack-related operations (such as subroutine callings and interrupts).

The stack is a "Last-In-First-Out" (LIFO) data structure and it is used to store the return addresses on subroutine calls. One good thing about the HCS08 stack is that it can also store any kind of data, allowing C functions to pass their parameters through the stack (in opposition to some RISC architectures that do not allow data storage on the stack).

This structure grows from the highest address (frequently the last RAM address) down to the start of the RAM (the stack can have any size, limited only by the quantity of RAM available in the chip). Each value stored into the stack decrements the SP by one (technically, this operation is known as pushing onto the stack or simply a PUSH operation). The reverse operation (pulling data from the stack) is called POP.

There are two assembly instructions for subroutine calling: BSR (branch to subroutine) and JSR (jump to subroutine). They both push the PC content into the stack and then branch (or jump) to the instruction destination address. The following listing demonstrates a simple program to exercise basic stack operations.

Address (hex)	Opcode (hex)	Assembly Instruction		Description
		ORG	$F000	the next instruction is located at address 0xF000 *
F000	A605	LDA	#5	load A with 5 decimal
F002	451234	LDHX	#$1234	load H with 0x12 and X with 0x34
F005	CDF200	JSR	my_subroutine	jump to my_subroutine
F008	C71800	STA	$1800	store A at address 0x1800 (clear COP counter)
F00B	20F3	BRA	*-$B	branch to PC-11 (0xF00B – 0xB = 0xF000)
...	...	...		
		ORG	$F200	the next instruction is located at address 0xF200
F200	87	PSHA		save A into stack
F201	4F	CLRA		clear A (A = 0)
F202	86	PULA		restore A from stack
F203	81	RTS		return from subroutine

* ORG is an assembler directive and not an assembly instruction!
Listing 2.1

Figure 2.2A depicts the stack content before the execution of the JSR instruction. The SP points to the top of stack (0xFF by default after a reset). When the JSR instruction is executed, the following actions are taken:

1. The current content of the PC (which already points to the next instruction) is pushed onto the stack. Note that at this time, PC = 0xF008 (BRA instruction) because that one is the next instruction to be executed in the natural program sequence.

2. The low byte of the address (0x08) is stored in the address pointed to by SP.

3. SP is decremented by one.

4. The high byte of the address (0xF0) is stored in the address pointed to by SP.

5. SP is decremented by one. Now the top of stack is 0x00FD (figure 2.2B).

	Address	Content
top of stack →	0x00FF	??
	0x00FE	??
	0x00FD	??
	0x00FC	??
	0x00FB	??

	Address	Content
	0x00FF	0x08
	0x00FE	0xF0
top of stack →	0x00FD	??
	0x00FC	??
	0x00FB	??

Figure 2.2A   **Figure 2.2B**

6. PC is loaded with the destination address (0xF200). Program flow continues from address 0xF200 (PSHA instruction).

There are also two instructions for returning from subroutines: RTS (return from subroutine) and RTI (return from interrupt). The RTS instruction pulls the PC from the stack and jumps to that address. It is used for returning from a subroutine. The RTI instruction pulls the PC, X, A and CCR from the stack and jumps to the new address pointed to by the PC.

We can look at our example again to see how the subroutine processing takes place: after processing the subroutine, the RTS instruction POPs the return address from the stack and stores it into PC. This makes the program flow jump to address 0xF008 (the address of BRA, the next instruction after JSR). After executing RTS, SP is equal to 0x00FF (the stack is empty). This is the classic implementation of a subroutine calling/returning.

 Notice that the POP operation is not destructive: it just increments the SP; the stack content is not changed and remains unchanged until a new PUSH operation takes place!

There are also three instructions to push registers A, H and X onto the stack (PSHA, PSHH and PSHX), and three others to pull them from the stack (PULA, PULH and PULX). In addition, most instructions can actually read/write to the stack through the stack-related addressing modes (this helps using high-level languages such as C).

As an example, we can again report to our previous example: within our small (and absolutely useless!) subroutine, two important operations take place:

1. The PSHA instruction pushes the accumulator (A) onto the stack. This effectively preserves A content (A can be changed within the subroutine and restored to its previous state before returning to the caller). The operation (PSHA) stores A into the address pointed to by SP and decrements SP (so it points to the next available stack position). This is shown in figure 2.3A.

2. The PULA instruction pulls the value of the top of the stack and stores it into the accumulator (A), effectively restoring the previous A content. The operation (PULA) increments SP by one and reads the content of the address pointed to by the new SP. This is shown in figure 2.3B (notice the operation is not destructive and does not change the content of address 0x00FD).

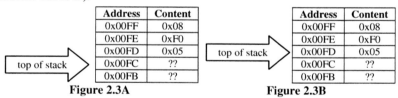

Figure 2.3A   **Figure 2.3B**

Note that on the HCS08 CPU, the stack can be located in any of the 65536 possible addresses (of course this address must be a RAM area).

The SP is loaded with 0x00FF after a a reset or a reset stack pointer (RSP) instruction. This was usually the top of the RAM area on the former HC05 devices, but on the newer HC08 and HCS08 devices with larger RAM memories, the top of the stack can be manually initialized using the LDHX and TXS instructions.

Note that on C applications the SP is automatically initialized to point to the top of the RAM; this is done by the startup code as we will see later.

 Remember that the stack can grow up to the maximum size of the amount of RAM available on the chip! This can lead to data overriding and in most cases to an application failure. It is a good practice to size it in a way it can not override the RAM area used to store the application data. The STACKSIZE linker parameter can be used to modify the stack size. The standard value is 0x50 (80 bytes).

2.1.5. Condition Code Register (CCR)

The condition code register stores the processor flags and the interrupt mask bit.

	BIT 7	BIT 6	BIT 5	BIT 4	BIT 3	BIT 2	BIT 1	BIT 0
Read	V	1	1	H	I	N	Z	C
Write								
Reset	x	1	1	x	1	x	x	x

Let us see a brief description of each processor flag as well as its basic operating characteristics:

V - Overflow flag:

The V flag indicates the occurrence of an overflow after an arithmetic operation with signed operands.

This flag is set when the last math operation (either addition, subtraction, increment, decrement, rotation or comparison operation) results in a value larger than +127 or smaller than -128. This indicates that the operation cannot fit an 8-bit register and that the result is not valid.

Some examples of operations:

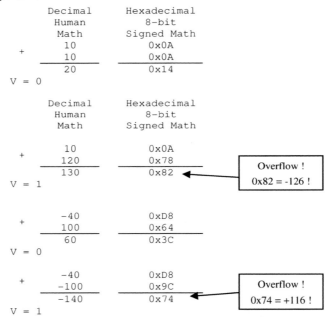

H - Half carry flag:

The H flag indicates the carry event from the third to the fourth bit of the result of any arithmetic operation.

This bit is also known as "digit carry" as it indicates a carry event from one hexadecimal digit to another.

This flag is especially useful on the BCD (binary coded decimal) arithmetic operations.

Examples:

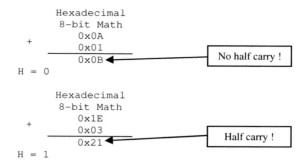

```
        Hexadecimal
        8-bit Math
           0x0A
    +      0x01
        ─────────
           0x0B  ◄────────  No half carry !
  H = 0
```

```
        Hexadecimal
        8-bit Math
           0x1E
    +      0x03
        ─────────
           0x21  ◄────────  Half carry !
  H = 1
```

N - Negative flag:

The N flag is set after any arithmetic, logic or data movement operation that results in a negative value (bit 7 set).

This bit is actually a copy of the most significant bit of the result. Remember that for signed values, the sign bit is located in bit 7. For unsigned operations, the N flag must be ignored.

Examples:

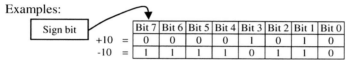

	Bit 7	Bit 6	Bit 5	Bit 4	Bit 3	Bit 2	Bit 1	Bit 0
+10 =	0	0	0	0	1	0	1	0
-10 =	1	1	1	1	0	1	1	0

(Sign bit → Bit 7)

Z - Zero flag:

The Z flag is set when any arithmetic, logic or data movement operation results in zero (all bits clear). If the operation results in a value different from zero, the Z flag is cleared.

C - Carry/borrow flag:

The C flag has four different uses:

* In add operations, the C flag is set when the result is greater than 255 decimal. This way, that flag behaves like a carry flag. **Remember that a carry is equal to 256 or 0x100!**

```
        Decimal        Hexadecimal
      Human Math        8-bit Math
          10               0x0A
    +    120          +    0x78
        ─────             ──────
         130               0x82
  C = 0
```

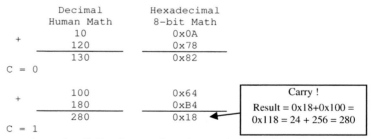

```
         100               0x64
    +    180          +    0xB4
        ─────             ──────
         280               0x18  ◄───
  C = 1
```

Carry !
Result = 0x18+0x100 =
0x118 = 24 + 256 = 280

* In subtraction operations, the C flag is set when the result is smaller than 0 (a negative value). This way, the flag behaves like a borrow flag, indicating the need for a borrow to complete the subtraction operation.

```
        Decimal        Hexadecimal
        Human           8-bit Math
        Math
          32               0x20
    -     15          -    0x0F
        ─────             ──────
          17               0x11
  C = 0
```

```
          32              0x20
    -     40              0x28
          ‾‾‾‾             ‾‾‾‾‾
          -8              0xF8
  C = 1
```

Notice that the subtraction operation is actually performed as an addition where the subtrahend is in form of 2's complement. In this case, the C flag always starts set and is cleared if there is a carry (result > 255). This way, the operation described above can be rewritten as:

```
        Decimal      Hexadecimal
        Human        8-bit Math
        Math
          32            0x20
    +    -40            0xD8
         ‾‾‾‾           ‾‾‾‾‾
          -8            0xF8
  C = 1
```

- In bit shift and bit rotation operations the C flag acts like a "ninetieth bit", receiving the value from the eightieth bit (bit 7) on left rotations or the value from the first bit (bit 0) on right rotations.

- In test-bit operations, the C flag receives the value of the tested bit.

I -   Interrupt mask:

This bit controls whether an interrupt request is processed by the CPU (I=0) or not (I=1). An interrupt requested while I=1, remains pending until I is cleared (enabling the interrupt processing) or the source interrupt flag is "manually" reset by the application.

2.1.6. Addressing Modes and Instruction Set

Despite the fact that the HCS08 CPU has almost the same instruction set as the HC08 CPU, the HCS08 CPU has some nice features that can significantly improve the speed while reducing the code size of the application.

In this topic we will learn about the instruction set and the addressing modes of the HCS08 CPU and how its improvements can make the difference comparing to the HC08 devices.

2.1.6.1. Addressing Modes

There are sixteen addressing modes, but only some of them are available to each instruction. By addressing mode we mean the way to represent how the CPU can reach the operands necessary to accomplish its job. Some instructions need only one operand, while others need two of them.

- **Inherent (INH):** the operand is part of the opcode itself. This is the addressing mode that provides access to the CPU registers like the accumulator (A), the stack pointer (SP), H:X and CCR. Some examples of this mode are:
  ```
  LDA #10      ; load A with 10 decimal (two operands, inherent (A) and immediate (10))
  INCX         ; increment A (one inherent operand (A))
  ```

- **Immediate (IMM):** the operand follows the opcode (the next byte after the opcode). This addressing mode is almost always used with the inherent mode. Some examples are:
  ```
  LDA #10      ; load A with 10 decimal (two operands, inherent (A) and immediate (10))
  ADD #5       ; add the accumulator with 5 decimal (two operands, inherent (A) and
               ; immediate (5))
  ```

- **Direct (DIR):** the direct mode uses one byte to represent the memory address to be used in the operation. This is the fastest way to address a memory position located in the first 256 memory address (this region is also known as **direct page**). The benefit of using this address mode is the saving of one byte in the address, thus saving memory space and speeding up the instruction fetch and execution. Some examples:

```
STA 10        ; store the accumulator onto address 10 of memory
ADD $10       ; add the accumulator to the content of the address 16
              ; decimal (10 hexadecimal) of the memory
```

 The bit manipulation and bit test instructions (BSET, BCLR, BRSET and BRCLR) can only use this addressing mode and thus are limited to operate only on direct page addresses (0x0000 to 0x00FF).

- **Extended (EXT):** the extended mode is the opposite of the direct mode. This mode allows access to the full address space (all the 65536 memory addresses) by using 2-byte (16-bit) addresses. It is also slower than the direct mode and uses more memory. Some examples:

```
STA 310       ; store the accumulator onto address 310 of memory
ADD $1000     ; add the accumulator to the content of the address
              ; 1000 hexadecimal of the memory
```

- **Indexed, no offset (IX):** the indexed mode uses the content of the H:X register as the address where the data is read or written. The indexed mode works exactly like a C language pointer: the content of the pointer (H:X) is the memory address of the data to be used in the operation. Examples:

```
CLR ,X        ; clear the content of the memory address pointed by H:X
AND ,X        ; logical AND operation between the accumulator and the
              ; data in the memory address pointed by H:X
```

- **Indexed, no offset with post-increment (IX+):** this is a variation of the indexed mode. In this mode, the content of the H:X register is automatically incremented by one after the operation. This is useful for data movement operations. There are only two instructions that can use this addressing mode CBEQ and MOV. Example:

```
MOV $20,X+    ; move the content from the memory address 20
              ; hexadecimal to the memory address pointed by the H:X
              ; register and add 1 to H:X after the operation
```

- **Indexed, with 8-bit offset (IX1):** this is another variation of the indexed mode. In this addressing mode, the content of the H:X register is added to an 8-bit constant (the offset) and the result is used to address the instruction operand. This addressing mode is useful when accessing data arrays: the 8-bit constant is the base address of the array and the H:X content is the number of the element to be accessed inside it. Example:

```
COM $10,X     ; complement the value stored at the address pointed by (H:X + 0x10)
```

- **Indexed, with 8-bit offset and post-increment (IX1+):** the IX1+ works in the same way as the IX1 mode, but it also increments the content of the H:X register by one at the end of the operation. This addressing mode is only supported by the CBEQ instruction. Example:

```
CBEQ ,X+,END  ; compare the content of the accumulator to the
              ; content of the memory address pointed by H:X and
              ; branch to the address represented by END if they are equal
```

- **Indexed, with 16-bit offset (IX2):** the IX2 mode works in the same way as the IX1 mode, but it uses a 16-bit offset instead of an 8-bit one, allowing the access to an array located in any memory address. Example:

```
COM $200,X    ; complement the value stored at the address pointed by (H:X + 0x0200)
```

- **SP-relative, with 8-bit offset (SP1):** the SP-relative addressing mode works in the same way as the indexed mode IX1, but using the stack pointer (SP) as the index register. This mode is useful for accessing parameters stored into the stack (up to 255 locations deep) like function parameters or local variables.

- **SP-relative, with 16-bit offset (SP2):** this addressing mode works in the same way as the indexed mode IX2, but again the index register is the stack pointer (SP). As the SP1, this mode is also used to access data stored into the software stack (farther than 255 levels into the stack). Example:

```
ADC $200,SP   ; add the value in the accumulator and the carry bit to
              ; the value stored at the address pointed by (SP + 0x0200)
```

- **Relative (REL):** in the relative addressing mode, the destination address is calculated by adding the program counter (PC) to the signed constant operand. This way, the destination address can be 127 bytes ahead or 128 bytes aback the current program position. This addressing mode is used only by the branch instructions. Examples:

```
BEQ *+5        ; compute the result of adding 5 to the actual value of the PC and branch
               ; to this address if the EQual condition (Z=1) is true
BEQ *-5        ; compute the result of adding -5 to the actual value of the PC and branch
               ; to this address if the EQual condition (Z=1) is true
```

- **Memory-to-memory:** this addressing mode is used only by the MOV instruction to transfer data to a memory address or between two memory addresses without using the accumulator. There are four different memory-to-memory transfer modes:

    o **Immediate to direct (IMM/DIR):** this mode can be used to initialize a memory address (a variable for example) with a constant (the immediate operand indicated by the # symbol) value. This addressing mode is faster and smaller than the same operation done through the accumulator (a LDA instruction followed by a STA). Example:

    ```
    MOV #$05,$20      ; copy the immediate value 0x05 to the address 0x20
    ```

    o **Direct to direct (DIR/DIR):** this mode can be used to copy data from one direct memory address to another. Example:

    ```
    MOV $05,$20       ; copy the value from the direct address 0x05 to the address 0x20
    ```

    o **Indexed to direct with post-increment (IX+/DIR):** this addressing mode was designed to transfer data blocks from a memory table (pointed by the H:X register) to a direct memory address (a variable for example), after each transfer the H:X register is incremented by one (so it points to the next position of the memory table).

    o **Direct to indexed with post-increment (DIR/IX+):** this mode works on the opposite way of the previous one: the contents of the direct memory address pointed by the first operand is transferred to the memory address pointed by the H:X register) and the H:X register is incremented by one after the operation (so it points to the next position of the memory table).

Now that we know about the addressing modes, let us take a look at the HCS08 instruction set. We divided the set into some categories depending on the instruction purpose: data transfer, arithmetic, logic, compare and test, branch and internal control instructions.

2.1.6.2. Data Transfer Instructions

The data transfer instructions are used to transfer data between the memory and the CPU internal registers or to initialize memory positions or CPU registers.

Actually, it is important to notice that the transfer operation is, in fact, just a copy operation: the data in the source operand is copied to the destination operand and left unchanged in the source.

LDA   load the accumulator with an immediate constant or memory value.

STA   store the accumulator into the memory.

LDX   load the X register with an immediate constant or memory value.

STX   store the X register into the memory.

LDHX   load the H:X register with 16-bit immediate constant or memory value.

STHX   store the H:X into the memory.

MOV   transfer (copy) data to/from memory without using the accumulator.

CLR   clear the content of the specified register or memory address.

| **TAP** | transfer (copy) the accumulator to the CCR register. |

TAP   transfer (copy) the accumulator to the CCR register.

TPA   transfer (copy) the content of the CCR register to the accumulator.

TAX   transfer (copy) the accumulator to the X register.

TXA   transfer (copy) the X register to the accumulator.

TSX   transfer (copy) the 16-bit contents of the SP+1 to the H:X register (H:X=SP+1).

TXS   transfer (copy) the 16-bit contents of the H:X register minus one to the SP (SP=H:X-1).

RSP   reset the stack pointer (SP = 0x00FF).

PSHA   push the accumulator onto the stack (store A on the stack).

PULA   pull the accumulator from the stack (restore A from the stack).

PSHH   push the H register onto the stack (store H on the stack).

PULH   pull the H register from the stack (restore H from the stack).

PSHX   push the X register onto the stack (store X on the stack).

PULX   pull the X register from the stack (restore X from the stack).

Some examples of instruction usage:

```
LDA #5              ; load A with 5 decimal
STA 128             ; store A into memory address 128 decimal
MOV 128,129         ; copy data from address 128 to address 129
MOV #8,130          ; move 8 decimal into memory address 130
```

After the execution of the instructions above, the content of the accumulator (A) will be 8 decimal and the memory will have the following contents:

Address:	127	128	129	130	131
Value:	Unknown	5	5	8	unknown

Figure 2.4

It is possible to give identifiers to the memory addresses making it easier to reference them:

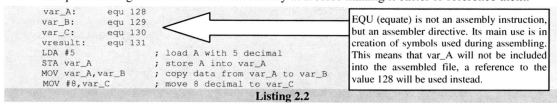

```
var_A:      equ 128
var_B:      equ 129
var_C:      equ 130
vresult:    equ 131
LDA #5              ; load A with 5 decimal
STA var_A           ; store A into var_A
MOV var_A,var_B     ; copy data from var_A to var_B
MOV #8,var_C        ; move 8 decimal to var_C
```

EQU (equate) is not an assembly instruction, but an assembler directive. Its main use is in creation of symbols used during assembling. This means that var_A will not be included into the assembled file, a reference to the value 128 will be used instead.

Listing 2.2

After the execution of the instructions above, the contents of the memory will be the same as the figure 2.4.

The following code snippet can be found on C startup files for stack initialization. This code initializes the stack pointer (SP) to the top of the RAM:

```
LDHX #RAMEnd+1      ; H:X = last RAM address plus one
TXS                 ; transfer (H:X-1) to SP (SP=H:X-1) (Now the stack is
                    ; initialized!)
```

Listing 2.3

On the HCS08 devices the LDHX and STHX instructions were expanded to support the following addressing modes: immediate, direct, extended, H:X indexed, H:X indexed with 8-bit offset, H:X indexed with 16-bit offset, SP indexed with 8-bit offset and SP indexed with 16-bit offset. On the HC08 devices only two modes are supported: immediate and direct.

The following code snippet shows the basics on stack operations:

```
; First we initialize A, H and X
LDA  #$12           ; A = $12
LDHX #$3456         ; H = $34 and X = $56
; Now we push A, H and X onto the stack (let us consider it is empty):
PSHA                ; push A onto stack
PSHH                ; push H onto stack
PSHX                ; push X onto stack
; Now we pull the registers from the stack in a different order, so that
; A will be $56, H will be $12 and X will be $34:
PULA                ; pull A from stack (A=$56)
PULX                ; pull X from stack (X=$34)
PULH                ; pull H from stack (H=$12)
```

Listing 2.4

Figure 2.3 shows four states of the stack: A – the stack is empty (before the PSHA instruction), B – after the PSHA instruction, C – after the PSHH instruction and D – after the PSHX instruction.

Address	Content		Address	Content		Address	Content		Address	Content
➤ $25F	??		$25F	$12		$25F	$12		$25F	$12
$25E	??		➤ $25E	??		$25E	$34		$25E	$34
$25D	??		$25D	??		➤ $25D	??		$25D	$56
$25C	??		$25C	??		$25C	??		➤ $25C	??
…	…		…	…		…	…		…	…
SP = $25F			SP = $25E			SP = $25D			SP = $25C	
A			B			C			D	

Figure 2.5

After the PSHX instruction, the PULA instruction will increment the SP and pull the value pointed by it, storing it in the accumulator. Note that the content of the memory address $25D is not changed by the PULA instruction.

After the three pull instructions, the SP will point to the address $25F, the accumulator will be equal to $56 and H:X equal to $1234.

2.1.6.3. Arithmetic Instructions

These instructions perform arithmetic operations like addition, subtraction, multiplication, division, increment, decrement and decimal adjustment (for binary-coded-decimal operands). Some instructions (ADC and SBC) can work with three operands: the accumulator, a memory or immediate operand and the carry from the previous operation, allowing easier multi-byte operations like 16 or 32-bit additions or subtractions.

ADD add the operand to the accumulator and store the result into the accumulator.

ADC add the operand plus the carry bit (C) to the accumulator and store the result into the accumulator.

AIS add the immediate signed operand to the SP register and store the result into the SP.

AIX add the immediate signed operand to the H:X register and store the result into the H:X register.

SUB subtract the operand from the accumulator and store the results into the accumulator.

SBC subtract the operand and the carry bit (C) from the accumulator and store the result into the accumulator.

MUL multiply the X register by the accumulator and the result into X (high byte) and A (low byte).

DIV divide a 16-bit value stored in H:A (H is the higher byte and A is the lower byte) by the value stored into the X . The result is stored into the accumulator and the remainder is stored into the X register.

ASL   arithmetic shift left (multiply by two).

ASR   arithmetic shift right (divide by two).

NEG   negate the accumulator, performing the 2's complement of the operand. This is the same of multiplying the value by -1.

INC   increment the register or memory by one.

DEC   decrement the register or memory by one.

DAA   decimal adjustment (adjust the content of the accumulator to the Binary Coded Decimal (BCD) format).

NSA   swap the nibbles of the accumulator (the lower nibble is exchanged with the higher nibble).

As a quick example, let us calculate the average value of the three variables shown in figure 2.4. We will need to add them all and then divide the result by 3. The following code could do this:

```
var_A:      equ 128
var_B:      equ 129
var_C:      equ 130
vresult:    equ 131
LDHX    #0x0003     ; H=0x00, X=0x03
LDA     var_A       ; load A with the value of var_A
ADD     var_B       ; add A to var_B and store the result in A
ADD     var_C       ; add A to var_C and store the result in A
DIV                 ; divide H:A / X -> place the result in A
STA     vresult     ; store the result in vresult
```
Listing 2.5

After the execution of the instructions above, the content of the accumulator (A) will be 6 decimal and the memory will have the following contents:

Address:	127	128	129	130	131
Value:	Unknown	5	5	8	6

Figure 2.6

The following code snippet shows how to add two 16-bit numbers:

```
; Add two 16-bit variables (v1 and v2) and store the result in v3
LDA   v1+1        ; load A with the low byte of v1
ADD   v2+1        ; add A with the low byte of v2
STA   v3+1        ; store the result into the low byte of v3
LDA   v1          ; load A with the high byte of v1 (A = v1)
ADC   v2          ; add A with the carry from the last ADD operation and
                  ; the high byte of v2 (A = A + C + v2)
STA   v3          ; store the result in v3
```
Listing 2.6

 The ADD instruction affects the carry bit, whereas LDA and STA do not. The carry bit is added with v2 and A by the ADC instruction.

Another code snippet to show how to subtract two 16-bit numbers:

```
; Subtract two 16-bit variables (v1 and v2) and store the result in v3
LDA   v1+1        ; load A with the low byte of v1
SUB   v2+1        ; subtract the low byte of v2 from A
STA   v3+1        ; store the result into the low byte of v3
LDA   v1          ; load A with the high byte of v1
SBC   v2          ; subtract the high byte of v2 and the carry bit from A
STA   v3          ; store the result in v3
```
Listing 2.7

 The SUB instruction affects the carry bit, whereas LDA and STA do not. The SBC instruction subtracts the carry bit and v2 from A.

2.1.6.4. Logic Instructions

These instructions can perform logic operations between its operands. The HCS08 ALU implements the following logical operations: logical AND, logical OR, logical Exclusive-OR, logical NOT, bit shift (to the right and to the left, through the carry or not). We also included the bit manipulation instructions BSET and BCLR is this category.

AND logical AND operation between the accumulator and the operand (result into the accumulator).

ORA logical OR operation between the accumulator and the operand (result into the accumulator).

EOR logical Exclusive-OR operation between the accumulator and the operand (result into the accumulator).

COM complement the operand (invert all bits). This is the logical NOT operation.

LSL logically shift the operand one bit to the left.

LSR logically shift the operand one bit to the right.

ROL logically shift the operand one bit to the left through the carry.

ROR logically shift the operand one bit to the right through the carry.

BSET set one bit in memory (only on the direct page area).

BCLR clear one bit in memory (only on the direct page area).

The logical operations can be used in mask operations. Let us take a look at a small example. Assume 0x0000 is the address of the port A data register (PTAD) and that through this register we can read or write to the external pins. Let us suppose we want to read only pins 2, 3 and 7. The following mask operation could be used:

Bit:	7	6	5	4	3	2	1	0
Mask:	1	0	0	0	1	1	0	0
Hex:	8				C			

Figure 2.7

```
LDA  #0x8C          ; load mask into A
AND  PTAD           ; perform logical AND between the mask in A and the PTAD register
```

Another example: let us say we have two variables (aux1 and aux2) and we want to mix them in the following way: the result will have the lower nibble of aux1 and the lower nibble of aux2 (see the figure below).

Bit:	7	6	5	4	3	2	1	0
aux1:	a	b	c	d	e	f	g	h
aux2:	i	j	k	l	m	n	o	p
result:	m	n	o	p	e	f	g	h

Figure 2.8

To achieve this, we could do:

```
LDA   aux1             ; load A with the value of aux1
AND   #0x0F            ; logical AND of A and 0x0F
; now the bits in the accumulator are equal to 0000efgh
STA   result          ; store A into result
; the bits in result are equal to 0000efgh
LDA   aux2            ; load A with the value of aux2
; the bits in the accumulator are equal to ijklmnop
NSA                   ; swap the nibbles of the accumulator
; the bits in the accumulator are equal to mnopijkl
```

```
AND     #0xF0                   ; logical AND of A and 0xF0
; the bits in the accumulator are equal to mnop0000
ORA     result                  ; logical OR of A and result
; the bits in the accumulator are equal to mnopefgh
STA     result                  ; store A into result
```
Listing 2.8

2.1.6.5. Compare and Test Instructions

These instructions are used to make decisions on the program based on the comparison between register and memory values. They are usually followed by change-of-flow instructions that change the program flow according to the result of the comparison. The comparison is made by the ALU by subtracting one operand from the other. None of the operands is changed in any way; only the CCR flags are changed by the operation.

CMP   compare the accumulator to the operand in memory and modify the flags accordingly.

CPX   compare the X register to the operand in memory and modify the flags accordingly.

CPHX   compare the H:X register to the 16-bit operand in memory and modify the flags accordingly.

BIT   compare the accumulator and the operand by using an AND operation and modify the flags accordingly.

TST   test the value in memory (compare to zero).

Examples:
```
LDA #9                  ; load A with 9
CMP #15                 ; compare A with 15
```

The CMP instruction above actually subtracts 15 from the accumulator and modifies the CCR flags accordingly. Let us take a look at the resulting flag states:

$$
\begin{array}{cc}
9 & 0x09 \\
-\;15 & 0x0F \\
\hline
-6 & 0xFA \\
\end{array}
$$
$$C = 1, \; Z = 0, \; N = 1 \text{ and } V = 0$$

This means that the compared value (A, which is equal to 9) is lower than the comparing value (15). The change-of-flow instructions we will see in the next topic can then be used to branch according to the result of the comparison.

 On the HCS08 devices the CPHX instruction is expanded and supports the following addressing modes: immediate, direct, extended, H:X indexed, H:X indexed with 8-bit offset, H:X indexed with 16-bit offset, SP relative with 8-bit offset and SP relative with 16-bit offset. On the HC08 devices only two modes are supported: immediate and direct.

2.1.6.6. Test and Change-of-flow Instructions

The change-of-flow instructions are used to change the sequential flow of the program by directly changing the PC value. There are two main types of change-of-flow instructions: branch/jump instructions and subroutine instructions. The main difference between them is that the subroutine instructions use the stack to store the return address, allowing the restoration of the original flow after the subroutine execution.

The branch instructions can also be classified into two categories: conditional branches and non-conditional branches. Conditional branches change the program flow if and only if the tested condition is true. In the non-conditional branches, no condition is analyzed to determinate if the branch is taken or not. All branch instructions use relative addressing mode and are limited to ranges from -128 to +127 bytes from the current PC. Relative addressing is characterized by adding a signed constant to the current PC. The resulting value is used as the new value for PC (the destination address). For farther branches, it is possible to use the JMP (jump) instruction.

BRA   branch always. Branch to relative address (-128 to +127 from the current PC).

BRN   branch never. This instruction is nearly useless. It solely provides a complement for the branch always (BRA) instruction. The BRN instruction can be used as a 2-byte and 3-cycle NOP.

There are conditional branches for all flags of the CCR register:

BEQ   branch if equal. The branch is taken if Z=1.

BNE   branch if not equal. The branch is taken if Z=0.

BPL   branch if plus. The branch is taken if the result of the last operation is positive (N=0).

BMI   branch if minus. The branch is taken if the result of the last operation is negative (N=1).

BCC   branch if the carry is clear. The branch is taken if C=0.

BCS   branch if the carry is set. The branch is taken if C=1.

BHCC branch if the half carry is clear. The branch is taken if H=0.

BHCS branch if the half carry is set. The branch is taken if H=1.

BMC   branch if the interrupt mask is clear (I=0). The branch is taken if interrupts are enabled.

BMS   branch if the interrupt mask is set (I=1). The branch is taken if interrupts are disabled.

The HCS08 instruction set also supports testing for signed and unsigned operation results. The usage of one or another depends entirely on the operands of the previous arithmetic operation (or the last operation that modified the flags).

For unsigned test operations the following instructions can be used:

BHI   branch if higher. The branch is taken if C and Z are both cleared (C | Z = 0).

BLO   branch if lower. The branch is taken if C=1.

BHS   branch if higher or same. The branch is taken if C=0. This is equivalent to BCC.

BLS   branch if lower or same. The branch is taken if C or Z are set (C | Z = 1).

For signed test operations the following instructions can be used:

BGT   branch if greater than. The branch is taken if Z=N=V=0.

BLT   branch if less than. The branch is taken if N≠V.

BGE   branch if greater than or equal to. The branch is taken if N=V.

BLE   branch if less than or equal to. The branch is taken if Z, N or V = 1.

Additionally, there are two branch instructions that test the IRQ pin. These instructions can only be used when the pin is configured for the IRQ function. We will see how to configure the pin later in this book.

BIL   branch if the IRQ pin is low (IRQ = 0).

BIH   branch if the IRQ pin is high (IRQ = 1).

The HCS08 instruction set also includes two instructions for testing bits on the direct page area.

BRCLR branch if the specified bit in the specified direct page memory address is clear.

BRSET branch if the specified bit in the specified direct page memory address is set.

The HCS08 CPU also includes a special decrement-and-branch instruction that can improve software loops.

DBNZ decrement the register or the content of the specified memory address and branch to the specified address if the result is different from zero.

There is also a jump instruction that can be used to branch to an absolute 16-bit address, enabling the program to jump anywhere within the 64 KiB memory area (notice that some addresses are illegal and can cause a CPU reset. We will see more on this topic later on).

JMP jump anywhere within the 65,536 possible memory addresses.

For subroutine handling there are four instructions (none of them in a conditional form). All subroutine calling instructions perform two operations: first the address of the next instruction (the one that immediately follows the branch/jump instruction) is pushed onto the stack and then the program branches or jumps to the new location. The return from the subroutine is done by simply pulling the return address from the stack and storing it into the PC register.

BSR branch to subroutine. This instruction branches to a subroutine located -128 to +127 positions away from the current PC. The return address is stored into the stack.

JSR jump to subroutine. This instruction jumps to a subroutine located anywhere within the addressable area (64 KiB).

RTS return from subroutine. This instruction performs the return to the original flow of the program. The return address is pulled off the stack and stored into the PC register. After completing the instruction, the flow restarts on the instruction following the BSR/JSR.

RTI return from interrupt. This instruction works like the RTS, but also restores the PC, X, A and CCR registers with their previous values stored into the stack. We will learn more about interrupts in topic 2.1.7.

Examples:

Implementing **if** statements:

```
; if (VAR==5) do_something();
LDA     VAR              ; load A with the VAR value
CMP     #5               ; compare A with 5 (set Z if they are equal)
; The CMP instruction subtracts 5 from the value of the accumulator
; (the current content of the accumulator is not changed) and set the V,
; N, Z and C flags accordingly
BNE     next             ; branch to SKIP if Z=0 (A not equal to 5)
BSR     do_something     ; call do_something (A is equal to 5)
SKIP:
```
Listing 2.9

```
; if (VA<VB) do_something();
LDA     VB               ; load A with the VB value
CMP     VA               ; compare A with the VA value from memory
BLS     SKIP             ; branch to SKIP if less (C=1) or same (Z=1)
BSR     do_something     ; call do_something (VA is lower than VB)
SKIP:
```
Listing 2.10

```
; if (VA>VB) do_something();
LDA     VB               ; load A with the VB value
CMP     VA               ; compare A with the VA value from memory
BCC     SKIP             ; branch to SKIP if greater than (C=0)
BSR     do_something     ; call do_something (VA is higher than VB)
SKIP:
```
Listing 2.11

2.1.6.7. Internal-Control Instructions

These instructions are used to control the CPU and the ALU.

SEC   set the C flag on the CCR register.

CLC   clear the C flag on the CCR register.

SEI   set the interrupt mask (I) disabling the interrupts.

CLI   clear the interrupt mask (I) enabling the interrupts.

BGND   stop the program and enter debug mode if bit ENBDM = 1 (register BDCSCR). If the background debug mode is disabled (ENBDM = 0) this instruction is treated as an illegal opcode.

 Trying to execute an illegal opcode causes a system reset!

NOP   no operation. This instruction does not perform any operation, spending one bus cycle of the CPU time.

SWI   software interrupt. This instruction changes the program flow branching to the address on the interrupt vector 1. More on this subject can be found in topic 2.1.7.

WAIT   clear I and enter wait mode. The wait mode is a low power mode that stops the CPU, while the other peripherals keep operating normally. All memory and register contents are preserved and the chip keeps waiting for an interrupt to restore its normal operation.

STOP   clear I and enter stop mode. The stop mode is a low power mode with lower power consumption than the wait mode. On the HCS08 there are up to three different stop modes: stop1, stop2 and stop3. The stop3 mode works on the same way as the stop mode on the HC08 devices. The stop2 and stop1 modes are two new modes exclusive to the HCS08 devices and allow lower power consumption (the stop1 mode is only available on low-voltage devices).

Further information about wait and stop modes will be presented in chapter 8.

 The STOP instruction has a unique feature: it can be disabled and treated as an illegal opcode, disabling the stop modes. The STOP instruction must be enabled prior to being used. This can be done by setting the bit STOPE in the SOPT1 register. In topic 5.5 we further discuss the system control registers .

2.1.6.8. Instruction Timing

The instruction timing is the amount of time needed to completely execute one instruction. This timing depends on the clock frequency and the number of clock cycles needed to fetch, decode and run one instruction (also known as clock per instruction or CPI).

On the HCS08, the CPU is clocked directly by the main oscillator output (OSCOUT). This signal is also divided by two, originating the bus clock (BUSCLK) signal, which clocks the internal HCS08 peripherals.

The HCS08 CISC architecture is based on a variable length instruction word approach, meaning that the instructions can have a variable size (from one and up to four bytes long) thus, depending on the instruction and the addressing mode, the HCS08 CPU may need at least one and up to eleven clock cycles to fully execute these instructions.

Despite being almost identical to the HC08 instruction set, the HCS08 instruction set presents some noticeable differences in the instruction timing. The upgrade from an 8 MHz bus clock to the 20 MHz one made it necessary to redesign the CPU microcode. To achieve the new timing constraints some changes were made on the decoding and execution mechanisms: the instruction queue was increased from 1-byte (on the HC08) to 2-bytes (on the HCS08) and the number of clock cycles for some instructions was changed (some of them increased by one whereas others decreased by one or more clock cycles).

On the HCS08 the instruction timing is measured in BUSCLK cycles. At full speed (OSCOUT = 50 MHz, BUSCLK = 25 MHz) the fastest instructions (single-byte ones, that only modify internal registers) need only 40 ns to get fully executed, whereas the slowest instruction (the single-byte/eleven-clocks software interrupt (SWI) instruction) needs 440 ns.

We can average the HCS08 instruction timing to something around 4 BUSCLK cycles per instruction. Thus, at full clock speed (BUSCLK = 25 MHz), the average time is 160 ns or 6.25 million instructions per second (6.25 MIPS). Of course, this average is application-dependent, varying from one application to another.

2.1.7. Interrupts

An interrupt is an external event that can cause a change in the program flow. Interrupts can be thought as subroutines as they are very similar to them. The main difference is that interrupts are called by the hardware whereas subroutines are called by the application.

On the HCS08, interrupts are divided into 32 different sources, each one with its own vector address. The vector address is the address of the interrupt service routine (ISR) written by the programmer. The ISR is automatically called by the hardware as soon as the event is detected. For this, two conditions must be met:

1. The global interrupt mask (I) must be clear.

2. The peripheral interrupt enable bit must be set.

 The SWI instruction is a special kind of interrupt as it always asserts an interrupt signal to the CPU, independently of the mask state (I)!

Figure 2.9 shows a simplified diagram of the HCS08 interrupt system. Each interrupt source comprises one enable bit and one flag bit. Those bits are connected to AND gates and ORed together into a large OR gate. The output of this OR gate goes to logic level "1" every time an interrupt is asserted. This signal is ANDed with the mask bit I and then it feeds the CPU. It is easy to understand that the CPU interrupt input is asserted only when I=0 and any group of flag/enable bits are set.

If multiple interrupts are pending, the one with the lowest vector number is serviced first, having priority over those with higher vectors. Once the higher priority interrupt is serviced, the other pending interrupts are serviced, according to the order of their interrupt vectors.

On the HCS08, vector 0 (RESET) has the highest priority, whereas vector 31 has the lowest priority.

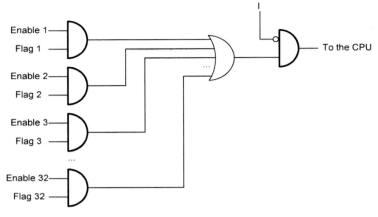

Figure 2.9 – Simplified interrupt structure

Once the conditions are met and an interrupt arrives into the CPU core, it starts some standard procedures:

1. The current instruction is completed;

2. The address of the next instruction (PC) is pushed onto the stack (the lower byte first and then the higher byte).

3. The index register X is pushed onto the stack.

4. The accumulator A is pushed onto the stack.

5. The CCR register is pushed onto the stack.

6. Interrupts are disabled (I=1).

7. The interrupt vector corresponding to the interrupt source that asserted the interrupt is fetched.

8. The program flow diverts to the address of the ISR.

 The H register is not automatically preserved by the hardware! This is to maintain software compatibility with the former HC05 devices. User is advised to store the H content into the stack at the beginning of the ISR and to restore it before returning from the interrupt.

Once the interrupt is serviced, the ISR code should issue an RTI instruction which returns the program flow to the point it had left when the interrupt happened.

The RTI instruction triggers the following sequence:

1. The CCR register is pulled from the stack.

2. The accumulator is pulled from the stack.

3. The index register X is pulled from the stack.

4. The return address is pulled from the stack (the higher byte first and then the lower byte) and stored into the PC.

5. The program flow returns to the point it had left when the interrupt was asserted.

The following tables show the available interrupt sources and their respective vector number for the some HCS08 devices.

Vector	Address	Source	C Vector Symbol	Flag	Enable Bit
0	0xFFFE	RESET	VectorNumber_Vreset	-	-
1	0xFFFC	SWI	VectorNumber_Vswi	-	-
2	0xFFFA	IRQ	VectorNumber_Virq	IRQSC:IRQF	IRQSC:IRQIE
3	0xFFF8	LVD	VectorNumber_Vlvd	SPMSC1:LVDF	SPMSC1:LVDIE
4	0xFFF6	ICG	VectorNumber_Vicg	ICGS1:ICGIF	ICGC2:LOLRE ICGC2:LOCRE
5	0xFFF4	TPM1 channel 0	VectorNumber_Vtpm1ch0	TPM1C0SC:CH0F	TPM1C0SC:CH0IE
6	0xFFF2	TPM1 channel 1	VectorNumber_Vtpm1ch1	TPM1C1SC:CH1F	TPM1C1SC:CH1IE
7	0xFFF0	TPM1 channel 2	VectorNumber_Vtpm1ch2	TPM1C2SC:CH2F	TPM1C2SC:CH2IE
8	0xFFEE	TPM1 channel 3	VectorNumber_Vtpm1ch3	TPM1C3SC:CH3F	TPM1C3SC:CH3IE
9	0xFFEC	TPM1 channel 4	VectorNumber_Vtpm1ch4	TPM1C4SC:CH4F	TPM1C4SC:CH4IE
10	0xFFEA	TPM1 channel 5	VectorNumber_Vtpm1ch5	TPM1C5SC:CH5F	TPM1C5SC:CH5IE
11	0xFFE8	TPM1 overflow	VectorNumber_Vtpm1ovf	TPM1SC:TOF	TPM1SC:TOIE
12	0xFFE6	TPM2 channel 0	VectorNumber_Vtpm2ch0	TPM2C0SC:CH0F	TPM2C0SC:CH0IE
13	0xFFE4	TPM2 channel 1	VectorNumber_Vtpm2ch1	TPM2C1SC:CH1F	TPM2C1SC:CH1IE
14	0xFFE2	TPM2 overflow	VectorNumber_Vtpm2ovf	TPM2SC:TOF	TPM2SC:TOIE
15	0xFFE0	SPI	VectorNumber_Vspi	SPI1S:SPRF SPI1S:MODF SPI1S:SPTEF	SPI1C1:SPIE SPI1C1:SPIE SPI1C1:SPTIE
16	0xFFDE	SCI1 error	VectorNumber_Vsci1err	SCI1S1:OR SCI1S1:NF SCI1S1:FE SCI1S1:PF	SCI1C3:ORIE SCI1C3:NFIE SCI1C3:FEIE SCI1C3:PFIE
17	0xFFDC	SCI1 RX	VectorNumber_Vsci1rx	SCI1S1:IDLE SCI1S1:RDRF	SCI1C2:ILIE SCI1C2:RIE
18	0xFFDA	SCI1 TX	VectorNumber_Vsci1tx	SCI1S1:TDRE SCI1S1:TC	SCI1C2:TIE SCI1C2:TCIE
19	0xFFD8	SCI2 error	VectorNumber_Vsci2err	SCI2S1:OR SCI2S1:NF SCI2S1:FE SCI2S1:PF	SCI2C3:ORIE SCI2C3:NFIE SCI2C3:FEIE SCI2C3:PFIE
20	0xFFD6	SCI2 RX	VectorNumber_Vsci2rx	SCI2S1:IDLE SCI2S1:RDRF	SCI2C2:ILIE SCI2C2:RIE
21	0xFFD4	SCI2 TX	VectorNumber_Vsci2tx	SCI2S1:TDRE SCI2S1:TC	SCI2C2:TIE SCI2C2:TCIE
22	0xFFD2	KBI1	VectorNumber_Vkeyboard1	KBI1SC:KBF	KBI1SC:KBIE
23	0xFFD0	ADC	VectorNumber_Vadc	ADCSC1:COCO	ADCSC1:AIEN
24	0xFFCE	I²C	VectorNumber_Viic	IIC1S:IICIF	IIC1C:IICIE
25	0xFFCC	RTI	VectorNumber_Vrti	SRTISC:RTIF	SRTISC:RTIE
26	0xFFCA	not implemented	-	-	-
27	0xFFC8	not implemented	-	-	-
28	0xFFC6	not implemented	-	-	-
29	0xFFC4	not implemented	-	-	-
30	0xFFC2	not implemented	-	-	-
31	0xFFC0	not implemented	-	-	-

Table 2.1 – AW interrupt vector table

Vector	Address	Source	C Vector Symbol	Flag	Enable Bit
0	0xFFFE	RESET	VectorNumber_Vreset	-	-
1	0xFFFC	SWI	VectorNumber_Vswi	-	-
2	0xFFFA	IRQ	VectorNumber_Virq	IRQSC:IRQF	IRQSC:IRQIE
3	0xFFF8	LVD	VectorNumber_Vlvd	SPMSC1:LVDF	SPMSC1:LVDIE
4	0xFFF6	MCG	VectorNumber_Vlol	MCGSC:LOLS	MCGC3:LOLIE
5	0xFFF4	TPM1 channel 0	VectorNumber_Vtpm1ch0	TPM1C0SC:CH0F	TPM1C0SC:CH0IE
6	0xFFF2	TPM1 channel 1	VectorNumber_Vtpm1ch1	TPM1C1SC:CH1F	TPM1C1SC:CH1IE
7	0xFFF0	TPM1 channel 2	VectorNumber_Vtpm1ch2	TPM1C2SC:CH2F	TPM1C2SC:CH2IE
8	0xFFEE	TPM1 channel 3	VectorNumber_Vtpm1ch3	TPM1C3SC:CH3F	TPM1C3SC:CH3IE
9	0xFFEC	TPM1 channel 4	VectorNumber_Vtpm1ch4	TPM1C4SC:CH4F	TPM1C4SC:CH4IE
10	0xFFEA	TPM1 channel 5	VectorNumber_Vtpm1ch5	TPM1C5SC:CH5F	TPM1C5SC:CH5IE
11	0xFFE8	TPM1 overflow	VectorNumber_Vtpm1ovf	TPM1SC:TOF	TPM1SC:TOIE
12	0xFFE6	TPM2 channel 0	VectorNumber_Vtpm2ch0	TPM2C0SC:CH0F	TPM2C0SC:CH0IE
13	0xFFE4	TPM2 channel 1	VectorNumber_Vtpm2ch1	TPM2C1SC:CH1F	TPM2C1SC:CH1IE
14	0xFFE2	TPM2 overflow	VectorNumber_Vtpm2ovf	TPM2SC:TOF	TPM2SC:TOIE
15	0xFFE0	SPI	VectorNumber_Vspi	SPI1S:SPRF SPI1S:MODF SPI1S:SPTEF	SPI1C1:SPIE SPI1C1:SPIE SPI1C1:SPTIE
16	0xFFDE	SCI1 error	VectorNumber_Vsci1err	SCI1S1:OR SCI1S1:NF SCI1S1:FE SCI1S1:PF	SCI1C3:ORIE SCI1C3:NFIE SCI1C3:FEIE SCI1C3:PFIE
17	0xFFDC	SCI1 RX	VectorNumber_Vsci1rx	SCI1S1:IDLE SCI1S1:RDRF SCI1S2:LBKIF SCI1S2:RXEDGIF	SCI1C2:ILIE SCI1C2:RIE SCI1BDH:LBKIE SCI1BDH:RXEDGIE
18	0xFFDA	SCI1 TX	VectorNumber_Vsci1tx	SCI1S1:TDRE SCI1S1:TC	SCI1C2:TIE SCI1C2:TCIE
19	0xFFD8	SCI2 error	VectorNumber_Vsci2err	SCI2S1:OR SCI2S1:NF SCI2S1:FE SCI2S1:PF	SCI2C3:ORIE SCI2C3:NFIE SCI2C3:FEIE SCI2C3:PFIE
20	0xFFD6	SCI2 RX	VectorNumber_Vsci2rx	SCI2S1:IDLE SCI2S1:RDRF SCI2S2:LBKIF SCI2S2:RXEDGIF	SCI2C2:ILIE SCI2C2:RIE SCI2BDH:LBKIE SCI2BDH:RXEDGIE
21	0xFFD4	SCI2 TX	VectorNumber_Vsci2tx	SCI2S1:TDRE SCI2S1:TC	SCI2C2:TIE SCI2C2:TCIE
22	0xFFD2	Port interrupt	VectorNumber_Vport	PTASC:PTAIF PTBSC:PTBIF PTCSC:PTCIF	PTASC:PTAIE PTBSC:PTBIE PTCSC:PTCIE
23	0xFFD0	ADC	VectorNumber_Vadc	ADCSC1:COCO	ADCSC1:AIEN
24	0xFFCE	I²C	VectorNumber_Viic	IIC1S:IICIF	IIC1C:IICIE
25	0xFFCC	RTC	VectorNumber_Vrti	RTCSC:RTIF	RTCSC:RTIE
26	0xFFCA	not implemented	-	-	-
…	…	…	…	…	…
31	0xFFC0	not implemented	-	-	-

Table 2.2 – DN, DV and DZ interrupt vector table

Vector	Address	Source	C Vector Symbol	Flag	Enable Bit
0	0xFFFE	RESET	VectorNumber_Vreset	-	-
1	0xFFFC	SWI	VectorNumber_Vswi	-	-
2	0xFFFA	IRQ	VectorNumber_Virq	IRQSC:IRQF	IRQSC:IRQIE
3	0xFFF8	LVD	VectorNumber_Vlvd	SPMSC1:LVDF	SPMSC1:LVDIE
4	0xFFF6	ICG	VectorNumber_Vicg	ICGS1:ICGIF	ICGC2:LOLRE ICGC2:LOCRE
5	0xFFF4	TPM1 channel 0	VectorNumber_Vtpm1ch0	TPM1C0SC:CH0F	TPM1C0SC:CH0IE
6	0xFFF2	TPM1 channel 1	VectorNumber_Vtpm1ch1	TPM1C1SC:CH1F	TPM1C1SC:CH1IE
7	0xFFF0	TPM1 channel 2	VectorNumber_Vtpm1ch2	TPM1C2SC:CH2F	TPM1C2SC:CH2IE
8	0xFFEE	TPM1 overflow	VectorNumber_Vtpm1ovf	TPM1SC:TOF	TPM1SC:TOIE
9	0xFFEC	TPM2 channel 0	VectorNumber_Vtpm2ch0	TPM2C0SC:CH0F	TPM2C0SC:CH0IE
10	0xFFEA	TPM2 channel 1	VectorNumber_Vtpm2ch1	TPM2C1SC:CH1F	TPM2C1SC:CH1IE
11	0xFFE8	TPM2 channel 2	VectorNumber_Vtpm2ch2	TPM2C2SC:CH2F	TPM2C2SC:CH2IE
12	0xFFE6	TPM2 channel 3	VectorNumber_Vtpm2ch3	TPM2C3SC:CH3F	TPM2C3SC:CH3IE
13	0xFFE4	TPM2 channel 4	VectorNumber_Vtpm2ch4	TPM2C4SC:CH4F	TPM2C4SC:CH4IE
14	0xFFE2	TPM2 overflow	VectorNumber_Vtpm2ovf	TPM2SC:TOF	TPM2SC:TOIE
15	0xFFE0	SPI	VectorNumber_Vspi	SPI1S:SPRF SPI1S:MODF SPI1S:SPTEF	SPI1C1:SPIE SPI1C1:SPIE SPI1C1:SPTIE
16	0xFFDE	SCI1 error	VectorNumber_Vsci1err	SCI1S1:OR SCI1S1:NF SCI1S1:FE SCI1S1:PF	SCI1C3:ORIE SCI1C3:NFIE SCI1C3:FEIE SCI1C3:PFIE
17	0xFFDC	SCI1 RX	VectorNumber_Vsci1rx	SCI1S1:IDLE SCI1S1:RDRF	SCI1C2:ILIE SCI1C2:RIE
18	0xFFDA	SCI1 TX	VectorNumber_Vsci1tx	SCI1S1:TDRE SCI1S1:TC	SCI1C2:TIE SCI1C2:TCIE
19	0xFFD8	SCI2 error	VectorNumber_Vsci2err	SCI2S1:OR SCI2S1:NF SCI2S1:FE SCI2S1:PF	SCI2C3:ORIE SCI2C3:NFIE SCI2C3:FEIE SCI2C3:PFIE
20	0xFFD6	SCI2 RX	VectorNumber_Vsci2rx	SCI2S1:IDLE SCI1S1:RDRF	SCI2C2:ILIE SCI1C2:RIE
21	0xFFD4	SCI2 TX	VectorNumber_Vsci2tx	SCI2S1:TDRE SCI2S1:TC	SCI2C2:TIE SCI2C2:TCIE
22	0xFFD2	KBI1	VectorNumber_Vkeyboard1	KBI1SC:KBF	KBI1SC:KBIE
23	0xFFD0	ATD	VectorNumber_Vatd1	ATD1SC:CCF	ATD1SC:ATDIE
24	0xFFCE	I^{2}C	VectorNumber_Viic	IIC1S:IICIF	IIC1C:IICIE
25	0xFFCC	RTI	VectorNumber_Vrti	SRTISC:RTIF	SRTISC:RTIE
26	0xFFCA	not implemented	-	-	-
27	0xFFC8	not implemented	-	-	-
28	0xFFC6	not implemented	-	-	-
29	0xFFC4	not implemented	-	-	-
30	0xFFC2	not implemented	-	-	-
31	0xFFC0	not implemented	-	-	-

Table 2.3 – GB and GT interrupt vector table

Vector	Address	Source	C Vector Symbol	Flag	Enable Bit
0	0xFFFE	RESET	VectorNumber_Vreset	-	-
1	0xFFFC	SWI	VectorNumber_Vswi	-	-
2	0xFFFA	LVD	VectorNumber_Vlvd	SPMSC1:LVDF	SPMSC1:LVDIE
3	0xFFF8	IRQ	VectorNumber_Virq	IRQSC:IRQF	IRQSC:IRQIE
4	0xFFF6	not implemented	-	-	-
5	0xFFF4	TPM1 channel 0	VectorNumber_Vtpm1ch0	TPM1C0SC:CH0F	TPM1C0SC:CH0IE
6	0xFFF2	TPM1 channel 1	VectorNumber_Vtpm1ch1	TPM1C1SC:CH1F	TPM1C1SC:CH1IE
7	0xFFF0	TPM1 overflow	VectorNumber_Vtpm1ovf	TPM1SC:TOF	TPM1SC:TOIE
8	0xFFEE	TPM2 channel 0	VectorNumber_Vtpm2ch0	TPM2C0SC:CH0F	TPM2C0SC:CH0IE
9	0xFFEC	not implemented	-	-	-
10	0xFFEA	TPM2 overflow	VectorNumber_Vtpm2ovf	TPM2SC:TOF	TPM2SC:TOIE
11	0xFFE8	not implemented	-	-	-
12	0xFFE6	not implemented	-	-	-
13	0xFFE4	not implemented	-	-	-
14	0xFFE2	not implemented	-	-	-
15	0xFFE0	not implemented	-	-	-
16	0xFFDE	not implemented	-	-	-
17	0xFFDC	not implemented	-	-	-
18	0xFFDA	KBI	VectorNumber_Vkeyboard	KBISC:KBF	KBISC:KBIE
19	0xFFD8	ADC	VectorNumber_Vadc	ADCSC1:COCO	ADCSC1:AIEN
20	0xFFD6	not implemented	-	-	-
21	0xFFD4	not implemented	-	-	-
22	0xFFD2	not implemented	-	-	-
23	0xFFD0	RTI	VectorNumber_Vrti	SRTISC:RTIF	SRTISC:RTIE
24	0xFFCE	not implemented	-	-	-
25	0xFFCC	not implemented	-	-	-
26	0xFFCA	not implemented	-	-	-
27	0xFFC8	not implemented	-	-	-
28	0xFFC6	not implemented	-	-	-
29	0xFFC4	not implemented	-	-	-
30	0xFFC2	not implemented	-	-	-
31	0xFFC0	not implemented	-	-	-

Table 2.4 – QD interrupt vector table

Vector	Address	Source	C Vector Symbol	Flag	Enable Bit
0	0xFFFE	RESET	VectorNumber_Vreset	-	-
1	0xFFFC	SWI	VectorNumber_Vswi	-	-
2	0xFFFA	IRQ	VectorNumber_Virq	IRQSC:IRQF	IRQSC:IRQIE
3	0xFFF8	LVD	VectorNumber_Vlvd	SPMSC1:LVDF	SPMSC1:LVDIE
4	0xFFF6	not implemented	-	-	-
5	0xFFF4	TPM channel 0	VectorNumber_Vtpmch0	TPMC0SC:CH0F	TPMC0SC:CH0IE
6	0xFFF2	TPM channel 1	VectorNumber_Vtpmch1	TPMC1SC:CH1F	TPMC1SC:CH1IE
7	0xFFF0	TPM overflow	VectorNumber_Vtpmovf	TPMSC:TOF	TPMSC:TOIE
8	0xFFEE	not implemented	-	-	-
9	0xFFEC	not implemented	-	-	-
10	0xFFEA	not implemented	-	-	-
11	0xFFE8	not implemented	-	-	-
12	0xFFE6	MTIM overflow	VectorNumber_Vmtim	MTIMSC:TOF	MTIMSC:TOIE
13	0xFFE4	SPI	VectorNumber_Vspi	SPIS:SPRF SPIS:MODF SPIS:SPTEF	SPIC1:SPIE SPIC1:SPIE SPIC1:SPTIE
14	0xFFE2	SCI error	VectorNumber_Vscierr	SCIS1:OR SCIS1:NF SCIS1:FE SCIS1:PF	SCIC3:ORIE SCIC3:NFIE SCIC3:FEIE SCIC3:PFIE
15	0xFFE0	SCI RX	VectorNumber_Vscirx	SCIS1:IDLE SCIS1:RDRF	SCIC2:ILIE SCIC2:RIE
16	0xFFDE	SCI TX	VectorNumber_Vscitx	SCIS1:TDRE SCIS1:TC	SCIC2:TIE SCIC2:TCIE
17	0xFFDC	I²C	VectorNumber_Viic	IICS:IICIF	IICC:IICIE
18	0xFFDA	KBI	VectorNumber_Vkeyboard	KBISC:KBF	KBISC:KBIE
19	0xFFD8	ADC	VectorNumber_Vadc	ADCSC1:COCO	ADCSC1:AIEN
20	0xFFD6	ACMP	VectorNumber_Vacmp	ACMPSC:ACF	ACMPSC:ACIE
21	0xFFD4	not implemented	-	-	-
22	0xFFD2	not implemented	-	-	-
23	0xFFD0	RTI	VectorNumber_Vrti	SRTISC:RTIF	SRTISC:RTIE
24	0xFFCE	not implemented	-	-	-
25	0xFFCC	not implemented	-	-	-
26	0xFFCA	not implemented	-	-	-
27	0xFFC8	not implemented	-	-	-
28	0xFFC6	not implemented	-	-	-
29	0xFFC4	not implemented	-	-	-
30	0xFFC2	not implemented	-	-	-
31	0xFFC0	not implemented	-	-	-

Table 2.5 – QG interrupt vector table

Vector	Address	Source	C Vector Symbol	Flag	Enable Bit
0	0xFFFE	RESET	VectorNumber_Vreset	-	-
1	0xFFFC	SWI	VectorNumber_Vswi	-	-
2	0xFFFA	IRQ	VectorNumber_Virq	IRQSC:IRQF	IRQSC:IRQIE
3	0xFFF8	LVD	VectorNumber_Vlvd	SPMSC1:LVDF	SPMSC1:LVDIE
4	0xFFF6	TPM1 channel 0	VectorNumber_Vtpm1ch0	TPM1C0SC:CH0F	TPM1C0SC:CH0IE
5	0xFFF4	TPM1 channel 1	VectorNumber_Vtpm1ch1	TPM1C1SC:CH1F	TPM1C1SC:CH1IE
6	0xFFF2	TPM1 channel 2	VectorNumber_Vtpm1ch2	TPM1C2SC:CH2F	TPM1C2SC:CH2IE
7	0xFFF0	TPM1 overflow	VectorNumber_Vtpm1ovf	TPM1SC:TOF	TPM1SC:TOIE
8	0xFFEE	TPM2 channel 0	VectorNumber_Vtpm2ch0	TPM2C0SC:CH0F	TPM2C0SC:CH0IE
9	0xFFEC	TPM2 channel 1	VectorNumber_Vtpm2ch1	TPM2C1SC:CH1F	TPM2C1SC:CH1IE
10	0xFFEA	TPM2 channel 2	VectorNumber_Vtpm2ch2	TPM2C2SC:CH2F	TPM2C2SC:CH2IE
11	0xFFE8	TPM2 overflow	VectorNumber_Vtpm2ovf	TPM2SC:TOF	TPM2SC:TOIE
12	0xFFE6	SPI2	VectorNumber_Vspi2	SPI2S:SPRF SPI2S:MODF SPI2S:SPTEF	SPI2C1:SPIE SPI2C1:SPIE SPI2C1:SPTIE
13	0xFFE4	SPI1	VectorNumber_Vspi1	SPI1IS:SPRF SPI1S:MODF SPI1S:SPTEF	SPI1C1:SPIE SPI1C1:SPIE SPI1C1:SPTIE
14	0xFFE2	SCI1 error	VectorNumber_Vsci1err	SCI1S1:OR SCI1S1:NF SCI1S1:FE SCI1S1:PF	SCI1C3:ORIE SCI1C3:NFIE SCI1C3:FEIE SCI1C3:PFIE
15	0xFFE0	SCI1 RX	VectorNumber_Vsci1rx	SCI1S1:IDLE SCI1S1:RDRF SCI1S2:LBKIF SCI1S2:RXEDGIF	SCI1C2:ILIE SCI1C2:RIE SCI1BDH:LBKIE SCI1BDH:RXEDGIE
16	0xFFDE	SCI1 TX	VectorNumber_Vsci1tx	SCI1S1:TDRE SCI1S1:TC	SCI1C2:TIE SCI1C2:TCIE
17	0xFFDC	I²C	VectorNumber_Viicx	IICS:IICIF	IICC:IICIE
18	0xFFDA	KBI	VectorNumber_Vkeyboard	KBISC:KBF	KBISC:KBIE
19	0xFFD8	ADC	VectorNumber_Vadc	ADCSC1:COCO	ADCSC1:AIEN
20	0xFFD6	ACMP	VectorNumber_Vacmpx	ACMPSC:ACF	ACMPSC:ACIE
21	0xFFD4	SCI2 error	VectorNumber_Vsci2err	SCI2S1:OR SCI2S1:NF SCI2S1:FE SCI2S1:PF	SCI2C3:ORIE SCI2C3:NFIE SCI2C3:FEIE SCI2C3:PFIE
22	0xFFD2	SCI2 RX	VectorNumber_Vsci2rx	SCI2S1:IDLE SCI2S1:RDRF SCI2S2:LBKIF SCI2S2:RXEDGIF	SCI2C2:ILIE SCI2C2:RIE SCI2BDH:LBKIE SCI2BDH:RXEDGIE
23	0xFFD0	SCI2 TX	VectorNumber_Vsci2tx	SCI2S1:TDRE SCI2S1:TC	SCI2C2:TIE SCI2C2:TCIE
24	0xFFCE	RTC	VectorNumber_Vrtc	RTCSC:RTIF	RTCSC:RTIE
25	0xFFCC	TPM3 channel 0	VectorNumber_Vtpm3ch0	TPM3C0SC:CH0F	TPM3C0SC:CH0IE
26	0xFFCA	TPM3 channel 1	VectorNumber_Vtpm3ch1	TPM3C1SC:CH1F	TPM3C1SC:CH1IE
27	0xFFC8	TPM3 channel 2	VectorNumber_Vtpm3ch2	TPM3C2SC:CH2F	TPM3C2SC:CH2IE
28	0xFFC6	TPM3 channel 3	VectorNumber_Vtpm3ch3	TPM3C3SC:CH3F	TPM3C3SC:CH3IE
29	0xFFC4	TPM3 channel 4	VectorNumber_Vtpm3ch4	TPM3C4SC:CH4F	TPM3C4SC:CH4IE
30	0xFFC2	TPM3 channel 5	VectorNumber_Vtpm3ch5	TPM3C5SC:CH5F	TPM3C5SC:CH5IE
31	0xFFC0	TPM3 overflow	VectorNumber_Vtpm3ovf	TPM3SC:TOF	TPM3SC:TOIE

Table 2.6 – QE interrupt vector table

2.1.7.1. Interrupt Latency

We call interrupt latency to the elapsed time between the interrupt event and the execution of the first instruction of the ISR.

This latency time is the amount of time necessary to complete the current instruction, recognize the interrupt, push registers PC, X, A and CCR onto the stack, decode the interrupt, fetch and execute the first instruction of the ISR.

On the HCS08 devices the minimum interrupt latency is equal to 11 BUSCLK cycles (for the SWI instruction) and the maximum latency is 22 BUSCLK cycles (for an interrupt arriving right at the beginning of a SWI instruction).

2.1.8. Memory Maps

Microcontrollers are composed not only by the CPU, but also by peripherals and memories connected to the CPU. As we said earlier, on the HCS08 microcontrollers, these peripherals and memories are connected to the CPU through common data and address buses (Von Neumann architecture).

The peripheral registers are addressed by the same bus that also addresses RAM and FLASH memories. Some addresses access these registers, while others access RAM or FLASH memory (some addresses may not be implemented on some devices).

Having a single address space simplifies the instruction set because any instruction that can access the memory can also access any peripheral register. A single address space also means that there is no distinction between the program-only and the data-only spaces: it is possible to store data onto FLASH addresses and run instructions onto RAM addresses without using special instructions or special addressing modes.

A single address space also contributes to lower the silicon costs because it has a simpler implementation and uses less silicon space.

That said, figure 2.10 shows the generic map for the whole HCS08 family. As we can see, the memory map is very simple, with some areas allocated for specific uses (registers, RAM, FLASH and vectors). The HCS08 memory map is very similar to that of the HC08, with a major difference: there is no ROM monitor (it was replaced by the BDM).

0x0000 to 0x007F	Direct page registers
0x0080 to 0x00FF	Direct page RAM
0x0100 to 0x17FF	RAM/FLASH/Unused
0x1800 to 0x186F	High page registers
0x1870 to 0xFFAF	FLASH/Unused
0xFFB0 to 0xFFBF	Nonvolatile registers
0xFFC0 to 0xFFFF	Reset and interrupt vectors

Figure 2.10

Starting on address 0x0000 and up to address 0x00FF we find the direct page area. This name is due to the addressable range of the direct addressing mode. Access to this area is faster than to any other areas of the memory map.

The direct page area is divided into two portions:

1. The direct page registers area, located from address 0x0000 up to 0x007F is where most peripheral registers are located.

2. The direct page memory area, located from address 0x0080 up to address 0x00FF is occupied by the direct page RAM area. User is advised to store frequently used variables into this area for faster access (the C compiler does this by default when using the tiny memory model).

Table 2.7 shows a more complete memory map for some of the HCS08 family devices.

Address	Model					
	QD4/QG4	QG8	AW60	GB/GT60A	LC60	RC/RD/RE/RG60
0x0000						Direct Page Registers
0x0045	Direct Page Registers	Direct Page Registers			Direct Page Registers	
0x0046			Direct Page Registers			
0x005F				Direct Page Registers		
0x0060						
0x006F						
0x0070	RAM (256 bytes)					
0x007F		RAM (512 bytes)				RAM (2,048 bytes)
0x0080						
0x015F			RAM (2,048 bytes)			
0x0160					RAM (4,096 bytes)	
0x025F						
0x0260						
0x0845				RAM (4,096 bytes)		
0x0846						
0x086F	Unused (5,792 bytes)					
0x0870		Unused (5,536 bytes)				
0x105F						FLASH (4,026 bytes)
0x1060			FLASH (3,984 bytes)			
0x107F					FLASH B (1,952 bytes)	
0x1080				FLASH (1,920 bytes)		
0x17FF						
0x1800				High page Registers		High page Registers
0x182B	High page Registers	High page Registers				
0x182C			High page Registers			
0x184F					High page Registers	
0x1850						
0x185F						
0x1860						
0x186F		Unused (51,120 bytes)				
0x1870					FLASH B (26,512 bytes)	
0x7FFF	Unused (55,216 bytes)			FLASH (59,268 bytes)		FLASH (59,268 bytes)
0x8000			FLASH (59,216 bytes)			
0xDFFF						
0xE000					FLASH (32,688 bytes)	
0xEFFF		FLASH (8,112 bytes)				
0xF000	FLASH (4,016 bytes)					
0xFFAF						
0xFFB0	Nonvolatile Registers					
0xFFBF						
0xFFC0	Vectors Area					
0xFFFF						

Table 2.7 – HCS08 memory maps

Two other register areas can be found on the HCS08 devices: the high page registers and the nonvolatile registers.

The high page registers area is mainly used by system control registers, but some devices (such as the LC60) also have peripheral registers on it.

The nonvolatile registers area stores the FLASH registers, which are programmed during the FLASH programming. Their values are not lost when the device is powered off.

2.1.9. Debug Module

The HCS08 microcontrollers also include an integrated debug system that allows in-circuit programming and debugging through just one pin.

The debug system comprises the following modules:

- **Background Debug Controller (BDC)** – responsible for accessing the internal resources of the CPU and controlling the communication with the host computer. The BDC is found on all HCS08 devices.

- **On-Chip Debug System (DBG)** – responsible for advanced debug features such as: two (three on QE devices) digital comparators for advanced triggers and breakpoints and an eight-level 16-bit first-in first-out (FIFO) buffer for CPU tracing and debug data storage. The DBG is found on almost all HCS08 devices (current exceptions are QA and QD devices).

The interface between the chip and the host computer is done through a special hardware: the BDM pod. The BDM can be actually a P&E Micro USB Multilink Interface, a P&E Micro Cyclone PRO or other BDM interface.

The connection between the BDM pod and the chip is done through a standard 6-pin header (shown in figure 2.11). The recommended electrical interface with the chip is shown in figure 2.12.

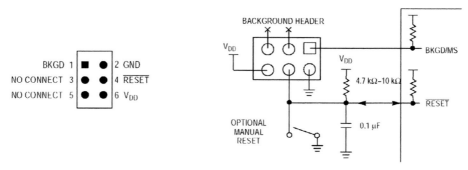

| Figure 2.11 | Figure 2.12 |

 On some devices the BKGD/MS signal is multiplexed with an output-only I/O pin and the RESET signal is multiplexed with an input-only pin. On such devices the function of the BKGD/MS pin is controlled by the BKGDPE bit and the reset function is controlled by the RSTPE bit, both located in the SOPT or SOPT1 register.

2.1.9.1. BKGD/MS Pin

As we said earlier, the communication between the BDM pod (outside the chip) and the BDC module (inside the chip) is carried out by using a pin called BKGD/MS.

This pin actually has two functions: the communication with the external debug hardware (BKGD function) and the selection of the operating mode of the chip (MS function).

While the chip is in reset, the pin acts in MS function and is used to select the mode in which the chip will operate: if debugging hardware is connected to the debug header (figure 2.11), it forces this pin to "0", thus forcing the CPU to the active background debug mode. When nothing is connected to the pin or the pin is pulled high by an external signal, the CPU enters the user program. The mode is selected when the chip leaves the reset state.

As we will see in the next topic, the active background mode implies that the user program execution is suspended.

Anyway, after reset, the BKGD/MS pin acts as the communication port with the external debug hardware, using a proprietary asynchronous half-duplex protocol (16 BDC clock cycles per bit or 1.25 Mbps with a 20 MHz bus clock). On devices with other multiplexed functions on the BKGD/MS pin, it defaults to the BKGD function after reset. The user is able to select the alternate function by writing a "0" onto the BKGDPE bit (SOPT or SOPT1 register).

2.1.9.2. Background Debug Controller (BDC)

The BDC module is responsible for managing the communication with the external BDM pod, processing the serial debug commands, accessing the CPU internal registers and controlling the hardware breakpoint.

There are two debug command categories: the non-intrusive and the intrusive commands.

The non-intrusive commands are mostly used to control the BDC and for reading/writing to the system memory. These commands can be issued even when the background debug mode is not active.

The intrusive commands are used to control the BDC and the program execution as well as to access the CPU registers and read/write into the system memory. These commands can only be issued when the BDC is in the active background debug mode (the program execution is stopped).

All these commands are automatically managed and sent by the debugging application running in the host computer (the Codewarrior IDE) and thus are transparent to the user. All you are required to do is to start a debugging session and use the buttons and windows to control the program flow and inspect the variables.

2.1.9.3. On-Chip Debug Module (DBG)

The higher-end devices include, in addition to the BDC module, an advanced debug module called DBG.

This module adds some important functionality to the BDC such as:

- Two trigger comparators (three on QE devices): can detect read/write operations on two different addresses or a read/write operation of a specific value into a specific memory address. Comparator A always compares its value against the current 16-bit CPU address, Comparator B can compare its value against the current 16-bit CPU address or 8-bit CPU data bus (depending on the selected trigger mode). On QE devices, Comparator C compares its value against the current 16-bit CPU address. Optionally it can be used to track the most recent change-of-flow address.

- 8-level 16-bit wide FIFO memory: allows storage of captured information such as change-of-flow addresses or data. By storing change-of-flow addresses, the DBG module enables the debug application (running on the host) to reconstruct the program flow, thus allowing a privileged view of the program flow before the breakpoint.

- Nine trigger modes: the high flexibility of the DBG module allows nine ways to trigger debug events, making it easy to locate even the hardest bugs. Table 2.8 describes the nine trigger modes.

Trigger Mode	Description
A-only	The debug event is triggered when the current address on the internal address bus matches the comparator A value while in a read or write operation.
A or B	The debug event is triggered when the current address on the internal address bus matches the comparator A or the comparator B value while in a read or write operation.
A then B	The debug event is triggered when the following events happen in order: 1- The current address on the internal address bus matches the comparator A while in a read or write operation. 2- The current address on the internal address bus matches the comparator B while in a read or write operation.
A and B data	The debug event is triggered when the current address on the internal address bus matches the comparator A value, and the current content of the data bus matches the comparator B value while in a read or write operation, all in the same bus cycle. This mode can be used to catch read or write operations into the peripheral registers or variables that match a specific value.
A and not B	The debug event is triggered when the current address on the internal address bus matches the comparator A value, and the current content of the data bus does not match the comparator B value while in a read or write operation, all in the same bus cycle. This mode can be used to catch read or write operations into the peripheral registers or variables, that are different then a specific value.
Event-only B	The debug event is triggered when the current address on the internal address bus matches the comparator A value while in a read or write operation. The debug event causes the value of the CPU data bus to be stored into the FIFO.
A then event-only B	The debug event is triggered when the following events happen in order: 1- The current address on the internal address bus matches the comparator A while in a read or write operation. 2- The current address on the internal address bus matches the comparator B while in a read or write operation. The debug event causes the value of the CPU data bus to be stored into the FIFO.
Inside range	The debug event is triggered when the current address on the internal address bus is greater than or equal to the comparator A value and smaller then or equal to the comparator B value. I.e.: A $\leq$ address $\leq$ B.
Outside range	The debug event is triggered when the current address on the internal address bus is smaller than the comparator A value and greater then the comparator B value. I.e.: address $<$ A or address $>$ B.

Table 2.8

We will learn how to use the advanced resources of the BDG module later in the topic 3.5.

2.2. Flexis Programmer's Model

The new Flexis devices (MC9S08AC96/128, MC9S08JM64/128 and MC9S08QE64/96/128) include some exclusive features. The major one is related to the new addressing scheme that allows accessing up to 4 MiB of memory.

The Flexis HCS08 devices include a memory management unit (MMU) that works much like the one present on HCS12 devices. The original 64 KiB memory map is divided into four 16 KiB pages. Page 2 (addresses between 0x8000 and 0xBFFF) is mapped to one of the MMU pages selected by the PPAGE register. When PPAGE=0, all accesses to address 0x8000 are mapped to 0x00000 on the FLASH memory, address 0x8001 is mapped to 0x00001 and so on.

Pages 0 to 3 are accessed directly by standard addressing modes, pages 4 to 255 can only be accessed through the paging window or by using paged access (through the linear address pointer registers LAP2 to LAP0).

	PPAGE	Paged Access	Direct Access
0x00000 to 0x0007F	0	FLASH (16,384 bytes)	Direct page registers
0x00080 to 0x000FF			Direct page RAM (128 bytes)
0x00100 to 0x017FF			RAM/Reserved (5,888 bytes)
0x01800 to 0x0187F			High page registers
0x01880 to 0x0207F			RAM/Reserved (2,048 bytes)
0x02080 to 0x03FFF			FLASH (8,064 bytes)
0x04000 to 0x07FFF	1		FLASH (16,384 bytes)
0x08000 to 0x0BFFF	2	FLASH (16,384 bytes)	Paging window: access the page pointed by the PPAGE register
0x0C000 to 0x0FFFF	3		FLASH (16,384 bytes)
0x10000 to 0x13FFF	4		FLASH (16,384 bytes)
0x14000 to 0x17FFF	5		FLASH (16,384 bytes)
0x18000 to 0x1BFFF	6		FLASH (16,384 bytes)
0x1C000 to 0x1FFFF	7		FLASH (16,384 bytes)
0x20000 to 0xFFFFF	8 to 255		Reserved

Figure 2.13

The registers used for memory addressing through the MMU are the following:

1. PPAGE – for selection of the active page accessed by the paging window. The page pointed by PPAGE is accessible through addresses 0x8000 to 0xBFFF on the standard programmer's model;

2. LAP0, LAP1 and LAP2 – these registers allow direct addressing of any 17-bit address:

Linear address (bits):	23	22	21	20	19	18	17	16	15	14	13	12	11	10	9	8	7	6	5	4	3	2	1	0
Register bits:	7	6	5	4	3	2	1	0	7	6	5	4	3	2	1	0	7	6	5	4	3	2	1	0
Register:	LAP2								LAP1								LAP0							

On QE64, QE96 and QE128 devices, only address bit 16 is implemented on LAP2 register (a maximum of 128 KiB of memory).

Access to the paged memory can be done through the paging window (addresses 0x8000 to 0xBFFF) or through the linear address pointer registers (LAP). The paging window is mainly used for program execution whereas the linear address pointer registers are used to read/write data into any address of the memory.

Once the variable address is loaded into the LAP registers, it is possible to read or write to the memory by using the LWP, LBP and LB registers:

- LWP – used for reading/writing words into the paged memory (at the address pointed by LAP registers). After each operation, the content of LAP registers is incremented by one;

- LBP – used for reading/writing words into the paged memory (at the address pointed by LAP registers). After each operation, the content of LAP registers is incremented by one;

The LWP and LBP registers are located in the direct address 0x007C and 0x007D respectively. 8-bit access can be done on any of them, whereas 16-bit access must use the LWP register. This enables LDHX and STHX instructions to do 16-bit access on addresses pointed by LAP registers.

- LB – used for reading/writing words into the paged memory (at the address pointed by LAP registers). This register does not post-increment the content of LAP registers;

- LAPAB – writing a 2's complement into this register adds the value to the current LAP registers (LAP = LAP + LAPAB). This is a write-only register and when read it returns all 0s. Each write onto it adds a value to the LAP registers.

Examples:

1. Accessing a variable stored at $10000 (in assembly):

```
ORG $10000
var: DS 1
ORG ROMStart
MOV #$01, LAP2      ; LAP2 = $01
CLR LAP1            ; LAP1 = $00
CLR LAP0            ; LAP0 = $00 (LAP = $10000)
LDA LB             ; load A with the byte value pointed by LAP
```

2. Accessing a variable stored at 0x10000 (in C):

```
LAP0 = 0; LAP1 = 0; LAP2 = 1; // LAP = 0x010000
temp = LB;                    // load temp with the byte value pointed by LAP
```

Or, using the special macros defined in the "mmu_lda.h" file:

```
__LOAD_LAP_ADDRESS(variable); // the address of variable is stored into LAP
temp = LB;                    // load temp with the byte value pointed by LAP
```

3. Manipulating the LAP registers:

```
CLR   LAP2         ; LAP2 = $00
MOV   #$12, LAP1   ; LAP1 = $12
MOV   #$34, LAP0   ; LAP0 = $34 (LAP = $01234)
LDHX LWP           ; load H with value pointed by LAP and X with
                   ; value pointed by LAP+1. After the instruction LAP = $01236
MOV #2, LAPAB      ; Add 2 to LAP (LAP = $01238)
MOV #-6,LAPAB      ; Add -6 to LAP (LAP = $01232)
```

In C:

```
LAP0 = 0x34; LAP1 = 0x12; LAP0 = 0;    // LAP = 0x001234
var16 = LWP;         // load a 16-bit value into var16
LAPAB = 2;           // add 2 to LAP
LAPAB = -6;          // add -6 to LAP
```

4.   Accessing a byte array on paged memory (include the "mmu_lda.h" file):

```
// For arrays up to 127 bytes
__LOAD_LAP_ADDRESS(array_name); // the address of the array is stored into LAP
LAPAB = index;       // for arrays with up to 127 bytes
temp = LB;           // temp = array[index] value

// For arrays larger than 127 bytes:
__LOAD_LAP_ADDRESS(array_name); // the address of the array is stored into LAP
__ADJUST_LAP_IMM_16BIT(index);
temp = LB;           // temp = array[index] value
```

Observe that the MMU registers are located on the direct page registers area, allowing direct page instructions to read/write data into any paged memory. This feature results in a performance improvement since the direct page addressing mode is faster than the extended mode.

The architecture also includes two new instructions:

CALL – for subroutine calling. This instruction acts pretty much like the BSR instruction, but it also pushes the PPAGE register onto the stack and modifies it to the instruction-supplied value, allowing subroutine calls to any valid address within the full 4 MiB space.

The syntax for the instruction is CALL page, 16-bit address

RTC – this is the complement for the CALL instruction. RTC returns from a subroutine in the same manner RTS would do, but it also pulls the PPAGE register from the stack.

2.3. Nomenclature

Freescale's nomenclature system observes the following convention:

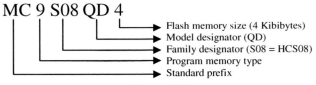

Figure 2.14

According to Freescale's naming system there are three program memory types:

7 – PROM memory (One-time programmable (OTP) devices)

8 – EEPROM memory

9 – FLASH memory

On the HCS08 family only FLASH memory devices are available.

The sub-model designator is used to distinguish specific devices among the family members, with the first letter indicating the general category as shown in the following table:

Model	Description
A	Automotive/industrial applications (2.7 – 5.5V devices): AC – high pin count (up to 80 pin package), improved security and peripherals AW – high pin count (up to 64 pin package)
D	Automotive/industrial applications (2.7 – 5.5V devices): DN – devices with EEPROM DV – devices with CAN DZ – devices with CAN and EEPROM

Model	Description
E	General automotive devices with LIN (2.7 to 5.5V): EL – automotive devices with LIN and EEPROM EN – same as EL but without EEPROM
G	General purpose (1.8 – 3.6V devices): GB – high pin count (up to 64 pin package) GT – lower pin count (from 32 up to 48 pins)
J	Other communication interfaces: JM – USB 2.0 full speed device JR – integrated 27 MHz transmitter
L	LCD driver: LC – low-voltage devices, up to 160 LCD segments LG – up to 296 LCD segments LL – low-voltage devices, up to 192 LCD segments
Q	General purpose devices: QA – 1.8 to 3.6V, lowest cost devices (8-pin only devices) QB – 1.8 to 3.6V, general purpose, pin-to-pin compatible with QE devices QE – 1.8 to 3.6V, general purpose, high speed devices QD – 2.7 to 5.5V, low cost HCS08 devices (8-pin only devices) QG – 1.8 to 3.6V, low cost devices
S	Small packages, general purpose (2.7 – 5.5V devices): SG – 8 to 28 pin packages, 4 to 32 KiB FLASH memory SH – 8 to 24 pin packages, improved 5V QG devices SL – automotive devices with LIN and EEPROM
R	Remote control: RC, RD, RE, RG – migration path from HC05 and HC08 devices

Table 2.9

2.4. Available Models

Below there is some information on the currently available HCS08 devices.

Model	V_{DD}	FLASH (KiB)	RAM (byte)	EEPROM (byte)	BUS CLK (MHz)	I/O	Timers / CCP channels	IR	SCI	USB	CAN	I^2C	SPI	ADC (channels/ bits)	AC	Package
AC8 AW8A	H	8	768	-	20	up to 38	3 – 16 bits / 4+2+2 CCP	-	2	-	-	1	1	8 / 10 bits	-	LQFP32, SDIP42, LQFP44, QFN48
AC16 AW16A	H	16	1,024	-	20	up to 38	3 – 16 bits / 4+2+2 CCP	-	2	-	-	1	1	8 / 10 bits	-	LQFP32, SDIP42, LQFP44, QFN48
AW16	H	16	1,024	-	20	up to 34	2 – 16 bits / 6+2 CCP	-	2	-	-	1	1	16 / 10 bits	-	QFN48, LQFP44
AC32	H	32	2,048	-	20	up to 54	3 – 16 bits / 6+2+2 CCP	-	2	-	-	1	1	16 / 10 bits	-	QFP64, LQFP64, QFN48, LQFP44, LQFP32
AW32	H	32	2,048	-	20	up to 54	2 – 16 bits / 6+2 CCP	-	2	-	-	1	1	16 / 10 bits	-	QFP64, LQFP64, QFN48, LQFP44
AC48	H	48	2,048	-	20	up to 54	3 – 16 bits / 6+2+2 CCP	-	2	-	-	1	1	16 / 10 bits	-	QFP64, LQFP64, QFN48, LQFP44, LQFP32
AW48	H	48	2,048	-	20	up to 54	2 – 16 bits / 6+2 CCP	-	2	-	-	1	1	16 / 10 bits	-	QFP64, LQFP64, QFN48, LQFP44
AC60	H	60	2,048	-	20	up to 54	3 – 16 bits / 6+2+2 CCP	-	2	-	-	1	1	16 / 10 bits	-	QFP64, LQFP64, QFN48, LQFP44, LQFP32
AW60	H	60	2,048	-	20	up to 54	2 – 16 bits / 6+2 CCP	-	2	-	-	1	1	16 / 10 bits	-	QFP64, LQFP64, QFN48, LQFP44
AC96	H	96	6,016	-	20	up to 70	3 – 16 bits / 6+2+2 CCP	-	2	-	-	1	1	16 / 10 bits	-	LQFP80, QFP64, LQFP44
AC128	H	128	8,192	-	20	up to 70	3 – 16 bits / 6+2+2 CCP	-	2	-	-	1	1	16 / 10 bits	-	LQFP80, QFP64, LQFP44

Table 2.10 (continued)

Model	V_{DD}	FLASH (KiB)	RAM (byte)	EE PROM (byte)	BUS CLK (MHz)	I/O	Timers / CCP channels	IR	SCI	USB	CAN	I²C	SPI	ADC (channels/ bits)	AC	Package
DN16	H	16	1,024	512	20	up to 54	2 – 16 bits / 6+2 CCP	-	2	-	-	1	1	16 / 10 bits	2	LQFP32, LQFP48
DN32	H	32	2,048	1,024	20	up to 54	2 – 16 bits / 6+2 CCP	-	2	-	-	1	1	24 / 10 bits	2	LQFP32, LQFP48, LQFP64, QFN64
DN48	H	48	3,072	1,536	20	up to 54	2 – 16 bits / 6+2 CCP	-	2	-	-	1	1	24 / 10 bits	2	LQFP32, LQFP48, LQFP64, QFN64
DN60	H	60	4,096	2,048	20	up to 54	2 – 16 bits / 6+2 CCP	-	2	-	-	1	1	24 / 10 bits	2	LQFP32, LQFP48, LQFP64, QFN64
DV16	H	16	1,024	-	20	up to 54	2 – 16 bits / 6+2 CCP	-	2	-	1	1	1	16 / 10 bits	2	LQFP32, LQFP48
DV32	H	32	2,048	-	20	up to 54	2 – 16 bits / 6+2 CCP	-	2	-	1	1	1	24 / 10 bits	2	LQFP32, LQFP48, LQFP64, QFN64
DV48	H	48	3,072	-	20	up to 54	2 – 16 bits / 6+2 CCP	-	2	-	1	1	1	24 / 10 bits	2	LQFP32, LQFP48, LQFP64, QFN64
DV60	H	60	4,096	-	20	up to 54	2 – 16 bits / 6+2 CCP	-	2	-	1	1	1	24 / 10 bits	2	LQFP32, LQFP48, LQFP64, QFN64
DZ16	H	16	1,024	512	20	up to 54	2 – 16 bits / 6+2 CCP	-	2	-	1	1	1	16 / 10 bits	2	LQFP32, LQFP48
DZ32	H	32	2,048	1,024	20	up to 54	2 – 16 bits / 6+2 CCP	-	2	-	1	1	1	24 / 10 bits	2	LQFP32, LQFP48, LQFP64, QFN64
DZ48	H	48	3,072	1,536	20	up to 54	2 – 16 bits / 6+2 CCP	-	2	-	1	1	1	24 / 10 bits	2	LQFP32, LQFP48, LQFP64, QFN64
DZ60	H	60	4,096	2,048	20	up to 54	2 – 16 bits / 6+2 CCP	-	2	-	1	1	1	24 / 10 bits	2	LQFP32, LQFP48, LQFP64, QFN64
EL16	H	16	1,024	512	20	up to 26	2 – 16 bits / 4+2 CCP	-	1	-	-	1	1	16 / 10 bits	2	TSSOP20, TSSOP28
EL32	H	32	1,024	512	20	up to 26	2 – 16 bits / 4+2 CCP	-	1	-	-	1	1	16 / 10 bits	2	TSSOP20, TSSOP28
EN16	H	16	512	-	20	up to 39	1 – 16 bits / 4 CCP	-	1	-	-	-	1	12 / 10 bits	1	LQFP32, LQFP48
EN32	H	32	1,024	-	20	up to 39	1 – 16 bits / 4 CCP	-	1	-	-	-	1	12 / 10 bits	1	LQFP32, LQFP48
GB32A	L	32	2,048	-	20	56	2 – 16 bits / 3+5 CCP	-	2	-	-	1	1	8 / 10 bits	-	QFP64
GB60A	L	60	4,096	-	20	56	2 – 16 bits / 3+5 CCP	-	2	-	-	1	1	8 / 10 bits	-	QFP64
GT8A	L	8	1,024	-	20	up to 39	2 – 16 bits / 3+2 CCP	-	1	-	-	1	1	8 / 10 bits	-	QFN32, SDIP42, QFP44, QFN48
GT16A	L	16	2,048	-	20	up to 39	2 – 16 bits / 3+2 CCP	-	1	-	-	1	1	8 / 10 bits	-	QFN32, SDIP42, QFP44, QFN48
GT32A	L	32	2,048	-	20	up to 39	2 – 16 bits / 2+2 CCP	-	2	-	-	1	1	8 / 10 bits	-	QFP44, QFN48
GT60A	L	60	4,096	-	20	up to 39	2 – 16 bits / 2+2 CCP	-	2	-	-	1	1	8 / 10 bits	-	QFP44, QFN48
JM8	H	8	1,024	-	24	up to 37	2 – 16 bits/ 6+2 CCP	-	2	1	-	1	2	8 / 12 bits	1	QFN48, LQFP44, LQFP32
JM16	H	16	1,024	-	24	up to 37	2 – 16 bits/ 6+2 CCP	-	2	1	-	1	2	8 / 12 bits	1	QFN48, LQFP44, LQFP32
JM32	H	32	2,048	-	24	up to 51	2 – 16 bits/ 6+2 CCP	-	2	1	-	1	2	12 / 12 bits	1	LQFP64, QFP64, QFN48, LQFP44
JM60	H	60	4,096	-	24	up to 51	2 – 16 bits/ 6+2 CCP	-	2	1	-	1	2	12 / 12 bits	1	LQFP64, QFP64, QFN48, LQFP44

Table 2.10 (continued)

Model	V_{DD}	FLASH (KiB)	RAM (byte)	EE PROM (byte)	BUS CLK (MHz)	I/O	Timers / CCP channels	IR	SCI	USB	CAN	I²C	SPI	ADC (channels/ bits)	AC	Package
LC36	L	24+12	4,096	-	20	up to 24	1 – 16 bits / 2 CCP	-	1	-	-	1	2	8 / 12 bits	1	LQFP80, LQFP64
LC60	L	32+28	4,096	-	20	up to 24	1 – 16 bits / 2 CCP	-	1	-	-	1	2	8 / 12 bits	1	LQFP80, LQFP64
LG16	H	16+2	2048	-	40	up to 69	1 – 8 bits + 2 – 16 bits/ 2+6 CCP	-	2	-	-	1	1	16 / 12 bits	-	LQFP80, LQFP64, LQFP48
LG32	H	16+16	2048	-	40	up to 69	1 – 8 bits + 2 – 16 bits/ 2+6 CCP	-	2	-	-	1	1	16 / 12 bits	-	LQFP64, LQFP48
LL8	L	8+2	2080	-	10	up to 38	2 – 16 bits/ 2+2 CCP	-	1	-	-	1	1	8 / 12 bits	1	LQFP48, QFN48
LL16	L	8+8	2080	-	10	up to 38	2 – 16 bits/ 2+2 CCP	-	1	-	-	1	1	8 / 12 bits	1	LQFP64, LQFP48, QFN48
QA2	L	2	160	-	10	up to 6	1 – 8 bits + 1 – 16 bits/ 1+0 CCP	-	-	-	-	-	-	4 / 10 bits	-	DIP8, SOIC8
QA4	L	4	256	-	10	up to 6	1 – 8 bits + 1 – 16 bits/ 1+0 CCP	-	-	-	-	-	-	4 / 10 bits	-	DIP8, SOIC8
QB4	L	4	256	-	10	up to 22	1 – 8 bits + 1 – 16 bits/ 1 CCP	-	1	-	-	1	1	8 / 12 bits	1	TSSOP16, QFN24, SOIC28
QB8	L	8	512	-	10	up to 22	1 – 8 bits + 1 – 16 bits/ 1 CCP	-	1	-	-	1	1	8 / 12 bits	1	TSSOP16, QFN24, SOIC28
QD2	H	2	128	-	10	up to 6	2 – 16 bits / 2+1 CCP	-	-	-	-	-	-	4 / 10 bits	-	DIP8, SOIC8
QD4	H	4	256	-	10	up to 6	2 – 16 bits / 2+1 CCP	-	-	-	-	-	-	4 / 10 bits	-	DIP8, SOIC8
QE4	L	4	256	-	20	up to 26	2 – 16 bits / 3+3 CCP	-	1	-	-	1	1	10 / 12 bits	2	DIP16, TSSOP16, SOIC20, SOIC28, LQFP32
QE8	L	8	512	-	20	up to 26	2 – 16 bits / 3+3 CCP	-	1	-	-	1	1	10 / 12 bits	2	DIP16, TSSOP16, SOIC20, SOIC28, LQFP32
QE16	L	16	1,024	-	25	up to 40	3 – 16 bits / 3+3+6 CCP	-	2	-	-	1	1	10 / 12 bits	2	SOIC28, LQFP32, LQFP44, QFN48
QE32	L	32	2,048	-	25	up to 40	3 – 16 bits / 3+3+6 CCP	-	2	-	-	1	1	10 / 12 bits	2	SOIC28, LQFP32, LQFP44, QFN48
QE64	L	64	4,096	-	25	up to 54	3 – 16 bits / 3+3+6 CCP	-	2	-	-	2	2	22 / 12 bits	2	QFP44, QFN48, LQFP64, LQFP80
QE96	L	96	6,016	-	25	up to 70	3 – 16 bits / 3+3+6 CCP	-	2	-	-	2	2	24 / 12 bits	2	QFP44, QFN48, LQFP64, LQFP80
QE128	L	128	8,064	-	25	up to 70	3 – 16 bits / 3+3+6 CCP	-	2	-	-	2	2	24 / 12 bits	2	QFP44, QFN48, LQFP64, LQFP80
QG4	L	4	256	-	10	up to 12	1 – 8 bits + 1 – 16 bits/ 2+0 CCP	-	1	-	-	1	1	8 / 10 bits	1	DIP8, DFN8, SOIC8, DIP16, QFN16, TSSOP16
QG8	L	8	512	-	10	up to 12	1 – 8 bits + 1 – 16 bits/ 2+0 CCP	-	1	-	-	1	1	8 / 10 bits	1	DIP8, DFN8, SOIC8, DIP16, QFN16, TSSOP16
RC8	H	8	1,024	-	8	up to 39	1 – 16 bits / 2 CCP	1	-	-	-	-	-	-	1	LQFP32, LQFP44
RC16	H	16	1,024	-	8	up to 39	1 – 16 bits / 2 CCP	1	-	-	-	-	-	-	1	LQFP32, LQFP44
RC32	H	32	2,048	-	8	up to 39	1 - 16 bits / 2 CCP	1	-	-	-	-	-	-	1	LQFP32, LQFP44

Table 2.10 (continued)

Model	V_{DD}	FLASH (KiB)	RAM (byte)	EE PROM (byte)	BUS CLK (MHz)	I/O	Timers / CCP channels	IR	SCI	USB	CAN	I²C	SPI	ADC (channels/ bits)	AC	Package
RC60	H	60	2,048	-	8	up to 39	1 - 16 bits / 2 CCP	1	-	-	-	-	-	-	1	LQFP32, LQFP44
RD8	H	8	1,024	-	8	up to 39	1 - 16 bits / 2 CCP	1	1	-	-	-	-	-	-	DIP28, SOIC28, LQFP32, LQFP44
RD16	H	16	1,024	-	8	up to 39	1 - 16 bits / 2 CCP	1	1	-	-	-	-	-	-	DIP28, SOIC28, LQFP32, LQFP44
RD32	H	32	2,048	-	8	up to 39	1 - 16 bits / 2 CCP	1	1	-	-	-	-	-	-	DIP28, SOIC28, LQFP32, LQFP44
RD60	H	60	2,048	-	8	up to 39	1 - 16 bits / 2 CCP	1	1	-	-	-	-	-	-	DIP28, SOIC28, LQFP32, LQFP44
RE8	H	8	1,024	-	8	up to 39	1 - 16 bits / 2 CCP	1	1	-	-	-	-	-	1	DIP28, SOIC28, LQFP32, LQFP44
RE16	H	16	1,024	-	8	up to 39	1 - 16 bits / 2 CCP	1	1	-	-	-	-	-	1	LQFP32, LQFP44, QFN48
RE32	H	32	2,048	-	8	up to 39	1 - 16 bits / 2 CCP	1	1	-	-	-	-	-	1	LQFP32, LQFP44
RE60	H	60	2,048	-	8	up to 39	1 - 16 bits / 2 CCP	1	1	-	-	-	-	-	1	LQFP32, LQFP44
RG32	H	32	1,024	-	8	up to 39	1 - 16 bits / 2 CCP	1	1	-	-	-	1	-	1	LQFP32, LQFP44
RG60	H	60	2,048	-	8	up to 39	1 - 16 bits / 2 CCP	1	1	-	-	-	1	-	1	LQFP32, LQFP44
SG4	H	4	256	-	20	up to 18	2 – 16 bits + 1 – 8 bits/ 2+2+0 CCP	-	1	-	-	1	1	12 / 10 bits	1	SOIC8, TSSOP16, TSSOP20
SG8	H	8	512	-	20	up to 18	2 – 16 bits + 1 – 8 bits/ 2+2+0 CCP	-	1	-	-	1	1	12 / 10 bits	1	SOIC8, TSSOP16, TSSOP20
SG16	H	16	1,024	-	20	up to 26	2 – 16 bits + 1 – 8 bits/ 2+2+0 CCP	-	1	-	-	1	1	16 / 10 bits	1	TSSOP16, 20, 28
SG32	H	32	1,024	-	20	up to 26	2 – 16 bits + 1 – 8 bits/ 2+2+0 CCP	-	1	-	-	1	1	16 / 10 bits	1	TSSOP16, 20, 28
SH4	H	4	256	-	20	up to 17	2 – 16 bits + 1 – 8 bits/ 2+2+0 CCP	-	1	-	-	1	1	12 / 10 bits	1	TSSOP16, 20, DIP20, QFN24
SH8	H	8	512	-	20	up to 17	2 – 16 bits + 1 – 8 bits/ 2+2+0 CCP	-	1	-	-	1	1	12 / 10 bits	1	TSSOP16, 20, DIP20, QFN24
SL8	H	8	512	256	20	up to 26	2 – 16 bits / 2+2 CCP	-	1	-	-	1	1	16 / 10 bits	1	TSSOP20, TSSOP28
SL16	H	16	512	256	20	up to 26	2 – 16 bits / 2+2 CCP	-	1	-	-	1	1	16 / 10 bits	1	TSSOP20, TSSOP28

Legend: V_{DD}: L = 1.8 to 3.6V, H = 2.7 to 5.5V, CCP – Capture/Compare/PWM Channel, IR – Infrared Modulator, CAN – Controller Area Network interface, SCI – Serial Communication Interface, USB – Universal Serial Bus interface (full speed device), I²C – Inter-Integrated Communication Interface, SPI – Serial Peripheral Interface, ADC – Analog to Digital Converter, AC – Analog Comparator.

Table 2.10

There are other chips that use the HCS08 CPU. They are the MC1321x devices (which integrate an HCS08GT device with a Zigbee-capable 2.4GHz transceiver) and the MPXY8300 device (an HCS08 CPU with integrated pressure sensor, 2-axis accelerometer and a 315/434 MHz RF transmitter), dedicated to tire pressure monitoring systems (TPMS).

2.5. Pinouts

Some of the pinouts available on the HCS08 family devices:

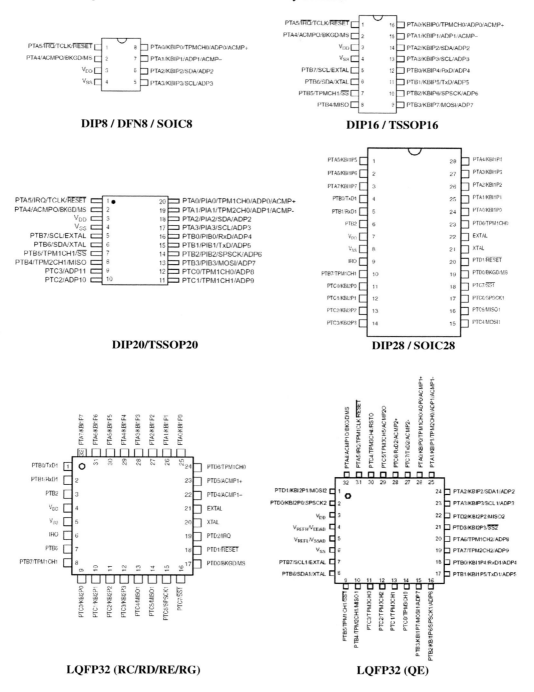

DIP8 / DFN8 / SOIC8

DIP16 / TSSOP16

DIP20/TSSOP20

DIP28 / SOIC28

LQFP32 (RC/RD/RE/RG)

LQFP32 (QE)

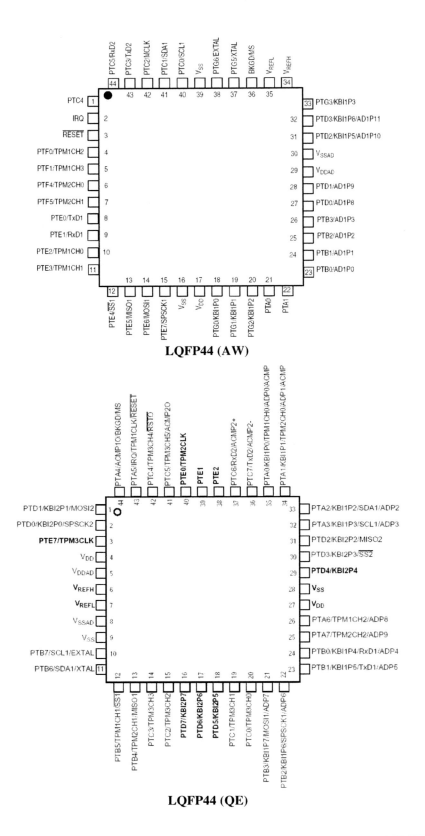

LQFP44 (AW)

LQFP44 (QE)

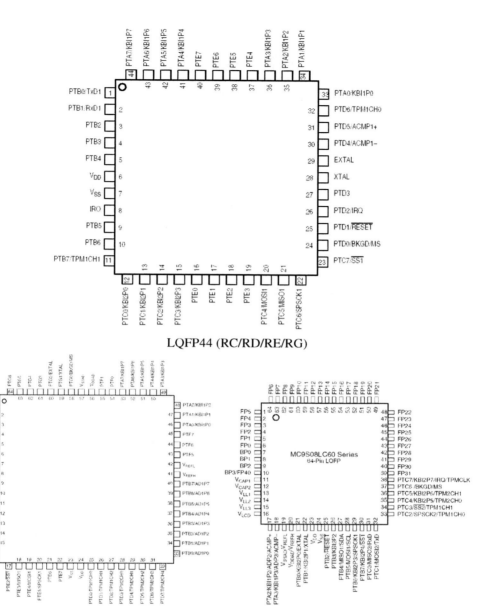

LQFP44 (RC/RD/RE/RG)

LQFP64
(GB/GT devices)

LQFP64
(LC devices)

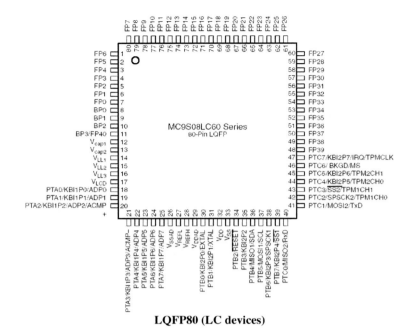

LQFP80 (LC devices)

3

The Codewarrior IDE

The Freescale's Codewarrior IDE is a complete system for editing, compiling, building, simulating, programming and debugging software developed for the Freescale's microcontrollers, microprocessors and digital signal controllers.

The Microcontrollers version 6.0 of the IDE is composed of an editor, assembler, optimizing C compiler, linker, simulator, programmer and a debugger.

This development studio supports the RS08, HCS08 and ColdFire V1 platforms and it is the first to support both 8 and 32-bit platforms in the same package.

Included in the package is the UNIS Processor Expert [TM] tool, which automates code generation for internal peripheral configuration and operation, external peripheral drivers and software algorithms. Another interesting tool is the UNIS Device Initialization tool, which can automatically generate CPU and peripheral initialization code. In this book, we are not going to use these tools, but focus on C code writing to understand how to configure and operate most peripherals found in the HCS08 devices.

3.1. IDE

The Integrated Development Environment (IDE) is an application software that integrates most programming tools into one unique piece of software.

On the following pages, we will learn how to configure the IDE and its options to create, simulate and debug projects for the HCS08 devices.

After starting the application, a window such as the one shown in figure 3.1 appears on the screen. This startup window allows the user to create a new project, run a tutorial demonstration, load a project or just start using the IDE.

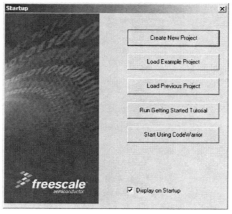

Figure 3.1

3.2. Creating a New Project

To start a new project in Codewarrior, press the Create New Project on the startup window or select the option FILE > NEW on the main menu, or press the icon on the tool bar. The new project wizard window opens, allowing the user to select the target device and the default connection for the debugger.

It is possible to select any available device from one of the three Freescale 8-bit families available: RS08, HC08 or HCS08. For the latest devices it may be necessary to install the service packs available through the Codewarrior updater tool.

In the default connection the user must select one of the available debug or simulation tools. Some of the options that may be available are:

- Full Chip Simulation: for software simulation, algorithm testing, execution timing measurements, and others, without using any real hardware.

- P&E Multilink/Cyclone Pro: for real-time in-circuit debugging, allowing the test of the application in the real circuit.

- Softec HCS08: this option should be used with the Softec's uDART08 BDM device (included in most Softec demonstration kits).

- HCS08 Serial Monitor: this option can be used when the target device was first programmed with the serial monitor software (refer to the AN2140 application note).

- HCS08 Open Source BDM: for using with the Open Source BDM pod.

Some of the options above may not be available for some devices.

> (i) *These options can be changed by the user at any time during the project development!*

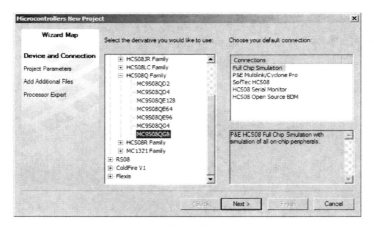

Figure 3.2

After selecting the device and the default connection and pressing the Next button the wizard window asks about some project options like the languages to be supported, the project name and its location (figure 3.3).

　　　　　　　　　　　　　　　　　　　　　　　　　　　　　HCS08 Unleashed

The Codewarrior IDE supports three different languages: assembly, C and C++. It is also possible to mix them within the same project.

The assembly language is supported both in the absolute and in the relocatable form. On the absolute assembler the variables are fixed in memory addresses specified by the programmer. The relocatable assembler uses special directives that allow the linker to choose the addresses for the variables.

 The selection of the absolute assembler is only possible when no other language is used in the project!

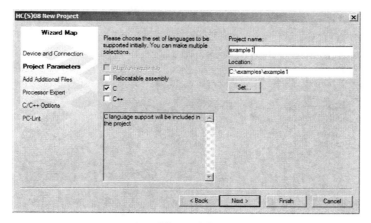

Figure 3.3

The C language is supported on all versions of the IDE (with a code limit of 32 KiB for the free special edition) and the C++ language is only available on the professional edition.

After selecting the project language (or languages, if more than one is selected), it is necessary to give it a name and set its location. It is possible to set a location to store the project files by directly typing it on the appropriate field or by pressing the Set button and choosing the desired folder.

Pressing the Next button opens the window shown in figure 3.4 allowing the user to add existing files (if any) to the project.

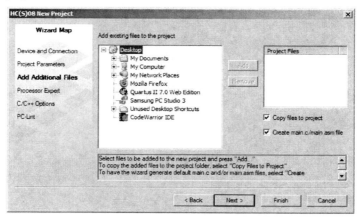

Figure 3.4

This window also allows the user to select two options:

- The "Copy files to project" option, when enabled, copies all selected files to the project folder.

- The "Create main.c/main.asm" option, when enabled, automatically creates these files in the "sources" folder within the project.

The next step is to select the rapid application development (RAD) options (figure 3.5). As we said earlier, the Codewarrior IDE has two options to help the designer speed up the application development: the device initialization tool and the Processor Expert tool.

The device initialization tool is an automatic code generation tool that simplifies the initial setup of the device: the user can select all the desired options and how each peripheral will be used, everything through an intuitive graphical interface.

The Processor Expert is a more complex bean-based code generation tool. It can be used to generate code for any internal peripheral and also for some external ones. In the Codewarrior Professional Edition the Processor Expert can also generate code for software algorithms.

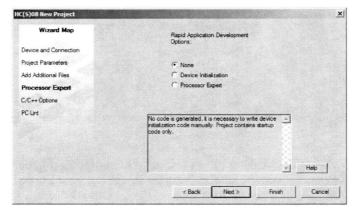

Figure 3.5

As we want to learn how to setup the chip and its peripherals we must select the None option and press the Next button to advance to the next window shown in figure 3.6.

In this window the user is able to select some C/C++ related options like: startup code generation, memory model and the floating point support.

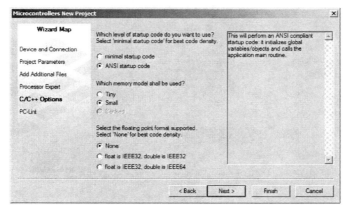

Figure 3.6

The startup code is an initialization code that is run before calling the main function. There are two options for the startup code generation:

1. Minimal startup code: selecting this option provides a faster and smaller startup code which only initializes the SP and calls the main function. No global or static variable is initialized. This option is not ANSI compliant and the user needs to initialize the variables himself. It should be preferred when using the stop2 low power mode.

2. ANSI startup code: selecting this option provides a full ANSI-compliant initialization code. This code initializes the SP and all global/static variables before calling the main function. The caveat is a bigger and slower startup code.

The memory model is related to how the compiler and the linker arrange code and data in memory:

1. On the tiny memory model, all data (including the stack) is stored in the direct page, unless it is declared by using **#pragmas** and the **__far** keyword. This model allows faster access to variables by using the direct addressing mode.

2. On the small memory model, data can reside anywhere within the 64 KiB address space by using 16-bit addresses (instead of the 8-bit addresses of the tiny memory model). This model is slower than the tiny one because it uses the extended addressing mode. On the other hand, it allows access to the full range of RAM memory available on the device. Using the direct memory area is only possible by the use of **#pragmas** (such as **#pragma** DATA_SEG __DIRECT_SEG MY_ZEROPAGE) and the **__near** keyword.

3. On the banked memory model, all data is stored outside the direct page unless declared by using **#pragmas** (like the small memory model). The CALL instruction is used to call functions on the banked memory. These functions return by the use of the RTC instruction. This mode is only available on the MMU-enabled devices (such as the MC9S08QE devices) and it is mandatory for accessing extended memory (addresses beyond the 64 KiB barrier).

The third selection field is where the user selects the type of floating point support. It is possible to disable the support for floating point variables (by selecting the None option), using the IEEE32 format to represent either floats and doubles or using the IEEE32 format to represent floats and the IEEE64 to represent doubles.

The next (and the last) window of the new project wizard is the one shown in figure 3.7. This window allows the user to use the PC-lint software to check the project source code. This option should only be enabled if the user has this application software installed on the computer.

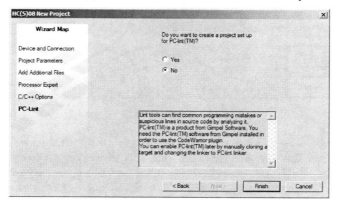

Figure 3.7

After pressing the Finish button the project is created.

3.2.1. Creating a New C Project

If the user selects the C language on the window shown in figure 3.3 and follows all the steps shown above, the IDE generates a new C language project as shown in figure 3.8 below. We can see the project window and the editor window.

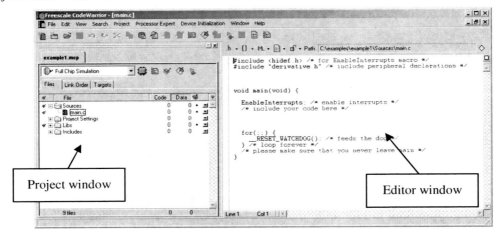

Figure 3.8

The project window shows the files included in the project, their status information and other useful information about the project itself. Figure 3.9 depicts all the elements of the project window.

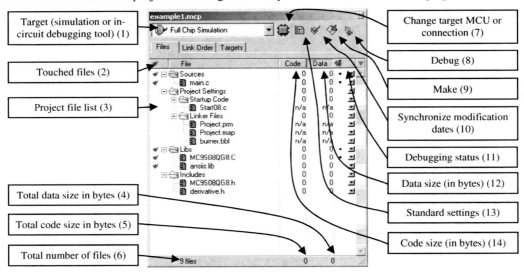

Figure 3.9

The project window consists of a file list containing all the files pertaining to the project (3). Some important files on this list are: Start08.c (the C startup file automatically generated by the compiler), Project.prm (the linker command file used to configure the operation of the linker), Project.map (the map file generated by the linker, which contains information on memory usage and allocation). We can also find the library files and include files automatically included by the compiler when the project was generated. We will see more on linker files later in this chapter.

In the file list, we can also see the amount of code memory (FLASH or ROM) (14) and data memory (RAM) (12) used by each file/module. The total amount of code and data memory is shown at the bottom of the window (4 and 5). At this time these values are all-zero: they will change only after the project is built.

On the top of the window, we find some buttons that control the project settings, command the build, and start the debug process.

The drop-down box (1) allows selecting the target connection for the debugging session: either the simulator or one of the in-circuit debugging tools can be selected with it.

The Change MCU/Connection button (7) allows the change of the target MCU on the fly. This button is very handy to change from one chip to another (do not forget to verify the peripherals and register usage!).

Another important control is the Standard settings button (13): pressing it opens the project settings window, allowing the user to select many compiler options (such as the optimization level), linker and many other options.

Finally, we have the Make (9) and the Debug (8) buttons: the Make button starts the making process where all modified or touched (2) files are compiled and linked. Pressing the Debug button does the same thing and also starts the simulator/debugger application.

 To successfully debug a file it is important that it is marked for debug (11) before the make/debug process is started!

Table 3.1 shows some of the file extension supported by Codewarrior.

File Extension	Description
*.ABS	Absolute file generated by the linker tool
*.ASM *.A08 *.INC	Assembler source file
*.BBL	Burner Batch File
*.BPT	Debugger breakpoint file
*.C *.CPP	C and C++ source files
*.CMD	Command file script
*.CWW	Workspace configuration file
*.ELF	Executable and linking format
*.H	Header file
*.HWL	Debugger layout file
*.HWC	Debugger configuration file
*.IO	I/O simulation file
*.PJT *.INI	Debugger configuration settings file
*.LST	Assembler listing file (automatically generated by the compiler)
*.MAK	Make file
*.MAP	Mapping file
*.MCP	Freescale Codewarrior IDE project file
*.MEM	Memory configuration file
*.O *.OBJ	Object file generated by the compiler or assembler
*.PRM	Linker parameter file
*.REC	Recorder file
*.REG	Register entries files
*.S *.S1 *.S19 *.SX	S-record file (for burning the microcontroller memory)
*.VTL	Visualization tool layout file

Table 3.1

Now that we know a little bit about the project window, let us try to modify the main.c file to create our first application: a led flasher!

At first, we need to open the main.c file in the editor. To do that, open the Sources group in the project window and double-click the main.c file. A window like the one shown in figure 3.10 appears on the screen.

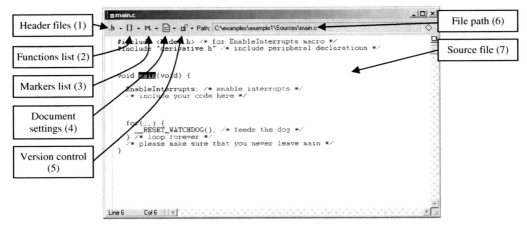

Figure 3.10

The editor window is simple yet powerful, with syntax coloring and options for code formatting such as auto indentation (for **switch** statement and opening/closing braces), format braces, code completion, etc. Unfortunately, there is no folding option for hiding portions of code (strangely, this option is only present on the simulator/debugger application).

On the top of the editor window, we can see five buttons: the Header list (1), the Functions list (2), the Markers list (3), the Document settings (4) and the Version control (5).

The header list shows a list of all the files within the present one (remember that this list is only updated after the project is built). It is possible to open any of these files by just selecting it from the list.

The functions list shows a list of all functions declared within the file. This is useful for navigating through the file: just select the desired function from the list and the editor jumps to the beginning of its code.

With the markers list it is possible to jump to specific positions marked previously. It is possible to add several markers through the source code, each one with a different name.

> *Markers can also be inserted by using the #pragma mark directive. Refer to topic 4.2.1 for further information on how to use this feature.*

The version control (CVS) is useful for controlling changes made to the source code. Prior to using it, it is necessary to enable the version control by selecting *Edit > Version Control Setting* on the main menu. Once the CVS is enabled, the padlock icon changes to a pencil icon and the options for version control are available.

Now, to understand how to work with the IDE, let us try a very simple program. Open the main.c file created by the project and type the following code in place of the one already there:

```
// A simple LED flasher
#include <hidef.h>        /* for EnableInterrupts macro */
#include "derivative.h"   /* include peripheral declarations */
#include "hcs08.h"        // This is our definition file !

void main(void)
{
  unsigned int temp;
  unsigned char counter1;
  SOPT1 = bBKGDPE;   // configure SOPT1 register, enable pin BKGD for BDM
  PTBDD = BIT_6;     // configure pin 6 of port B as an output
  counter1 = 0;
  while (1)
  {
    PTBD_PTBD6 = 1;                    // LED = off
    for (temp=10000; temp; temp--);   // wait for a while
    PTBD_PTBD6 = 0;                    // LED = on
    for (temp=10000; temp; temp--);   // wait for a while
    counter1++;
  }
}
```

Listing 3.1

After the program is correctly typed, we can proceed in building it. Just press the MAKE button (item 9 on figure 3.9) and we are done.

After the assemble/compile/link process, a window (figure 3.11) opens, showing the number of errors, warnings and hints.

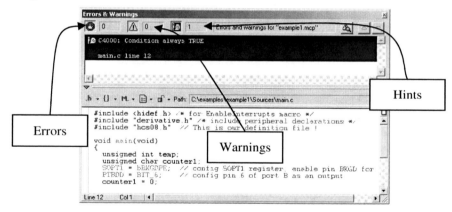

Figure 3.11

If errors are detected, no binary file is generated. The compiler also generates warnings and hints if any dubious code is found. Figure 3.11 shows the Errors & Warnings window for our example. There was only one hint telling us that line 12 has a condition that always evaluates true. That is not a big deal since it is exactly what we want it to do. It is possible to use a for statement (**for**(;;)) instead of **while**(1). Despite doing the same work, the **for** construction does not raise any hints.

If no errors are detected, an S19 file is generated and stored into the bin folder under the project folder.

Now, before starting with the simulation and debugging, let us take a look at the linker configuration files.

3.2.2. Linker Configuration File

The PRM file is the file that defines how the linker allocates memory for all pieces of code and data.

There are three important sections in the PRM file:

1. **Segments** : this section defines the valid addresses for each type of memory present on the chip:

   Z_RAM – RAM memory located into the direct page (variables allocated into this segment are accessed by using direct addressing mode).

   RAM – RAM memory located outside the direct page (variables allocated into this segment are accessed by using extended addressing mode).

   ROM – ROM (actually FLASH) memory.

   ROM1 – ROM (actually FLASH) memory located before the vector area.

   There are four segment qualifiers:

   READ_ONLY – segments of this type allow only reading operations. Read-only segments are initialized when the application is loaded into the flash memory.

   READ_WRITE – these segments allow reading and writing operations. These segments are usually initialized by application startup code.

   NO_INIT – these segments allow reading and writing operations and are not initialized by the startup code.

   PAGED – these segments allow reading and writing operations and are located into the paged memory. This qualifier is used only on Flexis devices.

2. **Placement**: this section assigns all predefined and user-defined segments into the valid areas defined in the "segments" section. Some of these segments are:

   DEFAULT_RAM – default segment for variables, usually placed into the RAM memory. All variables that do not belong to other segments are placed into this segment.

   DEFAULT_ROM – default segment for code (functions), usually placed into the ROM (FLASH) memory. All functions that do not belong to other segments are placed into this segment.

   _PRESTART – default segment for application entry point.

   STARTUP – default segment for startup code (used to initialize variables on the application initialization).

   COPY – initialization values stored into this segment (in the FLASH memory) are copied to the variables initialized in application startup.

   ROM_VAR – constant variables are stored into this segment by default.

   STRINGS – string literals are allocated into this segment.

3. **Stack**: defines the memory area reserved for the stack. The stack starts on the top of RAM and grows down to the number of bytes defined into the STACKSIZE option.

Listing 3.2 shows a typical PRM file (this one is for the MC9S08QG8 device).

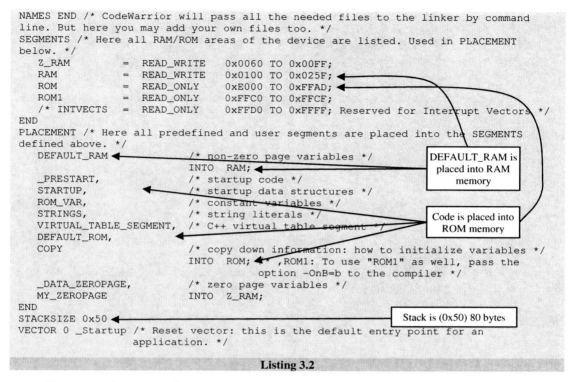

```
NAMES END /* CodeWarrior will pass all the needed files to the linker by command
line. But here you may add your own files too. */
SEGMENTS /* Here all RAM/ROM areas of the device are listed. Used in PLACEMENT
below. */
    Z_RAM        =  READ_WRITE    0x0060 TO 0x00FF;
    RAM          =  READ_WRITE    0x0100 TO 0x025F;
    ROM          =  READ_ONLY     0xE000 TO 0xFFAD;
    ROM1         =  READ_ONLY     0xFFC0 TO 0xFFCF;
    /* INTVECTS  =  READ_ONLY     0xFFD0 TO 0xFFFF; Reserved for Interrupt Vectors */
END
PLACEMENT /* Here all predefined and user segments are placed into the SEGMENTS
defined above. */
    DEFAULT_RAM                   /* non-zero page variables */
                                  INTO  RAM;
    _PRESTART,                    /* startup code */
    STARTUP,                      /* startup data structures */
    ROM_VAR,                      /* constant variables */
    STRINGS,                      /* string literals */
    VIRTUAL_TABLE_SEGMENT,        /* C++ virtual table segment */
    DEFAULT_ROM,
    COPY                          /* copy down information: how to initialize variables */
                                  INTO  ROM;   ,ROM1: To use "ROM1" as well, pass the
                                         option -OnB=b to the compiler */
    _DATA_ZEROPAGE,               /* zero page variables */
    MY_ZEROPAGE                   INTO  Z_RAM;
END
STACKSIZE 0x50
VECTOR 0 _Startup /* Reset vector: this is the default entry point for an
                application. */
```

DEFAULT_RAM is placed into RAM memory

Code is placed into ROM memory

Stack is (0x50) 80 bytes

Listing 3.2

Notice that the areas defined on the "SEGMENTS" section can be modified by the user to better suit a specific use. Typical situations are:

1. Modifying Z_RAM and RAM area to increase available area for extended addressing mode on small memory model:

```
// Reserves 0x20 bytes for direct addressing mode. The remaining RAM (0x0080
// to 0x025F) is accessed using extended mode:
Z_RAM = READ_WRITE   0x0060 TO 0x007F;
RAM   = READ_WRITE   0x0080 TO 0x025F;
```

2. Modifying ROM area to leave space (one or more pages) for storing non-volatile data. This is done by reducing the size of the ROM segment:

```
// Reserves the first 256 bytes of flash for non-volatile data (addresses
// from 0xE000 to 0xE0FF will not be used for code-storage):
ROM   = READ_ONLY   0xE100 TO 0xFFAD;
```

It is also possible to include the Z_RAM into the DEFAULT_RAM. This enables the linker to place variables into the RAM segment and once it is full, the Z_RAM segment is filled as needed:

```
DEFAULT_RAM      =  INTO RAM, Z_RAM;
```

The stack size can also be modified to fit an application (for simple applications, 80 bytes is too much memory, but more complex applications may need a larger stack).

The VECTOR command shown at the end of listing 3.2 is used to specify a function address or absolute memory address to be stored into the vector area.

The command: `VECTOR 0 _Startup` instructs the linker to store the address of _Startup at vector 0 (the reset vector).

The VECTOR ADDRESS command can be used to specify an absolute address:

```
VECTOR ADDRESS 0xFFFE _Startup // the reset vector points to _Startup
VECTOR ADDRESS 0xFFFE 0xF123   // the reset vector points to 0xF123
```

3.2.3. The MAP file

As a result of the linking process a .MAP file is also generated. This file shows how much RAM and flash memory are used by the application and where all the functions and variables are placed. The file is divided into sections; the most important ones are shown below:

The "file section" shows what files were linked together by the linker tool.

```
****************************************************************************
FILE SECTION
----------------------------------------------------------------------------
Start08.c.o                     Model: SMALL,        Lang: ANSI-C
MC9S08QG8.C.o                   Model: SMALL,        Lang: ANSI-C
main.c.o                        Model: SMALL,        Lang: ANSI-C
```

The "startup section" presents information on prestart code and values used on initialization.

```
****************************************************************************
STARTUP SECTION
----------------------------------------------------------------------------
Entry point: 0xE07B (_Startup)
_startupData is allocated at 0xE084 and uses 6 Bytes
extern struct _tagStartup {
  unsigned nofZeroOut      0
  _Copy    *toCopyDownBeg 0xE0CE
} _startupData;
```

The "section-allocation section" shows memory addresses and sizes of all code and data sections. We can see the size and placement of the initialization area (.init) and the size and placement of the stack (.stack). The final part of this section shows a summary of section sizes (READ_ONLY section refers to code placed into flash while the READ_WRITE section refers to data placed into RAM):

```
****************************************************************************
SECTION-ALLOCATION SECTION
Section Name            Size  Type   From      To        Segment
----------------------------------------------------------------------------
.init                   132   R      0xE000    0xE083    ROM
.startData               14   R      0xE084    0xE091    ROM
                               ...
.abs_section_ffaf         1   N/I    0xFFAF    0xFFAF    .absSeg76
.abs_section_ffae         1   N/I    0xFFAE    0xFFAE    .absSeg77
.stack                   80   R/W    0x100     0x14F     RAM
.vectSeg78_vect           2   R      0xFFFE    0xFFFF    .vectSeg78

Summary of section sizes per section type:
READ_ONLY  (R):       D2 (dec:      210)
READ_WRITE (R/W):     50 (dec:       80)
NO_INIT    (N/I):     57 (dec:       87)
```

The "vector-allocation section" shows the values stored for all interrupt vectors of the application (reset vector located at 0xFFFE is considered an interrupt):

```
****************************************************************************
VECTOR-ALLOCATION SECTION
    Address    InitValue    InitFunction
----------------------------------------------------------------------------
    0xFFFE     0xE07B       _Startup
```

The "object-allocation section" shows the address and size of all the objects used in the application. They are listed according to the module where they were declared. On this section, it is possible to see the address of the objects (Addr), its hexadecimal size (hSize), its decimal size (dSize) and the total number of times it is referenced within the application (Ref).

```
****************************************************************************
OBJECT-ALLOCATION SECTION
    Name          Module            Addr  hSize  dSize  Ref  Section     RLIB
----------------------------------------------------------------------------
MODULE:           -- Start08.c.o --
- PROCEDURES:
    loadByte                        E000    E     14     5   .init
    Init                            E00E   6D    109     1   .init
    _Startup                        E07B    9      9     0   .init
- VARIABLES:
    _startupData                    E084    6      6     4   .startData
- LABELS:
    __SEG_END_SSTACK                150     0      0     1
MODULE:           -- MC9S08QG8.C.o --
- PROCEDURES:
- VARIABLES:
    _PTAD                             0     1      1     0   .abs_section_0
```

```
                                      . . .
    _DBGF                                 1814       2       2       0   .abs_section_1814

MODULE:                  -- main.c.o --
- PROCEDURES:
    main                                  E092      3C      60       1   .text
- VARIABLES:
    OSC_TRIM                              FFAF       1       1       0   .abs_section_ffaf
    OSC_FTRIM                             FFAE       1       1       0   .abs_section_ffae
```

The "module statistic" shows how much memory is used by each application module:

```
*****************************************************************************
MODULE STATISTIC
  Name                              Data   Code  Const
-----------------------------------------------------------------------------
  Start08.c.o                          0    132      0
  MC9S08QG8.C.o                       85      0      0
  main.c.o                             2     60      0
  other                               80     16      2
```

Unused objects (declared in the application but never used) are listed on the following section:

```
*****************************************************************************
UNUSED-OBJECTS SECTION
-----------------------------------------------------------------------------
```

Dependencies can be found on these sections:

```
*****************************************************************************
OBJECT-DEPENDENCIES SECTION
-----------------------------------------------------------------------------
Init                     USES _startupData loadByte
_Startup                 USES __SEG_END_SSTACK Init main
main                     USES _SOPT1 _PTBDD _PTBD


*****************************************************************************
DEPENDENCY TREE
*****************************************************************************
 main and _Startup Group
 |
 +- main
 |
 +- _Startup
    |
    +- Init
    |  |
    |  +- loadByte
    |
    +- main              (see above)
```

Finally, we have the statistic section showing the total size for all blocks to be downloaded to the flash memory of the microcontroller (210 bytes in the example):

```
*****************************************************************************
STATISTIC SECTION
-----------------------------------------------------------------------------

ExeFile:
--------
Number of blocks to be downloaded: 3
Total size of all blocks to be downloaded: 210
```

3.3. Simulating and Debugging the Project

Once the program is successfully compiled, we can proceed to its simulation and debugging.

In a simulation session, the program is run in the host computer, which simulates the operation of the target CPU (the HCS08). This is a good choice for initial development and testing.

In a debug session, the program is run by the real target CPU in the application circuit. This allows checking the program operation directly into the real application, with interaction of external signals, etc.

The use of the BDM tool adds a great flexibility to the debugging process, allowing real-time visualization of variables and registers, insertion of breakpoints and special conditional breakpoints.

3.4. Simulation

The simulation tool can be started by pressing the debug button (item 8 on figure 3.9) or by pressing the F5 key. The True-Time Simulator & Real-Time Debugger (figure 3.12) is the standard simulation/debugging tool for the Codewarrior IDE.

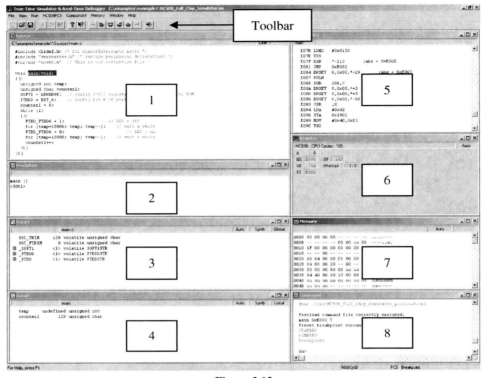

Figure 3.12

The application screen is divided into eight windows. These windows correspond to the internal visualization components. It is possible to select the desired components in the Component > Open menu item.

1- **Source code window**: this window shows the current source code being debugged. The highlighted line is the next one to be executed;

2- **Procedure**: shows the current function/sub-routine being executed;

3- **Data:1**: shows the global variables used by the program. It is possible to choose the data format by clicking with the right button on the name of the variable and selecting one of the options within the format option. It is also possible to change the content of the variable by double-clicking its value;

4- **Data:2**: this window works like the data:1 but shows the local variables declared by the current function;

5- **Assembly**: shows the assembly code being executed;

6- **Register**: shows the contents of the CPU registers (A, H:X, SP, PC and CCR) as well as the simulator cycle counter;

7- **Memory**: this window can be used to visualize the content of any memory address;

8- **Command**: shows the commands sent to and received from the target.

There is also a status bar at the bottom of the screen, which shows the current target CPU, the current connection and the execution status (figure 3.13). On the figure, the current CPU is an MC9S08QG8, the connection is FCS (Full Chip Simulation) and the program is stopped at a breakpoint.

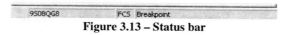

Figure 3.13 – Status bar

It is also important to know about the toolbar buttons and their function on controlling the simulator/debugger.

Start/Continue – by clicking this button or pressing the F5 key, the program execution is started or continued from the place where it was stopped;

Single Step – this button executes one instruction or C statement each time it is clicked (or the F11 key is pressed).

Step Over – this button, as the single step, executes one instruction or C statement each time it is clicked (or the F10 key is pressed). One major difference is that it treats function calls as single statements. The functions are executed and the simulator steps directly to the statement following the function call;

Step Out – this button executes the current function until it reaches the instruction immediately after the function call statement. This procedure can also be performed by pressing the SHIFT and F11 keys simultaneously;

Assembly Step – this button can be used to execute assembly instructions step by step. It is useful to inspect the execution of a C statement in "slow motion". This procedure can also be performed by pressing the CTRL and F11 keys simultaneously;

Halt – by clicking this button or pressing the F6 key, the program execution is halted and the contents of the registers and variables (shown in Register and Data windows) are refreshed;

Reset Target – use this button (or the CTRL and R keys simultaneously) to reset the CPU. The program flow diverts to the address pointed by the reset vector. When debugging in-circuit through BDM, the Codewarrior issues a BDM reset command, thus resetting the target hardware.

We can start program execution by clicking the ⬚ button or by pressing the F5 key. It is also possible to execute the program step by step (one C command at a time) by pressing the ⬚ button or by pressing the F10 key. Notice that the next line/command to be executed appears with a blue background.

The True-Time Simulator & Real-Time Debugger also offers some interesting features to simplify the simulation and debugging process:

- Breakpoints are points where the program execution must stop. This is usually done by specifying an address, which is compared to the current PC. When their content matches, the program execution is halted. There are two basic types of breakpoints: conditional and non-conditional. Conditional breakpoints halt the program execution when one or more conditions are met. Non-conditional breakpoints do not have any conditions; they halt the execution when a specific address is read or written.

- Markpoints are simple marks put into the memory (either code or data memory) and they enable easy access to the address marked. Notice that markpoints do not halt program execution. Their purpose is just making it easier to access specific points within the program.

- Watchpoints are special points marked to trigger debugger special actions (they are used on conditional breakpoints). Watchpoints can only be used when debugging code (they cannot be used in the simulation mode).

We will now take a closer look at the specific features of simulation/debugging application windows.

3.4.1. Source Window

This window displays the source code currently being run by the MCU. User can place breakpoints on points of interest and follow the execution through all modules of the application.

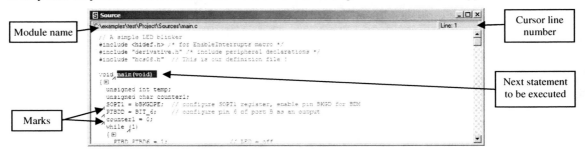

Figure 3.14

A right-click on the window content shows a menu with the following items:

- Set breakpoint: allows setting a non-conditional breakpoint at the current cursor position. A red arrow appears in front of the line marked with the breakpoint. A new right-click on the same place shows a slightly different menu with two new options: Delete breakpoint (to get rid of the breakpoint) and Disable breakpoint (to temporarily disable the breakpoint).

- Run to cursor: this marks a breakpoint at the current cursor position and starts program execution (until the program reaches the marked breakpoint).

- Show breakpoints: opens the "Controlpoints configuration" window, which shows data for all breakpoints, watchpoints and markpoints of the application. This window is an easy way for disabling or deleting undesired or unnecessary breakpoints.

Figure 3.15

- Show location: highlights the assembly instruction corresponding to the selected C source line or the C source line corresponding to the selected assembly instruction.

- Set Markpoint: places a markpoint into current cursor position.

- Show Markpoints: opens the "Controlpoints configuration" described above.

- Set Program Counter: changes the PC value so that it points to the current instruction (or the instruction corresponding to the selected C code).

- Open Source File: opens a different source-code file.

- Copy: copies the select text to the clipboard (same as CTRL + C).

- Go To Line…: jumps to a specific line in the source code.

- Find…: searches for a specific text into the current window.

- Find Procedure…: searches for a specific procedure into the current window.

- Folding: folds/unfolds blocks of the source file.

- Freeze: places source code window on freeze state (its content is not changed when the application is stopped).

- Marks: toggles the display of markings showing the positions where breakpoints can be set (these places are marked with small red triangles).

- Tooltips: shows a menu with options for activating/deactivating the tooltips (small balloons that appear when the cursor is placed over a variable for a while, showing the current variable value). It is also possible to select the desired format for the values.

 It is possible, at any time, to view the corresponding assembly code generated for a given C code. To do this, select the desired C code on the Source window, drag it to the Assembly window and drop it there. The application automatically selects the corresponding assembly code. The opposite is also true: selecting an assembly instruction, dragging and dropping it into the C source window shows the corresponding C line for that instruction.

3.4.2. Register Window

The register window allows viewing and editing the contents of all CPU registers.

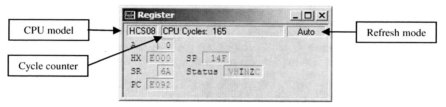

Figure 3.16

When the content of a register is changed, it is written in red (this also applies to the data and memory windows).

One important feature is the CPU cycle-counter, which counts the total number of cycles spent by the CPU. This is an easy way for measuring the time spent on sections of code or functions.

 To clear the cycle counter or to specify its value, type in the command window: cycles x (where x is the desired value for the cycle counter). I.e.: to clear the cycle counter type cycles 0

To measure the total time spent on one **for** statement in our example (listing 3.1), all we need to do is:

1. Mark two breakpoints (one at the beginning of the **for** statement and the other after the **for** statement). The source window will look like the one on figure 3.17.

2. Run the code (pressing F5): the program is run and halts on the first breakpoint.

3. Now type on the command window: cycles 0. This resets the cycle counter.

4. Restart the code (pressing F5): the program is run and halts on the second breakpoint. The value on the cycle counter is the total number of cycles spent on the **for** statement!

Figure 3.17

Figure 3.18

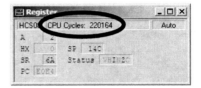

Figure 3.19

For a BUSCLK of 4 MHz, the delay time for a single **for** statement can be calculated by multiplying the number of cycles by the inverse of the BUSCLK frequency: 220164 * 1 / 4 MHz = 55.041 ms.

3.4.3. Memory Window

The memory window allows visualization of the memory contents. It is possible to select how the data is displayed and also to place markpoints and watchpoints (only in debug mode).

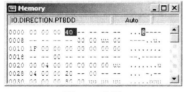

Figure 3.20

By selecting a position, the global symbol name corresponding to that address is shown in the field at the top of the window. With a double-click on the position, it is possible to edit and change its content.

Right-clicking on the window content opens the context sensitive menu.

Figure 3.21

This menu allows selecting some options such as word size (byte, word or long word), format (hexadecimal, octal, binary, signed decimal, unsigned decimal and bit-reversed), mode (automatic, periodical or frozen) and display options (view the address and ASCII code).

There are also options to fill addresses with a specific value, jump to an address, copy a memory block from one address to another, search memory for a specific pattern and refresh window content.

The content of the window is refreshed according to the selected mode: automatic – data is refreshed each time the application is halted; periodical – data is refreshed periodically (in units of 100ms); frozen – data is not refreshed when the application is halted.

3.4.4. Data Windows

Data windows are special components used to view variable-contents. They are very powerful and useful, especially when debugging C codes, supporting complex data types (such as arrays, structures, enumerations and unions) and allowing variable-content to be viewed in many different ways. Two data windows are opened by default: one for global and another for local symbols.

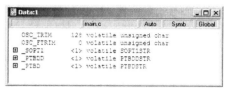

Figure 3.22

The global data window (figure 3.22) allows inspecting all global variables declared in the application (this also includes special registers), whereas the local data window (figure 3.23) allows inspecting the content of all local variables used within the current function. A field located at the right-top of the window shows the current scope.

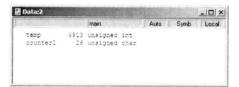

Figure 3.23

Besides the global and local scope, it is possible to have a user-defined scope where he is able to add any variables. Be aware of the fact that sometimes, local variables can be optimized by the compiler, being, therefore, inaccessible to the debugger.

New variables can be added by using the add expression on the context-sensitive menu or by simply selecting the desired variable on the source window, dragging and dropping it into the data window.

Figure 3.24

The context-sensitive menu for the data windows is shown in figure 3.24. It has many options regarding the way data is viewed in the window and also some options for controlling how their content is updated. Let us observe the meaning of the menu options:

- **Open Module…**: opens another source module for data visualization.

- **Add Expression…**: adds a new variable or expression to the data window. It is possible to add any C valid expression.

- **Set Watchpoint**: sets a watchpoint on the selected variable.

- **Show Watchpoints**: opens the "Controlpoints configuration" described above.

- **Set Markpoint**: sets a markpoint on the cursor's current position.

- **Show Markpoints…**: opens the "Controlpoints configuration" described above.

- **Show Location**: shows the current variable address and position in the memory window.

- **Zoom**: allows zooming in and out from structures (zooming in unfolds the structure and shows its fields, while zooming out folds the structure and hides its fields).

- **Scope**: selects the visualization scope for the variables within the current window.

- **Mode**: selects the refresh mode for the window content: this option works just like the one on the memory window.

- **Format**: selects the desired format for data shown in the window. It is possible to change just the format for the currently selected variable or for all variables within the window.

- **Options**: selects some window options:

- **Pointer as Array…**: views pointers as arrays (user can select the desired number of elements).

    - **Name width…**: maximum width (in characters) for variable-names.

- **Sort**: selects the sorting order for the variables within the window.

- **Refresh**: refreshes the contents of all variables.

3.4.5. Visualization Tool

The simulator also offers some advanced resources such as the virtual components: they allow simulating interaction of the microcontroller with many devices such as LEDs, keys, etc.

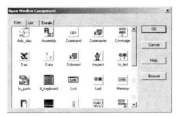

Figure 3.25

To demonstrate how to use this tool, let us add a virtual LED to our simulation. To do this, click on Component > Open option on the main menu.

A new window (figure 3.25) opens. Select the "VisualizationTool" at the bottom of the window.

After that, the "VisualizationTool" opens, showing a window such as the one on figure 3.26A.

The "VisualizationTool" is initially in "edit mode" as can be seen on the left-top of the window. While in "edit mode", it is possible to add new instruments and modify the properties of the existing instruments as well as the properties of the "VisualizationTool" itself. Notice that instruments are the many possible virtual components which can be added to the visualization.

To add a LED instrument, all we need to do is right-click on the work area of the window and then select "Add New Instrument > LED". A new LED component is placed on the work area. This component (or instrument) can be resized and moved at the programmer's will. After placing and sizing it, it is time to modify its properties. This can be done double-left-clicking it or by right-clicking it and selecting the properties option. A new window (figure 3.26B) opens.

Figure 3.26A

Figure 3.26B

To successfully configure our LED, it is important to set the "Port to Display" property to "_PTBD" and "Bitnumber to display" to "6". This window also allows selecting the active and inactive colors and other visual enhancements.

It is also important to modify one "VisualizationTool" property. Right-click on the work area and select "Properties". The "Properties of VisualizationTool" window opens. Change the refresh mode to "Periodical" and the "Refresh Time (100ms)" to "1". This redraws the components at a 100ms rate.

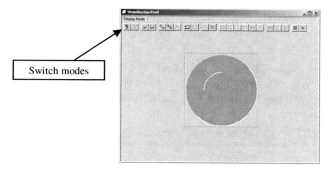

Figure 3.27

After configuring all parameters, right-click on the switch button to switch from "edit mode" to "display mode".

Now, press the F5 key and see our blinking led in action.

If desired, the visualization layout can be saved on disk by pressing CTRL+S or by clicking the disk icon.

3.5. Basic Debugging

As we said before, the HCS08 microcontrollers include a very powerful internal debugging hardware. On this topic we will learn about the basics on debugging software on HCS08 MCUs.

To start a debug session, the first step is to change the target from simulator to the desired debug tool (P&E Multilink/Cyclone PRO, Softec HCS08, HCS08 Serial Monitor or HCS08 Open Source BDM).

Pressing the F5 key starts the debug application (in fact, it is the same software application used for simulation). The first step is to establish contact with the target MCU. This is done through the "PEMICRO Connection Assistant" window shown in figure 3.28A. Owners of the Softec inDART-HCS08 see the "Target Connection" window (figure 3.28B).

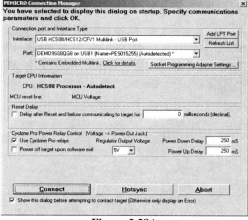

Figure 3.28A

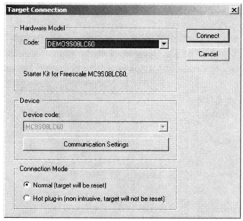

Figure 3.28B

When using the Softec inDART-HCS08, it is necessary to select the appropriate hardware model and device code (figure 3.28B shows the configuration for the DEMO9S08LC60 board).

Pressing the "Connect" button starts communication between the host PC and the target MCU and downloads the code to the target MCU. The "Hotsync" button can be used to connect to the target MCU without reflashing the code and without resetting the chip (this is useful for debugging an application already present in the MCU).

After pressing the "Connect" button, Multilink BDM and Cyclone Pro users see another window asking to erase and program the flash memory (figure 3.29). Answering "Yes" causes the target MCU flash memory to be erased and programmed with the new code. Answering "No" just resets the target MCU and starts the debug session.

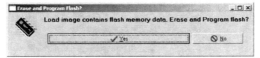

Figure 3.29

Figure 3.30A shows the progress dialog while the flash memory of an MC9S08QG8 is erased and programmed. After downloading the code to the target flash memory, the command window on the debugger looks like the one shown in figure 3.30B.

Figure 3.30A   **Figure 3.30B**

After downloading the code successfully, debugging proceeds exactly as in simulation. It is possible to run the program as well as halt the execution and inspect the state of registers, variables and memory.

 By using the "periodical" refresh mode, it is possible to see the variable contents changing on-the-fly while the program is running on the target MCU. It is also possible to change values in real-time: just double-click on the desired variable on the data or memory window, change its value and press ENTER!

Notice that after resetting the target MCU, the source window does not show the original application code, instead, a different code is shown (figure 3.31). This is the startup code necessary to initialize the C environment (the minimal version of the startup code only initializes the stack pointer).

Figure 3.31

As can be seen on figure 3.31, the "JMP main" instruction is responsible for jumping to the main function and starting the application.

3.6. Advanced Debugging Topics

Now that we are familiar with the Codewarrior simulation and debugging tools, it is time to explore some of the advanced debugging features of the HCS08 devices.

To test the debugging features we will use another version of our led flasher. This was tested on a DEMO9S08QG8 kit (with an MC9S08QG8 microcontroller):

```
// Another LED flasher
#include <hidef.h>          /* for EnableInterrupts macro */
#include "derivative.h"     /* include peripheral declarations */
#include "hcs08.h"          // This is our definition file !

#pragma profile on

unsigned char counter1;
unsigned int val;

void delay(unsigned int value)
{
  for (;value; value--);
}

void main(void)
{
  SOPT1 = bBKGDPE;    // configure SOPT1 register, enable pin BKGD for BDM
  PTBDD = BIT_6;      // configure pin 6 of port B as an output
  counter1 = 0;
  val = 10000;
  while (1)
  {
    PTBD_PTBD6 = 1;   // LED = off
    delay(val);       // wait for a while
    PTBD_PTBD6 = 0;   // LED = on
    delay(val);       // wait for a while
    counter1++;
    if (counter1>20) val = 30000;
  }
}
```

Listing 3.3

3.6.1. Periodical Refresh of Data Windows

This is one of the greatest features available on the HCS08 debug system and it enables the user to inspect the state of variables within an application in real-time.

Figure 3.32 shows how to setup this feature. After selecting the periodical mode, another window opens, asking for the desired update rate (in units of 100ms).

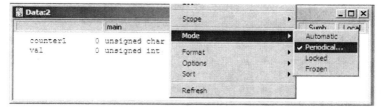

Figure 3.32

3.6.2. Conditional Breakpoints

Using listing 3.3, :we can demonstrate how to setup and use the triggers within DBG module to accomplish complex debugging tasks, which can be very useful while developing embedded applications.

 Before using the triggers be sure that they are configured to automatic mode (Trigger Module Usage > Automatic on the context sensitive menu)!

There are nine different trigger modes:

1. Memory access at address A: trigger occurs when the CPU accesses a specific address (the one stored in comparator A). This mode can be used to create a breakpoint (when comparator A points to an instruction) or a watchpoint (when the comparator A points to a RAM address).

2. Memory access at address A or address B: in this mode the trigger occurs if the CPU accesses an address which is equal to the one stored in comparator A or comparator B. This mode can be used to create breakpoints (when one or both comparators point to instructions) or watchpoints (when one or both comparators point to RAM addresses).

3. Memory access at address A and then at address B: in this mode the trigger occurs only when two conditions are met: first, the address pointed to by comparator A must be accessed by the CPU and then, the address pointed to by comparator B is accessed by the CPU. There is no timing constraint between the two accesses (they can be separated by microseconds to hours).

4. Memory access at address A with a value equal to B: in this mode the trigger occurs only when the CPU accesses a specific address (the one stored in comparator A) and the data bus is equal to a specific value (the one stored in comparator B). This mode is useful for catching bugs such as an unexpected write to a variable.

5. Memory access at address A with a value different from B: in this mode the trigger occurs only when the CPU accesses an address (the one stored in comparator A) and the data bus is different from a specific value (the one stored in comparator B).

6. Data capture on B match: this mode starts capturing data into the FIFO when a match condition for comparator B occurs.

7. Data capture on complex match: in this mode, the initial trigger occurs when the CPU accesses an address (the one stored in comparator A). After that, every time a match on comparator B occurs, the current content of the data bus is stored into the FIFO.

8. Memory access to an address within a range (A ≤ address ≤ B): in this mode the trigger occurs when the CPU accesses an address between a specific range (address is equal to or greater than comparator A and equal to or smaller than comparator B).

9. Memory access to an address outside a range (address < A or address > B): in this mode the trigger occurs when the CPU accesses an address outside a specific range (address must be smaller than comparator A and greater than comparator B).

Selection between these trigger modes is done on the "Trigger Module Settings" window (figure 3.33). This window also allows selecting the action performed on a trigger event:

1. Record data continuously and halt on trigger event.

2. Record data continuously and do not halt on trigger event.

3. Start recording after the trigger and halt when the FIFO is full.

4. Start recording after the trigger and do not halt when the FIFO is full.

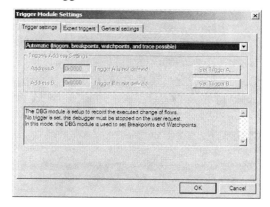

Figure 3.33

Using the DBG triggers is very straightforward. Let us see some examples on how to perform useful debugging tasks.

 There is a maximum number of three breakpoints on most HCS08 devices! Exceptions are QA and QD devices (one breakpoint) and QE devices (four breakpoints). This maximum number is reduced when using complex breakpoints.

3.6.2.1. Halting Execution on a Write Operation on a Variable

To demonstrate this feature we will set a trigger on the "counter1" variable. To do this, right-click on the name of the variable (in a data window) and select "Set Trigger Address A > Write Access", as shown in figure 3.34.

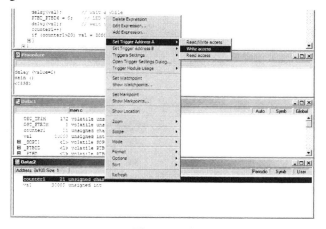

Figure 3.34

A variable marked on comparator A appears in the data window with a vertical red dashed line (figure 3.35). For comparator B the marking symbol is a blue dashed line.

Figure 3.35

That is it! Now, the program execution is halted after every write operation done into the variable.

3.6.2.2. Halting Execution When a Variable Reaches a Specific Value

In this example, we halt program execution when "counter1" reaches 10. To achieve this we follow two steps:

1. Set trigger A on "counter1" variable.

2. Modify trigger mode to mode 4: "Memory access at address A and value on data bus match". We also need to modify the match value to 0x0A (10 decimal). This is shown in figure 3.36.

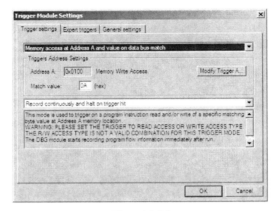

Figure 3.36

To test the trigger setting, just reset the debugger and restart program execution: after a brief time it is halted. The source window shows the line immediately after the command that wrote 10 to "counter1". Notice the command window reporting the trigger event (figure 3.37).

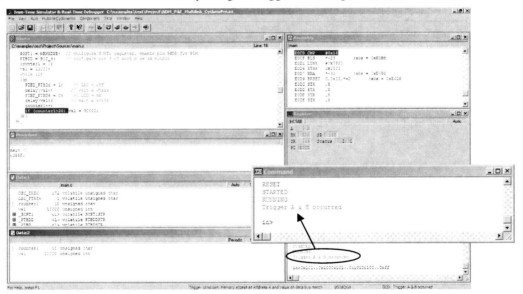

Figure 3.37

3.6.2.3. Halting Execution on Breakpoints Combination

This last example shows how to setup a complex breakpoint. Supposing we want to halt program execution after "val = 30000", but we also need to do that with the led unlit. How can we do that?

The answer is: using trigger mode 3 (memory access at address A and then at address B). All we need to do is set trigger A to the "val = 30000" expression and set trigger B to the first `delay(val)` function call (as shown in figure 3.38A). Notice that after setting up the triggers on the source window, two symbols appear in red: the A character indicates that trigger A is set on that position and the B character indicates that trigger B is setup on that position.

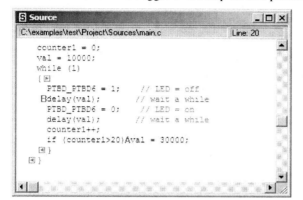
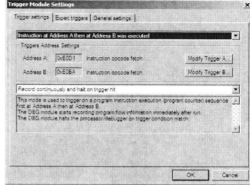

| **Figure 3.38A** | **Figure 3.38B** |

We also need to modify the trigger mode on the "Trigger Module Settings" window: select trigger "Instruction at Address A then at Address B was executed" as shown in figure 3.38B.

To test these trigger settings just reset the debugger and restart program execution: after a brief time it is halted on the `delay(val)` function call and the led on the DEMO9S08QG8 board is off.

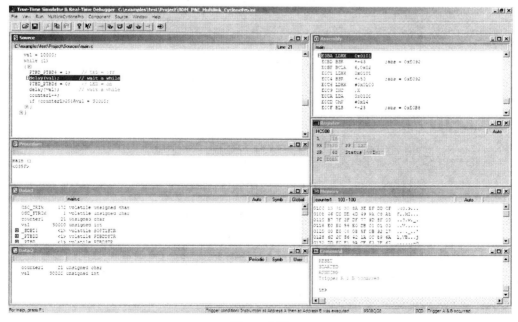

Figure 3.39

3.6.3. Profiling

The DBG FIFO memory can also be configured to profile the application code. This is a very interesting feature and it enables the programmer to know the amount of time spent on each part of the application code. This is often needed to determine critical sections of code that may need optimization.

To enable profiling, it is necessary to configure the BDM to the coverage/profiling mode. This can be done by selecting the "Trigger Module Usage > Profiling and coverage" option on the source window context-sensitive menu (figure 3.40).

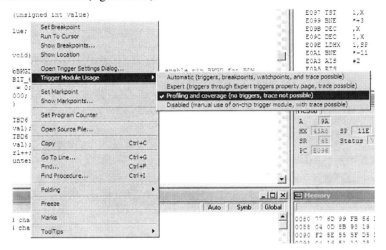

Figure 3.40

Once the debug system is configured to work in coverage/profiling mode, we can now open the profiling component by selecting the Component > Open… option on the main menu (this sequence is not mandatory). The profiler window (figure 3.41) opens.

Figure 3.41

Profiler window shows percentage numbers indicating the amount of time spent on each module (relative to the total execution time). The unlimited version (available with Codewarrior Standard and Professional versions) can also show values for each function within a module.

Initially, this window starts with all values zeroed. Once the program is run, the debugger starts collecting information about the currently instruction being executed and builds the application profiling map (figure 3.41 shows the percentage numbers after some minutes of execution).

It is also possible to configure the profiler component to show statistic information directly into source or assembly windows. This can be done by selecting Details > Source (or Details > Assembly) on the profiler menu. This introduces a split view in the selected window (Source or Assembly) as shown in figure 3.42.

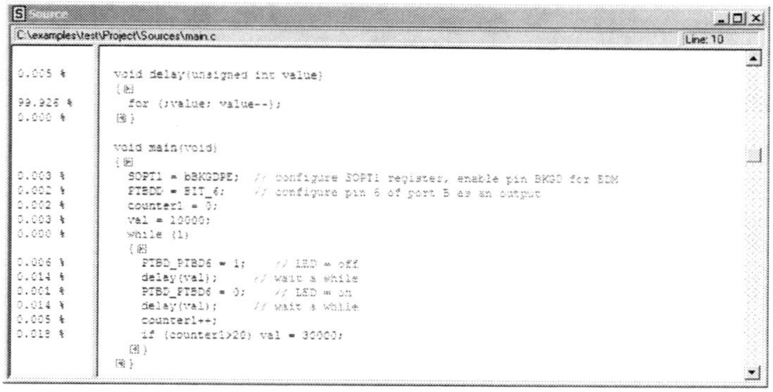

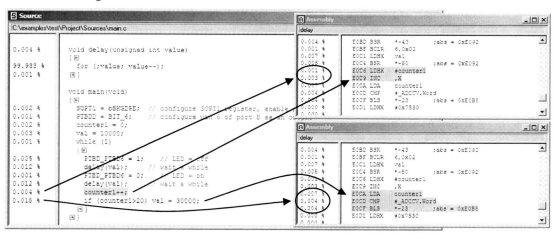

Figure 3.42

Values shown on the profiler window and on source/assembly split windows can also be updated in real-time. To enable this option select the "Timer Update" item on the profiler menu.

The mechanism to accomplish code profiling/coverage is very simple: when the debugger is not armed (triggers are not in use) the first position of the DBG FIFO stores the address of the current opcode being executed by the CPU. By periodically reading this FIFO memory, the debugger can count how many times that instruction was executed and then construct a profiling map for the application.

For being able to produce useful information, the profiling mechanism needs time: the more you let it capture information, the more accurate this information will be. Be aware that profiling C source has another issue: values presented in the source window are statement-relative while the profiling is done on an instruction basis. Thus, percentage values shown in the source window are resulting of the sum of all individual instruction percentages relative to that particular high-level command. This is demonstrated on figure 3.43.

Figure 3.43

3.6.4. Code Coverage

Another possible use of the DBG hardware is on measuring the code coverage. This feature enables the programmer to know the how much each part of the program was executed.

To enable code coverage it is necessary to configure the BDM to the coverage/profiling mode as demonstrated in the last topic.

The coverage component can be opened by selecting the Component > Open... option on the main menu (this sequence is not mandatory). The coverage window (figure 3.44) shows up.

Figure 3.44

This window shows module names and percentage numbers indicating how much of that module was executed on the current run.

The coverage mechanism uses the same principles of profiling and the options shown for profiling are also available for the coverage component.

3.6.5. Watchpoints

Watchpoints are another welcome feature to improve the debugging of complex applications. While the other debugging options seen until here are supported by specific hardware (the DBG module), watchpoints rely mainly on the BDC ability to inspect the microcontroller memory in real-time.

Despite that, watchpoints can introduce noticeable delays in program execution (depending on their configuration) and this is their worst drawback.

Watchpoints are available in three "flavors":

1. Non-conditionals: watchpoints that only check if the selected memory address was written or read.

2. Conditionals: watchpoints that check a selected memory address and compare its content against a value or expression.

3. Counting: watchpoints that allow specifying a number of times of condition matching before halting the application.

To setup a watchpoint, right-click on the desired variable (in a data window) or memory address (in a memory window) and select the "Set Watchpoint" option on the context sensitive menu (figure 3.45A). This opens the "Controlpoints Configuration" window (figure 3.45B).

Figure 3.45A

Figure 3.45B

In the "Controlpoints Configuration" window, the user can configure the desired options for the watchpoint: the kind of access (read/write, write or read operations), the optional condition to be matched and also the total number of events before halting execution.

As an example, we will demonstrate the setting up of a watchpoint on "counter1" variable to halt execution when it is greater than 10. To do that, right-click on the "counter1" variable in one of the data windows and select the option "Set Watchpoint" in the context sensitive menu.

After that, right-click on the same variable and select "Show Watchpoints" on the context sensitive menu. In the "Controlpoints Configuration" window, change the "Read/Write" field to "Write" and enter the following condition: "counter1>10" on the condition field (as shown in figure 3.46).

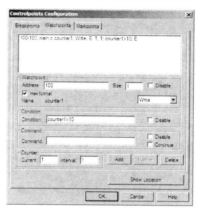

Figure 3.46

To test the watchpoint, just reset the debugger and restart program execution: after a brief time it is halted as shown in figure 3.47. Notice the slow down on the execution speed.

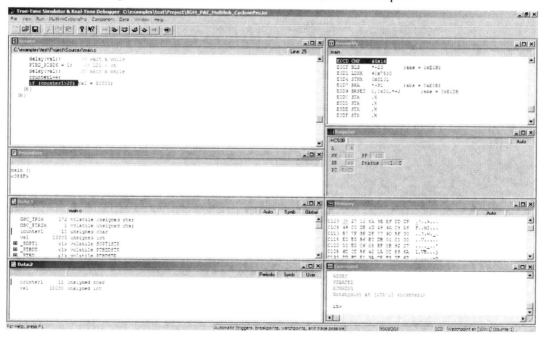

Figure 3.47

4

The C Programming Language

Dennis M. Ritchie from the AT&T Bell Laboratories originally created the C language in 1972. Its first usage was to allow portability of the UNIX operating system, written in PDP-11 assembly language, to other computer systems. C was born with the purpose of being easier and more portable than the assembly and much closer to the machine than other high-level languages.

Through the years, many flavors of the language appeared, causing some compatibility issues: programs written in one version of the language could not be compiled in another version.

This problem was only solved when the American National Standards Institute (ANSI) released the first standardized version of the C language, known as ANSI C, C89 or C90 (the ISO/IEC review) in 1989. In 1999 the ANSI committee reviewed the C89 specification, adding some new features and releasing the C99 review (the current one).

In this chapter we will present some characteristics of the C compiler integrated to the Codewarrior IDE, as well as some procedures to be adopted when using this language in the programming of the HCS08 devices.

4.1. C Language Review

Before studying the HCS08 C compiler we will briefly review some of the basics of the C language. If you are experienced with the C language, we suggest that you skip directly to the topic 4.2, where we focus on the details of the HCS08 C compiler.

4.1.1. Commands and Reserved Keywords

The C language has 32 reserved words which are the statements and declarations that form the basis of the language. These reserved words cannot be redefined anywhere within a program.

auto	break	case	char	const	continue
default	do	double	else	enum	extern
float	for	goto	if	int	long
register	return	short	signed	sizeof	static
struct	switch	typedef	union	unsigned	void
volatile	while				

Table 4.1 – C keywords

4.1.2. The Program Structure

A typical C program is composed of some remarkable parts:

1. Compiler directives.

2. Global variables declarations.

3. Functions prototypes declarations.

4. Functions bodies (the code of each function, including the **main** function).

From the four parts listed above, the only one we always find in any C program is the **main** function body. This is the first function that gets executed when the program is run.

Listing 4.1 shows a simple C program listing:

```
#include <stdlib.h> // This is a compiler directive
char va;               // This is a global variable declaration
void main(void)     // This is the main function declaration
{
    int vb;
    va = 10;
    vb = va + 5;
    /*
        beginning of comment 2
        end of comment 2
    */
}
```

Code block

Listing 4.1

The first line:

```
#include <stdlib.h>        // This is a compiler directive
```

It is a compiler directive that instructs the compiler to include another file in the compilation process (in this case the file called "stdlib.h"). The .h files are called "header files" and are commonly used to store function prototypes, global declarations, etc. The "stdlib.h" header is necessary for using the standard C library (although we are not really using any of its functions).

The filename is specified inside double quotes or angle brackets. The difference between them is the place where the compiler searches for the file.

Observe the presence of a comment in the same line: by using the // it is possible to add single-line comments to any line of the code. Everything after the // is ignored by the compiler until it reaches the end of the line.

The next line: `char va;` is a variable declaration. C needs that all variables be declared before being used. In the present case, "va" is being declared outside any function. All variables declared outside the functions of the program are called global variables, as they can be used by any function or code block. They last until the program reaches its end.

Following the "va" declaration we find the line:

```
void main(void)
```

This is the definition of the **main** function of the program. Just after that we find a code block delimited by curly braces: one for opening it "{" and another for closing it "}".

Code blocks are used to list a statement sequence to be executed by a function or by a conditional statement in C language.

 C statements can only be written inside a function! You cannot have statements outside any function!

The first line in the code block is a variable declaration:

```
int vb;
```

In C, all variable declarations are written in the form:

```
<data type> <variable name>;
```

The data type field specifies the type of the variable. In this case we are declaring a variable named "vb" of type **int**, which is a 16-bit integer type. We'll see more on data types in the next topic. This variable is also local to the function and can only be used within it. All variables declared within a code block are local to it and can only be used within it. Their content is lost as soon as the end of the block is reached (or the function exits).

Note that all C statements and variable declarations are ended by a semicolon ";". This allows writing multiple different statements in the same line of code.

The next line of code is:

```
va = 10; // comment 1
```

This line performs an attribution operation: the global variable "va" receives the value 10 decimal.

Numerical constants are, by default, represented in decimal format, but C also supports the hexadecimal format (a number preceded by 0x) and the octal format (a number preceded by 0). So, writing va = 0x0A and va = 012 produces the same result as va = 10.

The next line of code is:

```
vb = va + 5;
```

This is a typical expression. In this case the expression evaluates the addition of the current content of "va" plus five, storing the result at the local variable "vb".

It is important to see that "va" is a **char** variable which is added by 5. This results in an 8-bit value (a **char**) that must be stored into a 16-bit variable (the **int** variable "vb"). C completes this task by storing the result into the lowest byte of "vb", while clearing its highest byte. We will see more on this subject when we study expressions in topic 4.1.4.

The next lines are:

```
/*
  beginning of comment 2
  end of comment 2
*/
```

They are another kind of valid comment format supported by C. This format is used for long multiple-line comments. A multi-line comment is started by /* and finished by */. Everything within those symbols is ignored by the compiler.

4.1.3. Data Types

There are three basic data types in C: integers, floats and void. The integer types can be divided into signed (positive or negative) and unsigned (positive only). The float types can represent either integer numbers (rounded) or fractional numbers (represented using a special floating point format). The **void** type is mainly used to declare functions that do not receive or return any kind of data. It is also used to declare generic data pointers, as we will see later on.

The C language also comprises some type modifiers:

short   - for reducing the type representation range;

long   - for increasing the type representation range;

signed   - for specifying signalized data types (using the 2's complement format);

unsigned - for specifying non-signalized data types;

The ANSI standard defines the base type as being always **signed**, unless declared as **unsigned**. The exception is the **char** type, which was left undefined by the ANSI standard. For the HCS08 compiler, the standard format for **char** types is **unsigned**, unless declared as **signed**.

Considering the base types and the subtypes formed by using the type modifiers we have the following list of C supported data types:

Type	Range	Size in bytes
Unsigned char, char	0 a 255	1
signed char	−128 a 127	1
unsigned int, unsigned short int	0 a 65.535	2
int, signed int, short int, signed short int	−32.768 a 32.767	2
unsigned long int	0 a 4.294.967.295	4
long int, signed long int	−2.147.483.648 a 2.147.483.647	4
float	$1,18 \times 10^{-38}$ a $3,4 \times 10^{+38}$	4
double *	$1,18 \times 10^{-38}$ a $3,4 \times 10^{+38}$ or $2,23 \times 10^{-308}$ a $1,8 \times 10^{+308}$	4 or 8
long double *	$1,18 \times 10^{-38}$ a $3,4 \times 10^{+38}$ or $2,23 \times 10^{-308}$ a $1,8 \times 10^{+308}$	4 or 8

* On the Codewarrior C Compiler, double support can be set to 32 or 64-bit.

Table 4.2

The six first types listed on the table 4.2 are integer types, while the last three are real types (floating point numbers).

4.1.3.1. Variable Declaration

As we have seen, C variables must be declared before being used. This can be done in two different ways: at the beginning of a program or at the beginning of a function/code block.

The place where the variable is declared defines its visibility and duration, these are also called variables scope:

When the variable is declared outside any function (at the beginning of the program) it has global visibility, because it can be seen/changed by any function and lasts until the program finishes. This kind of variable is called global variable and has a global scope.

When the variable is declared inside any function or code block, it has local visibility, or, in other words, it is visible/usable only within the function or block it was declared and lasts until the function or block ends.

Therefore, a global variable can be modified or read by any function in the program, but a local variable can only be modified or read within the function or block it was declared. It is not possible to read or modify the local variable of a function from within another function.

Variable declaration follows the format:

```
type identifier <,comma-separated list of identifiers>;
```

The `type` is one of the supported data types listed on table 4.2. The identifier is the name by which the variable will be know through the program.

Identifiers are used to name not only variables but also functions. Valid C identifiers must follow some simple rules:

1. They must comprise only letters, numbers and the underline symbol "_". All other symbols including space are forbidden;

2. They must begin by a letter or underline;

3. Only the first 31 characters are valid (longer identifiers are still valid but can be misleading);

4. C differentiates capital letters from lowercase letters (TEST is different from Test that is different from test);

5. It is not possible to use C keywords as identifiers.

Let us see some examples of variable declarations:

```
int test;                 // declares a signed int variable called test
unsigned int my_var;      // declares an unsigned int variable called my_var
char first, second;       // declares two unsigned char variables
float _result;            // declares a float variable called _result;
```

4.1.3.2. Variable Initialization

It is possible to attribute a value to a variable at the moment it is declared. This can be done by simply attributing a constant value to it at the time it is declared:

```
char v1;          // declares a variable called v1
char v2 = 10;     // declares v2 and initializes with 10 decimal
```

It is also possible to use octal and hexadecimal numbers to initialize variables:

```
char v3 = 0x10;   // declares v3 and initializes with 10 hexadecimal (16
                  // decimal)
char v4 = 010;    // declares v4 and initializes with 10 octal (12 decimal)
```

Character constants can also be used to initialize variables:

```
char v5 = 'A';    // declares v5 and initializes with 'A' (the variable
                  // contents is 65 which is the ASCII code for 'A')
```

Finally, it is also possible to use constant strings to initialize variables:

```
char str[] = {"A string constant"};
```

In this case the compiler automatically reserves the memory amount necessary to store the whole string and its null terminator (C strings are always delimited by a null value (0x00)).

When it is necessary to include a non-printable character into the string, it is possible to use "inverted bar" constants:

Constant	Description
\b	Backspace (ASCII BS)
\f	Form feed (ASCII FF)
\n	New line (ASCII LF)
\r	Carriage return (ASCII CR)
\"	Inverted commas "
\'	Apostrophe '
\0	Null
\\	Inverted bar
\N	Octal constant (N is an octal value)
\xN	Hexadecimal constant (N is a hexadecimal value)

Table 4.3

Below we have an example of inverted bar usage:

```
char str1[] = {"Inverted bar:\\"};    // inserts an inverted bar after the :
Char str2[] = {"test\x20TEST"};       // inserts a space between test and TEST
```

4.1.3.3. Modifiers

In addition to the type modifiers, C language defines some modifiers called storage modifiers that can be used to change the way a variable is stored in the memory or the way the program can manipulate it. These modifiers are:

auto   - defines an automatic variable. Automatic variables can be used only within a function or code block. They are automatically destroyed when the function is exited or when the end of the code block is reached. This modifier is no longer used as ANSI stated that all local variables are **auto** by default.

const   - defines a constant variable (a memory position whose content cannot be modified by the program). These variables are usually stored in the FLASH memory.

extern - defines a global variable that is declared outside the program (in another module that is linked to it).

register - this modifier is used to request the compiler to store the variable into a CPU register instead of a memory address. This usually speeds up the program execution but is limited to local variables or functions formal parameters. This modifier is almost useless on the HCS08 architecture due to the lack of general purpose registers.

static - defines a variable that lasts for the whole program execution. Static variables can be defined locally or globally. Local static variables are not destroyed when exiting the function where they are declared. Also, they conserve their value between function calls. Local static variables can be thought as global variables with local visibility. On the other hand, global static variables act just like global variables with the side effect of being invisible to other modules external to the one where they are declared. Static variables are always initialized with zero at the start of the program.

volatile - this modifier tells the compiler not to apply any kind of optimization or assumption regarding the variable as it can be changed any time by the hardware without notice.

4.1.4. C Operators and Expressions

Operators are elements that can perform any kind of operation (arithmetical, logical and relational, are some examples of operations). These operations usually include one or two elements (the operands). Operators that need only one operand are called unary and operators that need two operands are called binary. There is a special kind of operator (the ? symbol) that needs three operands. This operator is called ternary.

The C language has a very complete set of operators, listed on the tables 4.4 and 4.5.

Operator	Operation	Function
=	Attribution	The left operand receives the value of the right operand
+	Addition	Add two operands
-	Subtraction or unary minus	Subtract two operands or make the operand negative
*	Multiplication	Multiply two operands
/	Division	Divide two operands
%	Integer division remainder	Return the remainder of the integer division of two operands
++	Increment	Add one to the operand
--	Decrement	Subtract one from the operand
&	Bitwise AND operation	Accomplish the bitwise AND operation between two operands
|	Bitwise OR operation	Accomplish the bitwise OR operation between two operands
^	Bitwise Exclusive-OR operation	Accomplish the bitwise Exclusive-OR operation between two operands
~	Negate	Negate the operand (invert all bits)
>>	Right shift	Shift the operand "n" bits to the right
<<	Left shift	Shift the operand "n" bits to the left

Table 4.4

Operator	Operation	Function
!	Boolean NOT operation	Return the opposite logical state of the operand
&&	Boolean AND operation	Return true if both operands are true
||	Boolean OR operation	Return true if any operand is true
>	Greater than (relational)	Return true if the left operand is greater than the right operand
>=	Greater than or equal (relational)	Return true if the left operand is greater than or equal to the right operand
<=	Less than or equal to (relational)	Return true if the left operand is less than or equal to the right operand
<	Less than (relational)	Return true if the left operand is less than the right operand
==	Equal to (relational)	Return true if both operands are equal
!=	Different to (relational)	Return true if the operands are not equal
&	Unary "address of" operator (pointers)	Return the address of the operand
*	Unary "contents of" operator (pointers)	Return the content of the pointer operand
.	Structure element reference operator	Return an element of a structure or union
->	Structure element reference pointer operator	Return an element of a structure or union
?	Conditional expression operator (ternary)	Evaluate the first expression and, if true, returns the result of the second expression. If false, returns the result of the third expression.
,	Sequence operator	Performs a series of operations within an expression
( )	Operation prioritizing operator	Prioritize an operation
[ ]	Array index operator	Points to an array element
sizeof	Compile-time data size operator	Return the size in bytes of the operand

Table 4.5

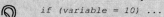

A very common mistake is to use the attribution operator (=) instead of the equality-test (==) in boolean expressions, something like:

`if (variable = 10) ...`

This line is wrong because instead of comparing the values of "variable" and 10, the expression actually attributes the value 10 to the "variable" and also passes the value 10 as the boolean result of the expression (always true in this case).

It is also possible to associate operators originating some reduced form operators:

Reduced Form	Standard Form
x += y	x = x + y
x -= y	x = x − y
x *= y	x = x * y
x /= y	x = x / y
x %= y	x = x % y
x &= y	x = x & y
x |= y	x = x | y
x ^= y	x = x ^ y
x <<= y	x = x << y
x >>= y	x = x >> y

Table 4.6

Now that we know something about the operators of the C language, let us see how to use them to build expressions.

As we know, expressions are a combination of operators, constants, variables and functions that are evaluated to give a result.

In C, expressions are always evaluated from left to right, with higher priority operators taking precedence over lower priority operators. The C language operator precedence is shown in table 4.7.

Precedence	Class	Operator
Higher	Select	( ) [ ] -> .
	Unary	! ~ ++ -- . (type) * & sizeof + −
	Arithmetical	* / %
	Arithmetical	+ -
	Shift	<< >>
	Comparison	< <= > >=
	Comparison	== !=
	Binary bitwise	&
	Binary bitwise	^
	Binary bitwise	|
	Binary Boolean	&&
	Binary Boolean	||
	Ternary	?
	Assignments	= += -= *= /= %= &= |= ^= <<= >>=
Lower	Sequence	,

Table 4.7

The evaluation order is especially important when dealing with increment and decrement unary operators. Consider two **int** variables (a and b), with "a" initially equal to 5. Writing:

```
b = a++;
```

actually attributes the current value of "a" to "b" and then increments "a" by one. The result is: a=6 and b=5!

By changing the order of the evaluation, we change the result of the expression. Considering "a" and "b", with "a" initially equal to 5. Writing:

```
b = ++a;
```

first increments the content of "a" and then attributes the new value of "a" to "b". The result is: a=6 and b=6!

Another interesting feature of the language is the automatic type conversion of different data types on the same expression. When multiple data types are mixed within an expression, the compiler promotes shorter types to the same level of the longer types until all expression operands have the same type. This is known as type promotion and follows some simple rules:

1. In expressions with two or more operands with different types, the shorter type is converted to the type of the longer type. The order followed by the compiler is: **char**, **short**, **int**, **long**, **float**, **double**.

2. In expressions with an **unsigned** type and a **signed** type, the **signed** type is converted to **unsigned**.

Remember that type promotion is done on an operand basis, according to the precedence rules we explained above.

Some examples of the automatic type promotion:

```
float a,b;
double c,d;
```

```
long int e,f;
int g,h;
char i,j;
unsigned int k,l;
...
g = h + i;          // i is promoted to signed int
e = g + j;          // j is promoted to signed int, the result is promoted to
                    // long int
e = a + h + j;      // a is promoted to long int (the value is truncated and
                    // the fractionary part), h and j are promoted to long
                    // int (contents are sign-extended)
```

When attributing a larger-type variable to a shorter-type variable, the result is truncated as shown below:

```
long va;
int vb;
char vc;
...
va = 0x12345678;
vb = va;       // vb = 0x5678;
vc = va;       // vc = 0x78;
```

C also admits explicit type conversions by using typecast operations. This can be done by using the type specifier before the variable name:

```
(signed) variable
(unsigned) variable
(char) variable
(int) variable
(long) variable
(float) variable
```

Notice that on expressions, numeric constants always default to the lowest possible type. This can lead to an unexpected behavior, for example:

```
unsigned long int a;
unsigned int b;
a = b * 1000;
```

Even considering that "a" can hold a **long** variable (in our case a 32-bit value), the result of b*1000 is 16-bit long (because "b" and "1000" are both 16-bit values). If b>65 the result overflows and only the lowest significant word is stored into "a".

To avoid this kind of problem, it is possible to force the constant to be a **long** value by adding the "l" or "L" suffix:

```
a = b * 1000L;
```

This forces the expression to be computed using 32-bit numbers leading to a correct answer. The same can be done by casting "b" to **long**:

```
a = (long)b * 1000;
```

4.1.5. Complex Types

Besides the basic types shown above, C also allows the construction of complex data types such as those we will see below.

4.1.5.1. Enumerations

Enumerations are a set of integer constants used to specify the possible values a variable can assume.

Enumerations are declared using the following syntax:

```
enum enumeration_name
{
  identifier_list
} variable_name
```

Elements are separated by commas. By default, the first element receives the zero value; the second receives one and so on. It is also possible to specify the order number of each element by simply attributing the desired value to the element.

Each element in the list receives the enumeration value of the previous element plus one. This means that by changing an element value, all the subsequent elements have their enumeration order changed accordingly.

Examples:

```
enum week { mon, tue, wed, thu, fri, sat, sun} days;
// mon=1, tue=2, wed=3, thu=4, fri=5, sat=6, sun=7
enum boolean { true=1, false=0};

// to create a enumerated variable using the boolean type:
enum boolean my_boolean_variable;
```

 Enumerations are especially useful for coding state machines: each state can be assigned a unique enumerated state.

4.1.5.2. Pointers

Pointers are variables that store addresses instead of data. They are one of the strengths that make the C language so powerful.

Figure 4.1 shows an example of pointers in action. There are two pointer variables: p1 and p2. Pointer p1 is stored at address 0x0100 and pointer p2 is stored at 0x0102.

The first pointer content is equal to 0x0104. This value is not data; it is an address. If we look at address 0x0104 we find the data1 variable. This means that p1 is pointing to data1 (which actually contains 0x32).

In the same way, p2 content is equal to 0x0106, which is the address of the aux variable. Therefore, p2 is pointing to aux (which actually contains 0x10).

Address	Variable	Content
0x0100	p1	0x0104
0x0102	p2	0x0106
0x0104	data1	0x32
0x0105	data2	0xF1
0x0106	aux	0x10

Figure 4.1

There are two pointer operators in C: & and *. The & operator returns the address of its operand while * returns the content of the variable located at the address specified by its operand.

Using the example of figure 4.1, we can write:

```
unsigned char data1, data2, aux;
unsigned char *p1, *p2;

main
{
  p1 = &data1;      // p1 gets the address of data1
  p2 = &aux;        // p2 gets the address of aux
  data1 = 0x32;
  aux = 0x10;
  data2 = *p1;      // data2 gets the value of the memory address pointed to
                    // by p1 (data2 = data1 = 0x32)
}
```

Listing 4.2

The listing above shows the usage of the & and the * operators. Even more, it also shows how to declare a pointer variable. All it is needed is to place the * before the name of the variable in the declaration statement:

```
unsigned char *p1, *p2;
```

This line declares two **char** pointer variables. By default, on the HCS08, each pointer occupies two bytes of memory (because addresses are 16-bit), that is why the variable p2 on figure 4.1 is located at address 0x0102 (two bytes after p1). Note that **__near** pointers use 1-byte address thus occupying only one byte; on the other hand, **__linear** pointers occupy three bytes (because they point to 24-bit addresses).

On the HCS08 CPUs, pointers make intensive use of the H:X register pair and indexed addressing mode. The simple attribution: `p1 = &data1;` compiles to:

```
LDHX #0x0104    ; load H:X with the address of data1
STHX 0x0100     ; store H:X at the address of p1
```

Using symbolic code:

```
LDHX #data1     ; load H:X with the address of data1
STHX p1         ; store H:X at p1
```

Another important topic on the pointer subject is the pointer arithmetic: pointers can only be added or subtracted.

Pointer arithmetic must also comply with the base pointer type: **char** pointers point to bytes of memory, **int** pointers point to words of memory, **long** pointers point to double-words and so on.

That said, let us see some examples on pointer arithmetic:

```
char *p1;   // pointer to byte data
int  *p2;   // pointer to word data (2 bytes)
long *p3;   // pointer to double-word data(4 bytes)

// Let us arbitrate the following addresses:
// p1 = 0x0100, p2 = 0x0110 and p3 = 0x0120
// Now let us do some pointer math:

p1++;       // p1 = 0x0101
p2++;       // p2 = 0x0112 (0x0110 + 2)
p3--;       // p3 = 0x011C (0x0120 - 4)

// It is also possible to attribute one pointer to another:
p1 = p2;    // p1 = p2 = 0x0112

// Another interesting example of pointer math:
p2 = p2 + 3;        // p2 = 0x0118 (0x0112 + (3 * 2))
```

Listing 4.3

C also admits pointers to functions: it is only necessary to attribute the name of the function to the pointer variable:

```
void *p4;

p4 = main;// p4 points to the main function
```

Notice the use of the **void** type for the pointer declaration: **void** pointers are generic pointers that can point to any data type.

 Performing any arithmetic operation on void pointers is not allowed!

Pointers can also point to another pointer (instead of pointing to a data variable); this is called indirection. Despite being possible, indirection is not recommended because it can lead to extremely hard-to-find bugs.

4.1.5.3. Arrays

Arrays are sequential and contiguous sets of elements of the same data type. The C language admits one-dimensional arrays (also called lists) or multi-dimensional arrays (also called tables) using any valid data types (including pointers and other complex data types).

Array declaration is done by specifying the maximum number of elements in any dimension using the following syntax:

```
data type array-name [max1] [max2] [maxN] ...;
```

In which:

- *data type* is the base data type for the array elements;

- *array-name* is the name of the variable;

- *max1* is the maximum number of elements in the first dimension;

- *max2* and *maxN* are the maximum number of elements in other dimensions.

Declaration examples:

```
unsigned char example1[10];// declares an array called example1 with
                           // 10 elements of unsigned char type
                           // This array occupies 10 bytes
long array_A[10][10];      // declares an array called array_A with
                           // 100 elements (10 * 10) of long type
                           // This array occupies 400 bytes
int test [3][4][5];        // declares a tri-dimensional array called
                           // test with 60 elements (3 * 4 * 5) of
                           // int type. This array occupies 120 bytes
                           // of memory
```

It is also possible to declare and initialize an array on the same statement:

```
// Declares an uni-dimensional array called temp with 5 elements: the
// first is 10, second is 20, third is 31, fourth is 44 and fifth is 59
char temp[5] = {10,20,31,44,59};
// Declares an uni-dimensional array called name with 5 elements: F, a, b, i, o
char name[5] = {'F', 'a', 'b', 'i', 'o'};
```

Multi-dimensional arrays can also be initialized using the same process:

```
int aux [2][3] =
{
  1,2,3,
  4,5,6
}
```

References to elements within an array are done by specifying the element index:

```
x = temp[0];            // x = 10
y = temp[2];            // y = 31
z = temp[4];            // z = 59
x1 = aux [0][2];        // x1 = 3 (first line, third column)
x2 = aux [1][1];        // x2 = 5 (second line, second column)
x3 = aux [1][0];        // x3 = 4 (second line, first column)
```

Notice that the fifth array element is the one with index 4. C arrays are always indexed by zero and the last element is the total number of elements in the dimension minus one.

Another interesting point on arrays is their proximity with pointers: in fact, a reference to the name of the array without any index is a pointer to the first element of the array. This means that:

```
x = temp[0];       // x = 10;
```

Is the same as:

```
x = *(temp);       // x = 10;
```

It is also possible to index array elements by using pointer arithmetic:

```
x = *(temp+2);       // x = temp[2] = 31
```

Finally, there is a special kind of array called string: strings are character-arrays dedicated to storing character sequences (texts). They are also null-terminated (their last character is a null (\0) character).

Strings are declared exactly like any other **char** array (remember to include one additional element for the null terminator):

```
char buffer[15] = {"This is a test"};
```

Notice that the null terminator is automatically inserted by the compiler when the string is initialized (string constants are specified by using inverted commas).

4.1.5.4. Structures

Structures are groups of individual variables of any type, referenced by a common name. Elements of structures are referenced as structure fields.

The declaration syntax is the following:

```
struct structure_name
{
  type field1;
  type field2;
  ...
} variables;
```

The declaration of the name of the structure is optional, as well as the immediate declaration of variables using the structure. It is possible to declare an unnamed structure, as long as the variables are declared immediately in the structure declaration.

On the other hand, it is also possible to declare a named structure and later declare the variables using it. As an example, let us see the structure declaration comprising of three **char** fields: hours, minutes and seconds.

```
struct stime
{
  char hours;
  char minutes;
  char seconds;
} time;
...
...
// Now, let us declare a new variable called alarm using the stime
// structure:
struct stime alarm;
```

Reference to a structure field is made by using the structure variable identifier followed by a dot and the field name. Therefore, to initialize the time variable with the value 8:45:30 all you have to do is:

```
time.hours = 8;
time.minutes = 45;
time.seconds = 30;
```

It is also possible to access a structure element by using a pointer to the structure. In this case, the structure element is pointed to by using the arrow operator "->":

```
structure pointer->field = value;
```

Based on the "stime" structure defined above, we could use pointers as in the following example:

```
struct stime *ptime;
...
...
ptime = &time;          // ptime is pointing to the time structure
ptime->minutes = 10;    // stime.minutes = 10
```

The C language also supports a special type of structure called bit fields. They are structures in which the field size can be defined by the programmer. Bit fields are especially useful to access specific bits within a variable or specific bits in a register.

Declaration of a bit field is done like the declaration of a standard field, but after the field identifier it is necessary to include ":" and the number of bits for that field. Referencing a bit field is done in the same way of standard fields.

Below we have a simple example of a bit field declaration and usage:

```
struct sbitfields
{
  char bit0        : 1;    // first bit field with 1 bit
  char fieldA      : 3;    // second bit field with 3 bits
  char fieldB      : 8;    // third bit field with 8 bits
  int  fieldC      : 12;   // fourth bit field with 12 bits
} bfield;
...
...
bfield.bit0 = 1;
bfield.fieldC = 5;
```

 *Notice that **char** bit fields can hold up to 8 bits, **int** bit fields can hold up to 16 bits and **long** bit fields can hold up to 32 bits! Note that signed bit fields hold one bit less than unsigned bit fields!*

It is important to notice that each field has a limited storage capacity. If you try to assign a larger value, the data actually stored will be truncated.

Trying to do:

```
bfield.fieldA = 10;        // 10 decimal = 1010 binary
```

actually stores 2 to "fieldA", because this bit field can only store three bits of data. Trying to store 10 decimal (1010 binary) into the field truncates the value to fit into the size of the field. As a result, the stored value is 010 (2 decimal).

 Be careful when using structures and bit fields because they can quickly increase the amount of code memory necessary to perform operations involving fields!

4.1.5.5. Unions

Unions are special data types used to share memory addresses among multiple identifiers of different types.

The compiler allocates memory to store the biggest type within the **union**, the other members are allocated within the same space.

For HCS08 devices, the compiler must follow the big endian model for storing the elements of the union.

 On the little endian format, variables larger than a byte are stored from the lowest byte to the highest byte (the highest address stores the highest significant byte); whereas on the big endian format, the storage is done from the highest to the lowest byte (the highest address stores the lowest significant byte). Little endian = LSB first, Big endian = MSB first!

Declaration of a **union** follows the same rules for structures:

```
union union-name
{
  type field1;
  type field2;
  ...
} variables;
```

The same declaration can specify the name of the union and the list of variables being declared. It is also possible to declare a **union** without any variable or without a name.

```
union u32
{
  unsigned long int var32;
  unsigned int var16[2];
  unsigned char var8[4];
} test;
```

By writing:

```
test.var32 = 0x12345678;
```

The **union** is stored into memory as shown in figure 4.2.

Field	Memory address			
	Base address	**Base address+1**	**Base address+2**	**Base address+3**
var32	0x12	0x34	0x56	0x78
var16[0]	0x12	0x34		
var16[1]			0x56	0x78
var8[0]	0x12			
var8[1]		0x34		
var8[2]			0x56	
var8[3]				0x78

Figure 4.2

Unions are useful for converting between different data types. Using the **union** above, we can easily access each individual byte or individual words of the **long int** var32 variable.

4.1.5.6. Type Definitions

Another interesting characteristic found on the C language is the ability to define new names for existing data types. This is done by using the **typedef** keyword.

The greatest use of this statement is to increase program portability between different platforms.

Let us say we need to port an application from platform A to platform B. On platform A, ints are 16-bit whereas on platform B they are 32-bit.

By using **typedef** to specify new platform-independent types, we can easily port the application:

```
// Platform A (ints are 16-bit)
typedef unsigned int u16;  // u16 is a 16-bit variable
typedef unsigned long u32; // u32 is a 32-bit variable


// Platform B (ints are 32-bit)
typedef unsigned short u16;    // u16 is a 16-bit variable
typedef unsigned int u32;      // u32 is a 32-bit variable
```

For the sake of portability all 16 and 32-bit variables must be declared using the u16 and u32 types:

```
u16 var16, test, aux;       // declares 16-bit variables
u32 var32, test32, temp;    // declares 32-bit variables
```

4.1.6. Statements

The C language includes a set of commands organized in the following categories:

- **Conditional statements**: for executing statements based on a condition being evaluated as true or false;

- **Iteration, loop or repetition**: for executing repetitive code;

- **Jump or branch**: for changing the program flow.

4.1.6.1. If

The **if** statement allows execution of a command or command-block based on the evaluation of a boolean expression.

The syntax is shown below:

```
if (expression) command1;
```

The expression must be specified between parenthesis and it is evaluated as a boolean expression with only two possible results: true, if it evaluates to a non-zero value, or false, if it evaluates to a zero value.

The "command1" is executed if and only if the expression evaluates true. If more than one command must be executed, they must be written within a block "{ }":

```
if (expression)
{
    command1;
    command2;
    ...
}
```

It is also possible to execute a command or command-block case the condition evaluates false. This is done by the **else** clause:

```
if (expression) command1; else command2;
```

In this case, "command2" is executed if and only if the expression evaluates false.

The C language also offers a powerful conditional operator: the "?". The syntax is:

```
(condition) ? command for true : command for false
```

The operation is very simple: the condition is evaluated. In case it is true, the first command is executed, in case it is false, the second command is executed.

The ? operator can be used for condition assignments:

```
variable = condition ? value if true : value if false
```

Typical uses for the ? operator are macro definitions like:

```
#define MAX(a, b) (((a)>(b))? (a): (b))
```

This macro returns the greatest value between "a" and "b". Example:

```
x = MAX(1,10);    // x = 10
a = MAX(b,c); // if b>c then a=b else a=c
```

4.1.6.2. Switch

The **switch** statement allows selecting one among many options, according to the content of a variable or expression.

The typical syntax is:

```
switch (expression)
{
    case constant_1:
        statement_1;
```

```
      statement_2;
      ...
      statement_n;
      break;
   case constant_2:
      statement_12;
      statement_22;
      ...
      statement_n;
      break;
   case constant_3:
   case constant_4:
      statement_14;
      statement_24;
      ...
      break;
   case constant_5:
      statement_15;
   case constant_6:
      statement_16;
      break;
   default:
      statement_17;
      statement_27;
      ...
      statement_n;
}
```

The "expression" can be any integer variable or an expression that results in an integer value. The statements associated with a specific **case** are executed when the "expression" results in a value equal to the constant value associated with the **case**.

The statements associated with a specific case are executed until a **break** statement is found. This allows the same portion of code to be executed for multiple values of the "expression" (constants 3 and 4 in the syntax example).

It is also possible to partially execute a portion of code. In the syntax example, the statement_16 is executed when "expression" is equal to "constant_6". When "expression" is equal to "constant_5", two statements are executed: statement_15 and statement_16. This behavior is called "fall through".

When no **case** statements meet the value of "expression", the statements associated with the optional **default** clause are executed.

 Each case statement must have a unique constant associated with it. One constant must not allowed to appear in more than one case!

 *Typical uses of **switch** statements are on coding finite state machines (FSMs).*

4.1.6.3. While

The **while** statement allows repetitive and conditional execution of a single or multiple statements.

The syntax is:

```
while (expression)
{
   code-block
}
```

If the "expression" is evaluated true, the statements/expressions within the code-block are executed one by one and when the end of the block is reached. The program flow returns to the expression-evaluation on the while statement. The process is repeated until the expression is evaluated false, when the program flow diverts to the statement immediately after the closing curly brace "}".

It is important to notice that if the expression is initially false, the statements within the **while** statement will never be executed!

The language also admits the use of **break** and **continue** clauses for controlling the current iteration of the **while** statement: the **break** clause causes the current iteration to be finished and the **while** statement exited, while the **continue** clause finishes the current iteration and returns to the expression-test on the **while** statement.

4.1.6.4. Do-while

The **do-while** construction works in a way that is similar to the **while** statement, but the expression evaluation is done at the end of the loop. The syntax for the **do** statement is:

```
do
{
   code-block
} while (expression)
```

As the expression evaluation is done at the end of the block, the statements within the **do while** constructions are executed at least once, even if the expression is initially false.

It is also possible to use of **break** and **continue** clauses for controlling the current iteration of the **do-while** construction.

4.1.6.5. For

This is a loop statement that allows conditional and repetitive execution of single or multiple statements.

The statement structure is the following:

```
for (initialization; expression; increment)
{
   code-block
}
```

The operation follows these procedures:

1. The "initialization" section is run;

2. The "expression" is evaluated and if the boolean result is true, the code-block is fully executed, otherwise the **for** structure is exited;

3. After the code-block, the "increment" section code is run;

4. After the "increment" section, the cycle repeats from step 2 above.

A progressive counter (0 to 10):

```
for (counter=0;counter<=10;counter++);
```

A regressive counter (10 down to 0):

```
for (counter=10;counter;counter--);
```

All sections within a **for** statement are optional. In fact, it is possible to have no sections at all:

```
for (;;);    // an infinite loop
```

As with **do** and **do-while** constructions, **for** allows using **break** and **continue** clauses for controlling the current iteration.

4.1.7. Functions

Functions are the building block of the C language. Basically a function is a set of statements and/or expressions used to perform a specific task.

Their main purpose is to allow reusing blocks of code among different parts of the application, allowing the reduction of the code size and code duplicity, at the same time it increases code readability by decomposing complex codes into simpler and smaller ones.

Functions are translated into assembly subroutines (called using the JSR/BSR instructions and returning through the RTS instruction) They can receive data (function parameters) from the caller and also send back data to it. The data interchange between the function and the caller is generally done through the stack.

Function declaration is done by specifying the return type, function name and its parameters. The generic syntax is:

```
type_specifier function_name (parameter list)
{
    function body ...
}
```

A function can receive (and return) any standard C data type.

Data can be passed to the function in two different ways:

1. By value: the actual parameter value is passed as the function argument;

2. By reference: the address of the parameter is passed as the function argument.

It is possible to use each one or both models in the same function. Notice that the function call must use the same model of the function declaration.

The main purpose of "by reference" parameter passing is allowing the function to modify the parameter value globally, instead of locally as in the "by value" parameter passing.

By default, function calls use the "by value" model. The only exception is when an array is used as an argument. In this case, the address of the first element of the array is passed to the function (instead of the whole array). Implementation of the "by reference" model is done by using pointers.

The next example shows the implementation of a function to calculate the average of two numbers. The resulting value is returned to the caller:

```
int average (int pa, int pb)
{
    long temp = ((long)pa + pb)/2;
    return (temp);
}
```

The first line declares the name of the function (average), the two parameters it receives ("pa" and "pb", both ints) and also states the return-type (**int**). If a function is declared as **void**, it does not return any value to the caller.

 When the function has multiple parameters, it is necessary to declare their type one-by-one.

Notice the "temp" variable declared at the beginning of the code of the function: this auxiliary variable is local to the function. It is created when the function is called and destroyed when the function returns. Parameters "pa" and "pb" are also considered local variables because they last until the function exits.

The **return** statement is not obligatory: it may be used to exit the function and return a specific value to the caller. The function also exits when all code within the body of the function is executed.

The function can be called as shown below:

```
result = average(variable_a,variable_b);
```

Or:

```
result = average(100,200); // calculate the average of 100 and 200
```

To make a "by reference" function call, it is necessary to declare the desired parameters as pointers. Moreover, the function call must also use pointers instead of simple variables.

To demonstrate this, let us take a look at a simple swap function: its purpose is to exchange the values of two variables. The first attempt to implement such function could be something like:

```
void swap (char v1, char v2)
{
    char temp;
    temp = v2;
    v2 = v1;
    v1 = temp;
}
```

Although the code looks correct, it is not: the values of "v1" and "v2" are exchanged as desired but this happens only locally. The caller variables are not exchanged at all.

The desired operation only works if we use the "by reference" function call. The second attempt to implement the swap function could be:

```
void swap (char *v1, char *v2)
{
    char temp;
    temp = *v2;
    *v2 = *v1;
    *v1 = temp;
}
```

The swap function can be called as shown below:

```
swap (&var1, &var2);        // exchange "var1" and "var2" contents
```

4.2. Compiler Characteristics

The HCS08 C compiler included in Codewarrior complies with the ANSI C89 standard and also implements some features of C99. There are also some language extensions to allow interrupt processing and a better code optimization.

4.2.1. Language Extensions

Besides the ANSI C reserved keywords, the HCS08 compiler includes a set of specific keywords and operators to allow a better code optimization. Table 4.8 shows the additional keywords/operators.

@	__far	__near	__linear	LINEAR
__alignof__	__va_sizeof__	__interrupt	__asm	

Table 4.8 – C language extensions

The operation of each keyword/operator is described below:

@ -   this operator can be used to allocate the variable into a specific memory segment. It is also possible to use it to specify an absolute memory address in which the (global) variable will be stored. Example:

```
// declare a global unsigned int variable and store it at address
// 0x0080
int my_variable @ 0x0080;
```

__far -   this keyword tells the compiler to use 16-bit addresses when accessing the variable, allowing access to variables located in any of the 65536 possible addresses. When used on pointer declarations this keyword acts as an access modifier, affecting the last variable declared on its left. The compiler also supports the **far** alias for **__far**. Examples:

```
// declare a far pointer to a far pointer to a char
char *__far *__far my_far_pointer;
// declare a pointer to a far pointer to an int
int *__far *my_far_pointer;
```

It is also possible to use this keyword with arrays. This can be done as follows:

```
// declare the function parameter as an element of the "param" array
void test_function (char param[2] __far);
// or by using a single pointer:
void test_function (char *__far param);
```

__near -   the near keyword is used to allow faster addressing on variables stored into the direct page. It works in the opposite way of the far keyword. The compiler also supports the **near** alias for **__near.**

__linear -   the __linear keyword is used as a pointer qualifier when accessing data in the extended (higher than 64KiB) memory on MMU-enabled devices.

LINEAR -   the linear keyword is used to specify linear addresses in extended (higher than 64KiB) memory on MMU-enabled devices. For example:

```
// declare a variable at linear address 0x010000
const char my_var @ LINEAR 0x010000 = 10;
```

__alignof__ -   returns the alignment of the specified type.

__va_sizeof__ -   returns the size (in bytes) of the specified data type (even after an eventual automatic casting).

__interrupt - this keyword tells the compiler that the function is called automatically by the hardware when an interrupt occurs. The keyword must be followed by a numeric constant which specifies the interrupt vector number. The compiler also supports the **interrupt** alias for **__interrupt.** Since version 6.0, the Codewarrior IDE includes pre-defined symbols for all interrupt vectors (refer to tables 2.1 to 2.6 for all related symbols):

```
// SWI interrupt servicing function
void interrupt VectorNumber_Vswi void swi_service(void);
// IRQ interrupt servicing function
void interrupt VectorNumber_Virq trata_irq(void);
// ADC interrupt servicing function
void interrupt VectorNumber_Vadc trata_adc(void);
```

__asm -   allows assembly instruction insertion into the C source code. We will discuss this operation in greater details later in this chapter. The compiler also supports the **asm** alias for **__asm.** Example:

```
// Insert a STOP instruction
asm stop;
```

There are also some useful pragmas which can be used to configure the way the compiler generates the code. The following options can be used following a **#pragma**:

CODE_SEG –   to specify a memory segment in which the functions will be stored. Example:

```
#pragma CODE_SEG MY_SEG
// functions following the pragma are all stored at MY_SEG segment
void my_func (void)
{
    ...
}
```

```
#pragma CODE_SEG DEFAULT
// functions following the pragma are all stored at the
// default segment (usually DEFAULT_ROM)
```

CONST_SEG –   to specify the memory segment in which the constant data will be stored. Example:

```
#pragma CONST_SEG MY_CONST_SEG
// const variables following the pragma are stored into
// MY_CONST_SEG memory segment
const char myconst1=50;
```

```
#pragma CONST_SEG DEFAULT
// const variables following the pragma are stored into
// the default const segment (usually ROM_VAR)
```

DATA_SEG –   this option works like the CONST_SEG but is valid for variables. All options valid for CONST_SEG are also valid for DATA_SEG. By default, data is stored into the DEFAULT_RAM segment.

STRING_SEG –   this option works like the CONST_SEG but is valid for strings. All options valid for CONST_SEG are also valid for STRING_SEG. By default, strings are stored into the STRINGS segment.

INLINE –   instructs the compiler to inline the next function. This embeds the full function code in place of the function call. Inline functions are faster but use more memory than standard functions. Example:

```
#pragma INLINE
void swap (char *v1, char *v2)
{
    char temp;
    temp = *v2;
    *v2 = *v1;
    *v1 = temp;
}
```

This option has the same effect of the –Oi compiler option. Notice that the –Oi option can also be used to force inlining of functions smaller than "n" bytes (the use of –Oi=c50 inlines all functions with code smaller than 50 bytes).

Some functions cannot be inlined despite using any of the inlining options:

- functions with default arguments;
- functions with labels inside;
- functions with an open parameter list;
- functions with inline assembly statements;
- functions using local static objects.

mark –   adds an entry into the Codewarrior editor function list. All text following the pragma is used as an entry name. These entries can be accessed through the "marks list" button on the editor window (refer to figure 3.10 for details). Examples:

```
#pragma mark Variable definition
char va, vb, vc;
#pragma mark test
void test (int subject)
{
    ...
}
```

NO_ENTRY –   suppresses the entry code generation (useful for assembler functions). The user must ensure the full context restoration before returning to the caller. Examples:

```
#pragma NO_ENTRY
void my_asm_func (void)
{
    ...
}
```

NO_EXIT –   suppresses the exit code generation (useful for assembler functions). The user must ensure the full context restoration before returning to the caller.

NO_FRAME –   suppresses the frame code generation (useful for assembler functions). The user must ensure the full context restoration before returning to the caller. Examples:

```
#pragma NO_ENTRY
#pragma NO_EXIT
#pragma NO_FRAME
void my_asm_func (void)
{
    ...
}
```

NO_RETURN –   instructs the compiler that the function does not return and thus, no return instruction (RTS, RTI or RTC) should be generated for the function. Example:

```
#pragma NO_RETURN
void special_func (void)
{
    ...
}
```

NO_STRING_CONSTR –

disables the special handling of the # as a string constructor. This is useful when embedding assembly instructions into C macros. Example:

```
#pragma NO_STRING_CONSTR
#define lda(x) asm lda #x
```

For other **#pragma** options refer to the Codewarrior help files and documentation.

4.2.2. Compiler Options

The configuration of the compiler options is done on the "P&E PEDebug FCS-ICS-ICD Settings" window that is shown by pressing the ALT-F7 keys, by clicking on button 13 (figure 3.9) or by selecting the "Edit > Standard settings" option on the main menu. After opening the window, select the "Compiler for HC08" item (under the target node) as shown in figure 4.3.

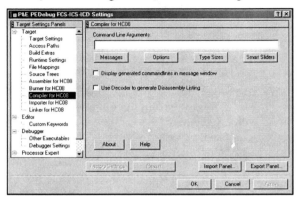

Figure 4.3

This window allows modifying many of the compiler parameters and the way the code is generated.

There are four main buttons. Each one opens a different configuration window:

Messages – on this window, it is possible to configure which messages will be generated by the compiler and also place them into different categories. It is also possible to disable unwanted messages;

Options – this window enables the configuration of the many compiler options divided into seven categories: optimizations, output, input, language, host, code generation and messages. We will discuss further about optimizations and code generation later in this chapter;

Type sizes – allows selecting the size (in bytes) for all standard C types. This is useful when porting code from other platforms (with different type sizes);

Smart Sliders – this button opens a window with sliders that allow a graphical selection of five main categories: code density, execution speed, debug complexity, compilation time and information level. By changing the sliders, the user can balance the variables for the desired compromise.

Notice that all changes done on these windows are valid for the whole project and affect all files within it.

4.2.3. Memory Models

The compiler uses some standard models for controlling how the data access is done. There are three memory models available:

- **Small**: this is the default model. All pointers and functions are addressed using 16-bit addresses, allowing code and data to be located anywhere within the 65536 possible addresses. It is possible to place variables into the direct page (by using **#pragma** DATA_SEG directive) and use the **__near** modifier to instruct the compiler to use the DIR addressing mode (which is faster than the default EXT addressing mode used in the small memory model). This model uses one of the following standard ANSI libraries:

o ANSIIS – no floating point support;

o ANSIS – floating point support (floats and doubles are 32-bit);

o ANSIFS – floating point support (floats are 32 and doubles are 64-bit).

- **Tiny**: on this model, all variables and the stack are placed into the direct page (first 256 memory addresses). Non-qualified pointers use 8-bit addressing mode (DIR). This memory model is faster than the small model but with a limited memory space (it is still possible to use the __**far** keyword to place variables outside the direct page). This model uses one of the following standard ANSI libraries:

o ANSITIS – no floating point support;

o ANSITS – floating point support (floats and doubles are 32-bit);

o ANSITFS – floating point support (floats are 32 and doubles are 64-bit).

- **Banked**: the banked memory model is suited for devices with large memories (such as the AC and QE devices). This model uses one of the following standard ANSI libraries:

o ANSIBIM – no floating point support;

o ANSIBM – floating point support (floats and doubles are 32-bit);

o ANSIBFM – floating point support (floats are 32 and doubles are 64-bit).

4.2.4. Supported Data Types

The HCS08 compiler supports all ANSI-C standard data types with a big endian alignment for data larger than eight bits (the lowest memory address stores the highest significant byte of the variable).

The floating point variables are stored according to the IEEE754 standard (**float** and **double** sizes can be changed by the user).

Floats are stored according to the IEEE32 format, using the following formula:

$$-1^{\text{sign}} * 2^{(\text{exponent}-127)} * 1.\text{mantissa} \, .$$

Doubles are stored according to the IEEE64 format, using the following formula:

$$-1^{\text{sign}} * 2^{(\text{exponent}-1023)} * 1.\text{mantissa} \, .$$

Figure 4.4 shows how the floating point data is stored into memory.

IEEE32	Sign	exponent (8-bits)	mantissa (23-bits)	
IEEE64	Sign	exponent (11-bits)		mantissa (52-bits)

Figure 4.4

4.2.4.1. Bit Fields Allocation

On this topic we will discuss more about bit fields and how they are stored into the microcontroller memory.

The first thing to know is that the memory allocation is done on a byte basis. By default, the first structure element is stored in the least significant byte. The remainder bits are filled following the declaration order of the structure. The compiler option "Bit Field Byte Allocation" allows selecting between two allocation models: the default model starts by the least significant bit (-BfaBLS option). It is also possible to instruct the compiler to start by the most significant bit (-BfaBMS option).

Let us see some examples on bit fields allocation. Consider the following structure:

```
struct
{
    unsigned char first    : 1;
    unsigned char second   : 1;
    unsigned char third    : 6;
} my_bit_field;
```

This structure occupies only one byte of memory, filled as shown in figure 4.5.

Bit 7	Bit 6	Bit 5	Bit 4	Bit 3	Bit 2	Bit 1	Bit 0
third						second	first

Figure 4.5

We can set a variable with the following expression:

```
my_bit_field.first = 1;
```

This generates the following C code:

```
LDHX    my_bit_field.first
LDA     ,X
ORA     #0x01
STA     ,X
```

However, this code can be substantially improved by storing the structure into the direct page memory:

```
#pragma DATA_SEG __DIRECT_SEG MY_ZEROPAGE
volatile struct
{
    unsigned char first    : 1;
    unsigned char second   : 1;
    unsigned char third    : 6;
} near my_bit_field;
```

Now, the same attribution:

```
my_bit_field.first = 1;
```

Leads to the following code:

```
BSET  0, my_bit_field.first
```

The reason for the improvement is very simple: by forcing the structure to be stored into the direct page, the compiler can use the bit manipulation instructions (such as BSET and BCLR) to set/clear bits into the structure fields.

We can also force another alignment for the structure. By using the "-BfaBMS" option, the memory allocation for the structure is the one shown in figure 4.6. In this case, code efficiency is even better because the field "third" is stored into the least significant bits of the memory (less code is needed to store/retrieve data from field "third").

Bit 7	Bit 6	Bit 5	Bit 4	Bit 3	Bit 2	Bit 1	Bit 0
first	second	third					

Figure 4.6

If the sum of all fields is greater than eight, the compiler allocates more bytes to store the remaining fields. The allocation strategy is very simple: by default, the compiler allocates memory starting by the least significant bit (or the most significant one, depending on the selected compiler option) and fills the memory in sequence, avoiding crossing byte boundaries whenever possible.

Moreover, the compiler avoids leaving a byte with two or more bits unused. In this case, it distributes the fields in a non-sequential fashion, for better memory space usage.

To demonstrate the explained above, let us show a more complex bit field structure occupying three bytes:

```
struct
{
    unsigned char first     : 3;
    unsigned char second    : 3;
    unsigned char third     : 3;
    unsigned char fourth    : 2;
    unsigned char fifth     : 4;
    unsigned char sixth     : 5;
} my_struct2;
```

This structure, when compiled with the "-BfaBLS" option (structure alignment starting from left), results in the allocation shown below:

Address								Address + 1								Address + 2							
7	6	5	4	3	2	1	0	7	6	5	4	3	2	1	0	7	6	5	4	3	2	1	0
fifth				fourth		third			second			first							sixth				

Figure 4.7

Using the "-BfaBMS" option (structure alignment starting from right), you will obtain the following allocation:

Address								Address + 1								Address + 2							
7	6	5	4	3	2	1	0	7	6	5	4	3	2	1	0	7	6	5	4	3	2	1	0
first			second			third		fourth			fifth					sixth							

Figure 4.8

Care should be taken when using bit fields: accessing non-unitary fields is frequently slower than accessing a standard variable. The compiler generates more code to store data into such variables.

4.2.5. Calling and Returning Conventions

The compiler follows some simple conventions for parameter passing on function calls:

1. Parameters are passed using the stack and are pushed from the left to the right (when possible, the compiler uses CPU registers for parameter passing, as described below);

2. If the function has a fixed number of arguments and the last one is one-byte long, it is passed in A. If the parameter next to the last is one byte, it is passed in X. If the parameter next to the last is two bytes, it is passed in H and X;

3. If the function has a fixed number of arguments and the last one is two bytes, it is passed using registers H and X;

The return value is passed according to the table 4.9:

Type	Register Used
char	A
int	H:X
Pointers or Arrays	H:X

Table 4.9

For larger types the compiler returns the address to a copy of the value using registers (H:X).

4.2.6. CPU-Flags Access Functions

To enable accessing the CPU flags (located in the CCR register), the compiler includes four intrinsic functions:

`__isflag_carry()` – return true if C = 1.

`__isflag_half_carry()` – return true if H = 1.

`__isflag_overflow()` – return true if O = 1.

`__isflag_int_enabled()` – return true if I = 0.

The use of these functions is quite straightforward. The following example tests the I flag and executes the attribution a = 0 if I = 0.

```
if (__isflag_int_enabled() ) aux = 0;
```

The greatest advantages of using these intrinsic functions rely on their high efficiency and reduced code size. For the last example, the generated code is just three-instructions long:

```
BMS     NEXT
CLRA
STA     aux
NEXT:
...
```

4.3. Embedding Assembly Into C Code

The compiler allows inserting assembly code directly into C source code (by using the keyword **asm**); this is called HLI (**H**igh **L**evel **I**nline) assembler.

While using HLI code inside a C source code, the programmer must know some important compiler characteristics:

- Assembly constants can be defined by using the directives: DCB (defines an 8-bit integer constant), DCW (defines a 16-bit integer constant), DCL (defines a 32-bit integer constant), DCF (defines a 32-bit **float** constant), DCD (defines a 64-bit **double** constant).

- Global variables are stored at absolute addresses, whereas local variables and function formal parameters are stored in the stack. Accessing C variables within the assembly code can be done by simply writing the name of the variable as the assembly instruction operand. All addresses are assumed to be **char***. For example:
  ```
  __asm lda aux      // load A with the value of the variable "aux"
  __asm sta result   // store A in result
  ```

- Accessing the higher part of a 16-bit address can be done by using the :MSB suffix after the name of the variable.

- The compiler automatically preserves and restores registers changed within an assembly block. This feature can be disabled by using the option "-Asr"

Examples:

```
// The following function demonstrates how to manipulate formal parameters.
// The function receives two parameters: a char pointer "my_pointer"
// and a char value "newvalue". The code simply writes the "newvalue" at
// the address pointed by "my_pointer".
void store_at_pointer (char *my_pointer, char newvalue)
{
   __asm
   {
     lda     newvalue
     ldhx    my_pointer
     sta     ,x
   }
}
```

Listing 4.4

```
// The next function divides the value of "v1" by the value of "v2". The
// result is returned in the accumulator
char my_asm_divide (char v1, char v2)
{
    __asm
    {
        lda     v1
        ldx     v2
        clrh
        div
    }
}
```

Listing 4.5

Calling the previous function is quite straightforward:

```
My_result = my_asm_divide (50,5);   // divides 50 by 5
```

4.4. Optimizing C Code

Before finishing our review of the C language, let us see some techniques and compiler options that can help produce smaller and faster code.

4.4.1. Compiler Options

The HCS08 compiler has some specific options that can help reduce the code size or improve the code-speed.

Let us see some of them:

-Cni   Disables the integer promotion. ANSI C states that all integer constants are **int** by default. When comparing a **char** variable to a numeric constant, the variable is first converted to **int** and then the comparison takes place. This operation consumes time and memory. By using the –Cni option, these automatic promotions are disabled, increasing code efficiency (be aware that this is not ANSI compliant).

-Cc   Allocates constants in ROM, thus reducing RAM usage. The linker configuration file (.PRM) must have a ROM_VAR entry to enable using this option.

-Oi   Enables the inline code generation. When combined with the #PRAGMA INLINE directive, it instructs the compiler to include the function code instead of a function call. This option can produce a faster code but it also increases code size. It is also possible to specify the maximum size for function inlining.

-Or   CPU-registers allocation for local variables. This option instructs the compiler to try to store local variables into CPU-registers. While reducing RAM usage and increasing code-speed this option is hardly useful due to the lack of spare CPU-registers on the HCS08 architecture.

-Cu   Loop unrolling: instructs the compiler to unroll for loops into sequential operations. This tends to speed up operations but it also increases code size. Example:

```
int a,b;
a = 1;
for (b=0; b<=3; b++) a += b;
```

Compile to:

```
a += 1;
a += 2;
a += 3;
b = 3;
```

Optionally it is possible to specify the maximum number of iterations to be unrolled.

-Mt   Selects the tiny memory model (refer to topic 4.2.3).

-Ms   Selects the small memory model (refer to topic 4.2.3).

-Mb   Selects the banked memory model (refer to topic 4.2.3).

4.4.2. Variables

In this topic we will review some good practices which can help reduce code size and increase code efficiency:

1. Always use the shortest possible variable: if the data can fit in a **char**, there is no reason for using an **int** or **long int**;

2. Avoid using **signed** variables: they can increase code size when compared to **unsigned** variables, especially on arithmetic operations;

3. **float** and **double** variables should be used only when strictly necessary. Sometimes it is possible to use fixed point math instead of floating point math. This reduces code size and increases code speed;

4. **near** variables produce more optimized code because they are stored in the direct page and use direct addressing mode, which is more efficient and smaller than other addressing modes;

5. When using variables as boolean flags, **char** variables should be preferred instead of larger types (such as **int** and **long**). In boolean expressions, avoid comparing variables to explicit values ("0" or "1"), instead, use the intrinsic evaluation available on C (a variable is evaluated true when its content is nonzero and false when zero):

   Instead of: Use:
   ```
   if (variable==0) ...        if (!variable) ...
   if (variable==1) ...        if (variable) ...
   ```

   > *Note that "if (variable==1)..." is not the same thing that "if (variable)...". The first expression evaluates true only if "variable" is equal to "1", while the second expression evaluates true if "variable" is equal to anything other than "0"!*

6. Care should be taken with variable scope: local variables should be preferred instead of global ones. Local variables use memory only during execution of the code block where they were declared, allowing memory reusing.

5

System Modules

Before studying the several peripherals found on the HCS08 devices, let us observe some internal modules which play a major rule on supporting CPU operation.

These modules are common to all HCS08 devices and knowledge about their operation is very important for designing successful embedded systems.

5.1. Reset

The reset module is responsible for placing the CPU and peripherals in a known initial state. The reset signal can be originated from one of the following sources:

- Power-on (POR) signal: triggered when the supply voltage on V_{DD} rises above 1.4 Volts.

- External signal on reset pin: triggered when the reset pin is in low state for a minimum period of 100ns. On some devices the reset pin must be previously configured to the reset function before being able to reset the system. On other devices, the reset pin also works as a reset output. Refer to topic 5.1.1 for more information.

- Low-voltage detection (LVD): triggered when the supply voltage falls down below a threshold value. Refer to topic 5.2 for more information.

- Watchdog (COP) timeout: triggered when the internal watchdog timer (COP – Computer Operating Properly as called by Freescale) times out. Refer to topic 5.2.1 for more information.

- Loss of clock (only on ICG-enabled devices): triggered when the internal clock generator (ICG) loses its external reference. Refer to topic 7.3 for more information.

- Illegal instruction: triggered when the CPU tries to execute an illegal (unimplemented) opcode. Notice that the STOP instruction is considered illegal when it is not previously enabled before being used. The BGND instruction is also considered illegal when executed when the BDM is not enabled.

- Illegal address: triggered when the CPU tries to access an illegal (unimplemented) memory address.

- BDM reset: triggered by the BDM interface.

Following a reset, a sequence of internal operations takes place:

1. Interrupts are disabled (CCR:I = 1).

2. The stack pointer (SP) is initialized with 0x00FF.

3. PC is loaded with the content of the reset vector located at addresses 0xFFFE and 0xFFFF.

4. All internal peripherals are disabled.

5. I/O ports are configured as inputs with pull up devices disabled.

6. On devices with ICG module, the oscillator switches to internal clock mode (self-clocked mode), operating at a standard frequency of 8 MHz. On devices with ICS module, the oscillator switches to internal clock mode (FLL engaged internal), operating at a standard frequency of 8 MHz.

5.1.1. The Reset Pin

Some HCS08 devices implement a dedicated reset pin while others implement the reset function multiplexed with other I/O functions.

Devices with a dedicated reset pin include a special feature built in this pin: whenever any reset is initiated, the RESET pin is actively driven low for 34 BUSCLK cycles and then it is released and the state of the pin is sampled for 38 BUSCLK cycles. This enables the MCU to reset external devices connected to the same reset signal and it also enables the identification of the reset source.

This special feature is also present on some devices with multiplexed functions on the reset pin (such as QB, QE4, QE8, QE16, QE32, LC, LG, LL, SH, RC, RD, RE and RG devices).

LC, LG and LL devices share the reset pin with generic I/O (PTB2 on LC and LL devices, PTC6 on LG devices), but the pin defaults to the reset function after a POR.

RC, RD, RE and RG devices share the reset pin with PTD1 function. On these devices the pin defaults to the PTD1 function. To enable the reset function it is necessary to set the SOPT:RSTPE bit.

QA, QD and QG devices do not implement the reset output function. On these devices the reset pin always defaults to the alternate multiplexed function (general purpose input). To enable the reset function it is necessary to set the SOPT1:RSTPE bit.

QE64/96/128 devices offer an independent reset output pin: RSTO. On these devices the reset pin is multiplexed with PTA5, IRQ and TCLK functions and after a POR it defaults to PTA5 function. To enable the reset function it is necessary to set the SOPT1:RSTPE bit. The RSTO function is multiplexed with PTC4 and TPM3CH4 functions and after a POR it defaults to the PTC4 function. To enable RSTO function it is necessary to set the SOPT1:RSTOPE bit.

Notice that when the pin is working as a reset input, an internal pull up device is automatically enabled.

> (i) *On noisy ambient and/or EMC-sensitive applications it may be necessary to add an external pull up resistor (4.7 to 10kΩ) and a small capacitor (100nF) to the reset pin (as shown in figure 5.1).This may not be necessary if this pin is not configured to the reset function.*

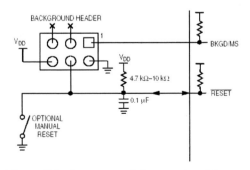

Figure 5.1 – Recommended reset pin wiring

Table 5.1 shows many configurations available on the HCS08 family.

Family	Multiplexed Function	Reset Output	Defaults to Reset
AC/AW	none	yes	yes
DN/DV/DZ	none	yes	yes
EL/SL	none	yes	yes
GB/GT	none	yes	yes
JM	none	yes	yes
LC/LL	PTB2	yes	yes
LG	PTC6	yes	yes
QD	PTA5/IRQ/TPM2CH0I	not available	no
QE64/96/128	PTA5/IRQ/TCLK	yes (RSTO pin)	no
QB, QE4, QE8, QE16, QE32, SH	PTA5/IRQ/TCLK	yes	no
QA/QG	PTA5/IRQ/TCLK	not available	no
SG	none	yes	yes
RC/RD/RE/RG	PTD1	yes	no

Table 5.1 – Reset functions

5.1.2. Software Reset

There are many ways to initiate a reset sequence under software control:

◆ The reset pin can be connected to another I/O pin. When the I/O pin is configured as an output and driven low, the chip is reset. The main disadvantage of this approach is the need of an I/O pin dedicated to the reset function (along with the reset pin itself).

◆ If the application uses the watchdog, a simple way to start a reset is by entering a loop without resetting the watchdog count: the chip will be reset once the watchdog times out.

◆ Another way to cause a reset is by attempting to execute an illegal instruction: the HCS08 CPU is automatically reset when an illegal opcode is decoded. If the application is not using low power modes (STOP instruction is disabled), the STOP instruction is considered illegal and resets the chip if decoded. The BGND instruction is also considered illegal when the system is not in debug mode.

Based on the illegal opcode resetting feature we can implement a software reset function that can be used to deliberately reset the application on a software command.

Our software reset function is based on the 0x9E opcode (the escaping code for 2-byte opcodes). We choose 0x9E00 as our illegal opcode. Listing 5.1 shows our software reset function.

```
void reset(void)
{
  unsigned int temp;
  temp = 0x9E00;
  __asm
  {
    LDHX   @temp      // load the address of temp into H:X
    JMP    ,X         // jump to the address pointed to by H:X
  }
}
```

Listing 5.1

To initiate a reset, all you have to do it is a call to the reset function:

```
reset();      // resets the application
```

 This kind of reset is different from simply jumping to the reset vector: a true reset (such as the software reset function presented in this topic) initializes the CPU state and peripheral registers, while a simple jump to the reset vector does not.

5.1.3. SRS Register

The System Reset Source (SRS) register stores the flags indicating the source of the last reset.

Name		BIT 7	BIT 6	BIT 5	BIT 4	BIT 3	BIT 2	BIT 1	BIT 0
	Read	POR	PIN	COP	ILOP	ILAD	ICG/LOC	LVD	0
SRS	Write				Writing on SRS clears COP count				
	Reset	?	?	?	?	?	?	?	?

Bit Name	Description	C Symbol
POR	When set, indicates that the last reset was a power-on reset (POR). It is also set by the low-voltage reset circuit.	bPOR
PIN	When set, indicates that the last reset was originated by the reset pin.	bPIN
COP	When set, indicates that the last reset was originated by a watchdog timer overflow.	bCOP
ILOP	When set, indicates that the last reset originated by an illegal opcode.	bILOP
ILAD	When set, indicates that the last reset originated by accessing an illegal address.	bILAD
ICG/LOC	When set, indicates that the last reset originated by ICG circuitry (loss of reference clock).	bICG bLOC
LVD	When set, indicates that the last reset originated by the low-voltage detection circuit (LVD). This flag is also set on a power-on reset.	bLVD

The procedure for evaluating the reset source is very simple (it should be done preferably at the beginning of the program). Below, we show some simple if statements that could call specific functions to deal with the reset cause:

```
...
if (SRS_PIN) external_reset();
if (SRS_COP) cop_timeout();
if (SRS_ILAD) illegal_address_reset();
if (SRS_ILOP) illegal_opcode_reset();
if (SRS_ICG) clock_failure();
if (SRS_LVD) low_voltage_reset();
...
```

 ILAD and ILOP resets are frequently caused by stack overflow and stack underflow errors. Testing these flags at the beginning of your code can help debug faulty applications!

The next example is another version of our led flasher. This time, it uses our software reset function to reset the application every time "counter1" reaches a counting of 25. A simple test, at the beginning of the program, lights LED2 (on the DEMO9S08QG8 board) if the reset source is due to the execution of an illegal opcode (ILOP).

```
// DEMO9S08QG8 - LED flasher and reset testing
#include <hidef.h>          /* for EnableInterrupts macro */
#include "derivative.h"     /* include peripheral declarations */
#include "hcs08.h"          // This is our definition file!

#define LED1  PTBD_PTBD6     // LED1 is connected to PTB6
#define LED2  PTBD_PTBD7     // LED2 is connected to PTB7

void delay(unsigned int value)
{
  for (;value; value--);
}

void reset(void)
{
  unsigned int temp = 0x9E00;
  __asm
  {
    LDHX  @temp             // load the address of temp into H:X
    JMP   ,X                // jump to the address pointer to by H:X
  }
}
```

```
void main(void)
{
  unsigned char counter1 = 0;
  SOPT1 = bBKGDPE;              // configure SOPT1 register, enable pin BKGD for BDM
  PTBDD = BIT_7 | BIT_6;        // configure pins 6 and 7 of port B as outputs
  // Now we test SRS to detect if the reset source was an ILOP
  // if ILOP = 1 then we turn LED2 on, if ILOP = 0 then we turn LED2 off
  if (SRS_ILOP) LED2 = 0; else LED2 = 1;
  while (1)
  {
    PTBD_PTBD6 = 1;            // LED = off
    delay(30000);             // wait for a while
    PTBD_PTBD6 = 0;           // LED = on
    delay(3000);              // wait for a while
    counter1++;
    if (counter1>=25) reset();
  }
}
```

Example 5.1

5.2. Low-Voltage Detection (LVD)

The low-voltage detection circuit protects the integrity of the application by resetting/interrupting the CPU on power shortages or brownout detection.

There are two protection levels: low-voltage detection (LVD) and low-voltage warning (LVW). The LVD can be configured to generate an interrupt or to reset the system when the supply voltage falls down below a specific value (the V_{LVD} parameter) for a least 10μs. On the other hand, the LVW is not capable of generating any kind of reset or interrupt: it must be read periodically by the application to detect when the supply voltage has fallen below the specified threshold (the V_{LVW} parameter).

 Newer devices (such as the LL family) include an interrupt feature for the LVW circuitry. This feature is enabled by the LVWIE bit on the SPMSC3 register!

The module is based on an analog comparator, which presents a hysteresis of approximately 100mV for a 5V supply (this hysteresis is represented by the rising and falling values on table 5.2).

Figure 5.2 shows a graph for the supply voltage over time. If we select the high range for V_{LVW} and low range for V_{LVD}, the LVWF bit is set when the supply voltage drops below the V_{LVW} falling value. It remains set until the supply voltage rises above V_{LVW} rising value. The same is true for the LVDF bit. Notice that LVWF and LVDF bits are not automatically cleared by hardware. The application needs to clear them by writing on the appropriate acknowledgement bit.

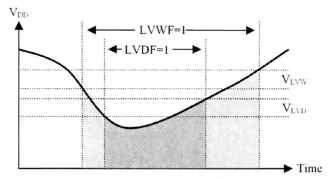

Figure 5.2 – LVD and LVW voltage levels

The LVD module also allows selecting different threshold levels for warning and detection levels. Table 5.2 shows the possible values for some HCS08 devices.

Family	V_{LVD}				V_{LVW}			
	Low Range (V_{LVDL})		High Range (V_{LVDH})		Low Range (V_{LVWL})		High Range (V_{LVWH})	
	Falling	Rising	Falling	Rising	Falling	Rising	Falling	Rising
AW	2.56	2.62	4.30	4.40	2.56	2.62	4.30	4.40
GB/GT	1.82	1.90	2.10	2.19	2.10	2.19	2.40	2.40
LC	1.88	1.93	2.15	2.23	2.15	2.21	2.45	2.48
QD	2.56	2.62	4.30	4.40	2.56	2.62	4.30	4.40
QE	1.82	1.90	2.10	2.19	2.10	2.19	2.46	2.46
QG	1.82	1.90	2.10	2.19	2.10	2.19	2.40	2.40
RC, RD, RE, RG	1.88	1.96	-	-	2.13	2.21	-	-

Table 5.2 – LVD and LVW voltages for some HCS08 devices

The LVD is enabled/disabled by the SPMSC1:LVDE bit (1 = enabled, 0 = disabled). This bit can be modified only once after reset. Once modified, all other writes onto it are ignored and produce no effect.

 Out of reset, the LVD module is always enabled and operates in reset mode!

5.2.1. Warning Event

When the supply voltage drops below the falling value of V_{LVW}, the LVWF bit is set by the hardware. An application monitoring the supply voltage should periodically read this bit and respond appropriately when it is set. On LL devices, the LVWF bit can optionally generate an interrupt request if the LVWIE bit is set.

To clear the LVWF bit, the application must write "1" into the LVWACK bit. This process is known as "acknowledgement of the LVWF bit". Notice that LVWF is a read-only bit and cannot be directly modified by the application. It is set only by the hardware and the only way to clear it is by acknowledging it as described.

There are two different ranges for the V_{LVW} value. They are selected by the LVWV bit: when LVWV=1, the high range is selected (V_{LVWH}) and when LVWV=0, the low range is selected (V_{LVWL}).

A sample code showing the monitoring and acknowledgement of LVWF is shown below (code designed to the MC9S08QG8):

```
...
if (SPMSC3_LVWF)
{
  // This code is executed when the supply voltage is below VLVW
  SPMSC3_LVWACK = 1; // Acknowledge the LVWF bit (clear it)
  ...
}
...
```

Listing 5.2

Notice that the LVWF bit is not cleared if the warning condition stands. This condition only ceases when the supply voltage rises above the rising V_{LVW} value.

A typical use of the low-voltage warning is, for example, to notify the user about a low battery condition. In this case, the V_{LVW} level should be configured to a value greater than the V_{LVD} level.

 Registers associated with LVWF, LVWACK and LVWV bits vary from one device to another. For more information check topic 5.2.5.

5.2.2. Detection Event

When the supply voltage drops below the falling value of V_{LVD}, the SPMSC1:LVDF bit is set by the LVD module and a low-voltage detection event is triggered.

The module can work in two distinct modes: reset or interrupt mode. The selection between these two modes is made by the SPMSC1:LVDRE bit (1 = reset mode, 0 = interrupt mode).

On reset mode, the system is reset when LVDF is set by the LVD module. Following a LVD reset, the SRS:LVD bit is set. To configure the LVD module to this mode use:

```
SPMSC1 = bLVDE | bLVDRE;
```

On interrupt mode, an interrupt is generated if it was previously enabled by the SPMSC1:LVDIE bit. When CCR:I = 0, LVDIE = 1, LVDRE = 0 and LVDF = 1, an interrupt signal is generated to the HCS08 CPU. To configure the LVD module to operate in this mode use:

```
SPMSC1 = bLVDE | bLVDIE;
```

Once LVDF is set, the only way to clear it is by acknowledging it. This can be done by writing "1" into the SPMSC1:LVDACK bit.

The function to service the LVD interrupt can be declared as follows:

```
void interrupt VectorNumber_Vlvd my_lvd_interrupt(void)
{
    SPMSC1_LVDACK = 1;       // acknowledge and clear LVDF
    // the code for the ISR starts below
    ...
}
```

 Bits LVDRE and LVDE in register SPMSC1 can be modified only one time after reset. This is also valid for bits PDC and PPDC in register SPMSC2 (these bits are related to the low power modes and will be studied in chapter 8.

There are two different ranges for the V_{LVF} value. They are selected by the LVDV bit: when LVDV=1 the high range is selected (V_{LVDH}) and when LVDV=0, the low range is selected (V_{LVDL}).

 The register which stores the LVDV bit varies from one device to another. For more information check topic 5.2.5.

5.2.3. LVD Module on RC, RD, RE and RG Devices

Operation of the LVD module on Rx devices is slightly different from other HCS08 devices:

1. There is no LVDE bit (the LVD module is always on).

2. LVWF and LVWACK bits are located on the SPMSC2 register (instead of SPMSC3). In fact, there is no SPMSC3 register on Rx devices.

3. There is no threshold selection for the V_{LVD} and V_{LVW} trip points.

Another difference is the safe mode, a preventive mode present on RC, RD, RE and RG devices. This mode is intended to avoid exiting stop mode in a low-voltage event (that can otherwise could wake up the device, increasing current consumption, thus decreasing even more the battery voltage).

The operation of the safe mode is very straightforward: when SPMSC1:SAFE bit is set, the LVD module inhibits all interrupt, reset and wake up events in a stop mode while a low-voltage condition is present and the supply voltage is under V_{REARM} parameter (3.0 V maximum). After the supply voltage rises above V_{REARM}, the SAFE bit is ignored and interrupt, reset and wake up events are treated as expected. Once the low-voltage condition is not present anymore (LVDF is not set), the application must clear the SAFE bit.

Notice that in SAFE mode, as long as the supply voltage remains above the POR trip voltage, all RAM contents (unless in stop1 mode) and register contents (unless in stop2 mode) are preserved.

5.2.4. Operation in Stop Modes

The LVD module continues operating in stop3 mode if the SPMSC1:LVDSE bit is set before entering stop mode.

If the module is enabled (LVDE=1) and configured to continue operating in stop modes (LVDSE=1) then, it is not possible to enter either stop1 or stop2 mode. **Attempts to enter one of these modes will enter stop3 instead!**

RC, RD, RE and RG devices do not allow LVD operation in stop modes. So far, they do not have the LVDSE bit!

Out of reset, the LVD module operation in stop modes is always enabled (LVDSE = 1)!

5.2.5. LVD Registers

The LVD module has three associate registers:

- System Power Management Status and Control 1 (SPMSC1) register.
- System Power Management Status and Control 2 (SPMSC2) register.
- System Power Management Status and Control 3 (SPMSC3) register.

5.2.5.1. SPMSC1 Register

Name	Model		BIT 7	BIT 6	BIT 5	BIT 4	BIT 3	BIT 2	BIT 1	BIT 0
SPMSC1	Ax, Dx, Ex, JM, Lx, Qx, Sx	Read	LVDF	0	LVDIE	LVDRE*	LVDSE	LVDE*	0	BGBE
		Write	-	LVDACK					-	
		Reset	0	0	0	1	1	1	0	0
	GB/GT	Read	LVDF	0	LVDIE	LVDRE*	LVDSE	LVDE*	0	0
		Write	-	LVDACK					-	-
		Reset	0	0	0	1	1	1	0	0
	RC RD RE RG	Read	LVDF	0	LVDIE	SAFE	LVDRE*	0	0	0
		Write	-	LVDACK				-	-	-
		Reset	0	0	0	0	1	0	0	0

* These are write-once bits which can be modified only once after a reset. All other writes onto them are ignored and produce no effect!

Bit Name	Description	C Symbol
LVDF	This flag indicates a low-voltage detect event. It is a read-only bit and is only valid when LVDE = 1.	bLVDF
LVDACK	Acknowledge bit for LVDF. Writing "1" to LVDACK clears LVDF.	bLVDACK
LVDIE	Interrupt mask for the low-voltage interrupt. This bit enables low-voltage interrupts when LVDRE = 0: 0 – LVD interrupt disabled. 1 – LVD interrupt enabled.	bLVDIE
LVDRE	Low-voltage reset enable: 0 – An LVD event triggers an interrupt. 1 – An LVD event causes a reset. **This is a write-once bit and can be modified only once after a reset. Subsequent writes produce no effect!**	bLVDRE

LVDSE	Enable the LVD module to operate in stop mode: 0 – LVD disabled in stop mode. 1 – LVD enabled in stop mode.	**bLVDSE**
LVDE	Enable the low-voltage detect module: 0 – LVD module disabled. 1 – LVD module enabled. **This is a write-once bit and can be modified only once after a reset. Subsequent writes produce no effect!**	**bLVDE**
SAFE	This bit controls the SAFE state on Rx devices: 0 – Resets, interrupts and wake up events are allowed in stop mode and when the supply voltage is above V_{REARM}. 1 – Resets, interrupts and wake up events are **not** allowed in stop mode or when the supply voltage is above V_{REARM}.	**bSAFE**

5.2.5.2. SPMSC2 Register

Name	Model		BIT 7	BIT 6	BIT 5	BIT 4	BIT 3	BIT 2	BIT 1	BIT 0
SPMSC2	AC, AW, QD	Read	LVWF	0	LVDV	LVWV	PPDF	0	-	PPDC*
		Write	-	LVWACK			-	PPDACK		
		Reset	0	0	0	0	0	0	0	0
	DN, DV, DZ, EL, EN, LG, SG, SH, SL	Read	0	0	LVDV	LVWV	PPDF	0	0	PPDC*
		Write	-	-			-	PPDACK	-	
		Reset	0	0	0	0	0	0	0	0
	LL, QB, QE	Read	LPR	LPRS	LPWUI	0	PPDF		PPDE*	PPDC
		Write		-			-	PPDACK		
		Reset	0	0	0	0	0	0	0	0
	LC, QA, QG	Read	0	0	0	PDF	PPDF	0	PDC*	PPDC*
		Write	-	-	-	-	-	PPDACK		
		Reset	0	0	0	0	0	0	0	0
	GB, GT	Read	LVWF	0	LVDV	LVWV	PPDF	0	PDC*	PPDC*
		Write	-	LVWACK			-	PPDACK		
		Reset	0	0	0	0	0	0	0	0
	RC, RD, RE, RG	Read	LVWF	0	0	0	PPDF	0	PDC*	PPDC*
		Write	-	LVWACK	-	-	-	PPDACK		
		Reset	0	0	0	0	0	0	0	0

These are write-once bits and can be modified only once after a reset. All other writes into them are ignored and produce no effect!

Bit Name	Description	C Symbol
LVWF	Low-voltage warning flag. This read-only bit is set when the supply voltage drops below the V_{LVW} threshold.	**bLVWF**
LVWACK	Acknowledge bit for LVWF. Writing "1" into LVWACK clears LVWF (if the low-voltage event is still present, LVWF is not cleared)	**bLVWACK**
LVDV	This bit allows selection between two threshold values for the low-voltage detection (refer to table 5.2): 0 – low range selected. 1 – high range selected.	**bLVDV**
LVWV	This bit allows selection between two threshold values for the low-voltage warning (refer to table 5.2): 0 – low range selected. 1 – high range selected.	**bLVWV**

The remaining bits (PDC, PPDC, LPR, LPRS and PPDE) are related to the low power modes and will be seen in chapter 8.

5.2.5.3. SPMSC3 Register

Name	Model		BIT 7	BIT 6	BIT 5	BIT 4	BIT 3	BIT 2	BIT 1	BIT 0
SPMSC3	LC, QA, QE, QG	Read	LVWF	0	LVDV	LVWV	0	0	0	0
		Write	-	LVWACK			-	-	-	-
		Reset	0	0	0	0	0	0	0	0
SPMSC3	QB, LL	Read	LVWF	0	-	-	LVWIE	0	0	0
		Write	-	LVWACK	-	-		-	-	-
		Reset	0	0	0	0	0	0	0	0

> ⓘ *The SPMSC3 register is present only on LC, LL, QA, QB, QE and QG devices!*

Bit Name	Description	C Symbol
LVWF	Low-voltage warning flag. This read-only bit is set when the supply voltage drops below the V_{LVW} threshold.	bLVWF
LVWACK	Acknowledge bit for LVWF. Writing "1" into LVWACK clears LVWF (if the low-voltage event is still present, LVWF is not cleared)	bLVWACK
LVDV	This bit allows selection between two threshold values for the low-voltage detection (refer to table 5.2 for device specific ranges): 0 – low range selected. 1 – high range selected.	bLVDV
LVWV	This bit allows selection between two threshold values for the low-voltage warning (refer to table 5.2 for device specific ranges): 0 – low range selected. 1 – high range selected.	bLVWV
LVWIE	This bit enables interrupt requests from the LVD circuitry (only on LL devices): 0 – LVW interrupt disabled. 1 – LVW interrupt enabled.	bLVWIE

5.3. Watchdog (COP)

The "Computer Operating Properly" or simply COP is a watchdog timer designed to protect the application against unexpected situations.

This protection is based on a simple mechanism: the watchdog timer counts up until it overflows. When it overflows, the CPU and peripherals are reset. Application software must periodically service the watchdog counting so it does not reach its overflow counting.

A well-designed application services the watchdog as little as possible (ideally, this operation should be done in just one place within the program). Placing too many watchdog service instructions reduces watchdog effectiveness.

On HCS08 devices the COP is built as a binary counter with two overflow values: 8,192 or 262,144 counts. This counter is enabled by the COPE bit and runs from the BUSCLK. Selection between the two timeout counts is done through the COPT bit. These bits are located in the SOPT register (SOPT1 on some devices), refer to topic 5.5 for further details.

Servicing watchdog counting is done by performing a write operation into SRS register (this is just a dummy write, the value written actually does not matter):

```
SRS = 0;    // clear watchdog timer
```

Some HCS08 devices (LC, QD, QE and QG families) include an alternate COP clock source: a 1 kHz independent internal oscillator. This internal oscillator is selected by default on these devices. Selection of BUSCLK as the COP clock source is done by writing "1" to the SOPT2:COPCLKS bit.

When the internal oscillator is the clock source for the COP timer, the timer overflow values change. Table 5.3 shows the four timeouts for the COP overflow.

COPCLKS	COPT	Clock Source	Timeout Period
0	0	Internal 1 kHz oscillator	32 counts (32 ms)
0	1	Internal 1 kHz oscillator	256 counts (256 ms)
1	0	BUSCLK	8,192 counts
1	1	BUSCLK	262,144 counts

Table 5.3 – COP timeouts

Following a reset, the COP is enabled (COPE=1) and the timeout period is set to 261,144 BUSCLK cycles (on LC, QA, QB, QD, QE and QG devices the clock source is the internal 1kHz oscillator and the timeout period is near to 256ms)!

COP can be disabled by writing "0" to COPE:

```
SOPT = 0;  // disables COP and BKGD debug pin
```

Be aware that SOPT is a write-once register. Refer to topic 5.6 for further information.

Following a stop mode recovery, the COP counting is always cleared.

The next example shows a possible implementation of our led flasher, this time including COP protection. Once the program is initially started, LED2 is turned off and LED1 blinks. The watchdog is enabled with an overflow interval of approximately 256ms. If the watchdog counting is not serviced each 256ms, it overflows and resets the system.

Our demonstration application services the watchdog within the **for** loops. If, for any reason, the main program gets lost, it probably stops servicing the watchdog, which resets the system.

For didactic purposes, our application also stops servicing the watchdog when the SW1 key is pressed and remains pressed. Without servicing, the watchdog overflows after ~256ms, resetting the chip.

The test procedure at the beginning of the program tests the SRS_COP bit and turns LED2 on if it is set. This way, it is possible to differentiate between a POR and a COP reset.

```
// LED flasher and reset testing
#include <hidef.h>          /* for EnableInterrupts macro */
#include "derivative.h"     /* include peripheral declarations */
#include "hcs08.h"          // This is our definition file!

#define LED1   PTBD_PTBD6    // LED1 is connected to PTB6
#define LED2   PTBD_PTBD7    // LED2 is connected to PTB7
#define SW1    PTAD_PTAD2    // SW1 is connected to PTA2
#define CLRCOP {SRS = 0;}    // Clear Watchdog Macro

void main(void)
{
  unsigned int temp;
  // configure SOPT1 register, enable COP (BKGD pin is disabled)
  SOPT1 = bCOPT | bCOPE;
  PTBDD = BIT_7 | BIT_6;     // configure pins 6 and 7 of port B as outputs
  PTAPE = BIT_2;             // enable internal pull up on pin PTA2
  // Now we test SRS to detect if the reset source was a COP timeout
  // if SRS_COP = 1 then we turn LED2 on, if COP_COP = 0 then we turn LED2 off
  if (SRS_COP) LED2 = 0; else LED2 = 1;
  while (1)
  {
    PTBD_PTBD6 = 1;                      // LED = off
    for (temp=30000; temp; temp--) CLRCOP;  // wait for a while
    PTBD_PTBD6 = 0;                      // LED = on
    for (temp=3000; temp; temp--) CLRCOP;   // wait for a while
    while (!SW1);                        // while SW1 is pressed, stay here
  }
}
```

Example 5.2

Avoid placing COP reset instructions within an interrupt servicing routine (ISR), as this is not recommended (the main application can fail, but the interrupts can still be serviced as expected).

A good programming practice is to use the fewest possible COP reset instructions (optimally, only one instruction should be placed on the whole application).

5.3.1. Enhanced Watchdog

Newer devices (such as DN, DV, DZ, EL, EN, JM, SG and SH) include an enhanced COP watchdog with increased security features such as:

- New reset sequence: the COP watchdog timer is only reset by writing the sequence 0x55, 0xAA into the SRS register. This sequence must be written within the selected timeout period or the COP times out and resets the system. Writing another value or an out-of-order sequence also resets the system.

  The reset sequence is showed below:

  ```
  SRS = 0x55;
  SRS = 0xAA;
  ```

  Writing the sequence in reverse order resets the system:

  ```
  SRS = 0xAA;
  SRS = 0x55;
  ```

  Writing any other values also resets the system:

  ```
  SRS = 0x10;   // resets the system
  ```

- New COP timeouts: the enhanced COP includes four selectable timeout periods and two clock sources. The timeout selection is enabled by the two COPT bits as shown in table 5.4.

COPCLKS	COPT1:COPT0	Clock Source	Timeout Period
x	00	COP disabled	COP disabled
0	01	Internal 1 kHz oscillator	32 counts (32 ms)
0	10	Internal 1 kHz oscillator	256 counts (256 ms)
0	11	Internal 1 kHz oscillator	1024 counts (1.024s)
1	01	BUSCLK	8,192 counts
1	10	BUSCLK	65,536 counts
1	11	BUSCLK	1,048,576 counts

Table 5.4 – Enhanced COP timeouts

- Windowed timeout: this new mode (available only when the COP is operating with the BUSCLK as the clock source) increases system security by only accepting COP resets within the last 25% of the timeout period. Any attempts to reset the COP (through the 0x55, 0xAA sequence) outside the allowed window causes the system to reset. The windowed option is controlled by the COPW bit (usually located in the SOPT2 register).

COPCLKS	COPT1:COPT0	COP Window Opens	COP Window Closes
1	01	after 6,144 counts	after 8,192 counts
1	10	after 49,152 counts	after 65,536 counts
1	11	after 196,608 counts	after 262,144 counts

Table 5.5 – Enhanced COP timeouts (windowed mode)

Notice that the window feature highly increases system security because it can catch a program locked on a close loop such as:

```
...
// this loop keeps servicing COP (unless the window option is enabled!)
while (1)
{
    SRS = 0xAA;
    SRS = 0x55;
}
```

5.4. Device ID

HCS08 devices include two special registers for device identification and revision information. This kind of information helps debugging tools identify the chip.

The identification code is a 12-bit binary number and the revision code is 4-bit binary number. Table 5.6 shows the identification numbers for current HCS08 devices.

Device Family	Identification Code
AC8,AC16, AW8A, AW16A	0x012
AC32, AC48, AC60	0x01D
AC96,AC128	0x01B
AW	0x008
DN, DV, DZ	0x00E
EL, SL	0x013
EN	0x00E
GB/GT 32/60	0x002
GT 8/16	0x00D
JM	0x016
LC	0x00C
LG	0x02A
LL	0x021
QA/QG	0x009
QB	0x023
QD	0x011
QE4, QE8	0x01C
QE16, QE32	0x01F
QE64/96/128	0x015
RC/RD/RE 8/16	0x003
RC/RD/RE/RG 32/60	0x004
SG, SH	0x014

Table 5.6 – HCS08 device IDs

5.4.1. Identification Registers

There are two identification registers:

- System Device IDentification High (SDIDH) register.
- System Device IDentification Low (SDIDL) register.

5.4.1.1. SDIDH Register

Name	Model		BIT 7	BIT 6	BIT 5	BIT 4	BIT 3	BIT 2	BIT 1	BIT 0
SDIDH	All	Read	REV3	REV2	REV1	REV0	ID11	ID10	ID9	ID8
		Write	-	-	-	-	-	-	-	-
		Reset	?	?	?	?	Refer to table 5.6			

5.4.1.2. SDIDL Register

Name	Model		BIT 7	BIT 6	BIT 5	BIT 4	BIT 3	BIT 2	BIT 1	BIT 0
SDIDL	All	Read	ID7	ID6	ID5	ID4	ID3	ID2	ID1	ID0
		Write	-	-	-	-	-	-	-	-
		Reset				Refer to table 5.6				

Revision codes refer to the current silicon mask revision and can be used to identify chips with possible mask errata. Refer to Freescale's website for errata documents and specific revision codes applicable to devices.

5.5. System Option Registers

The system option registers store special settings for device configuration (such as enabling the COP watchdog timer, enabling the STOP instruction, enabling the RESET and BKGD functions, etc.).

5.5.1. SOPT/SOPT1 Register

Name	Model		BIT 7	BIT 6	BIT 5	BIT 4	BIT 3	BIT 2	BIT 1	BIT 0
SOPT	AC, AW	Read	COPE	COPT	STOPE	-	0	0	-	-
		Write					-	-		
		Reset	1	1	0	1	0	0	1	1
SOPT1	DN, DV, DZ	Read	COPT*		STOPE	SCI2PS	IICPS		-	-
		Write								
		Reset	1	1	0	0	0	0	0	0
SOPT1	EL, SL	Read	COPT*		STOPE	SCIPS	IICPS		-	-
		Write								
		Reset	1	1	0	0	0	0	0	0
SOPT1	EN, JM	Read	COPT*		STOPE	-	-	-	-	-
		Write								
		Reset	1	1	0	0	0	0	0	0
SOPT	GB, GT	Read	COPE	COPT	STOPE	-	0	0	BKGDPE	-
		Write					-	-		
		Reset	1	1	0	1	0	0	1	1
SOPT1	Lx, Qx, Rx	Read	COPE	COPT	STOPE	-	0	0	BKGDPE	RSTPE
		Write					-	-		
		Reset	1	1	0	1	0	0	1	0
SOPT1	SG	Read	COPT*		STOPE	0	0	IICPS	0	0
		Write				-	-		-	-
		Reset	1	1	0	0	0	0	0	0
SOPT1	SH	Read	COPT*		STOPE	0	0	IICPS	BKGDPE	RSTPE
		Write				-	-			
		Reset	1	1	0	0	0	0	0	0

> ℹ️ *SOPT/SOPT1 is a write-once register: after a reset, this register can be written only once. Subsequent writes are ignored and produce no effect!*

Bit Name	Description	C Symbol
COPE	This bit enables the operation of the COP watchdog timer: 0 – COP disabled. 1 – COP enabled.	bCOPE
COPT	COP timeout selection (refer to table 5.3): 0 – short period. 1 – long timeout period.	bCOPT
COPT*	Enhanced COP timeout selection (refer to table 5.4): 00 – COP disabled.	COP_DISABLED
	01 – COP fast timeout (32ms or 8,192 counts).	COP_FAST
	10 – COP medium timeout (256ms or 65,536 counts).	COP_MEDIUM
	11 – COP slow timeout (1,024s or 262,144 counts).	COP_SLOW
STOPE	STOP instruction enable bit: 0 – STOP instruction disabled (STOP is considered an illegal instruction). 1 – STOP instruction enabled.	bSTOPE
SCIPS	Alternate pins for the SCI module: 0 – RxD on PTB0 and TxD on PTB1. 1 – RxD on PTA2 and TxD on PTA3.	bSCIPS
SCI2PS	Alternate pins for the SCI module: 0 – RxD2 on PTF1 and TxD2 on PTF0. 1 – RxD2 on PTE7 and TxD2 on PTE6.	bSCI2PS

	Alternate pins for the I²C module. On DN, DV and DZ devices: 0 – SDA on PTF3 and SCL on PTF2. 1 – SDA on PTE5 and SCL on PTE4.	**bIICPS_DZ**
IICPS	On EL and SL devices: 00 – SDA on PTA2 and SCL on PTA3.	**bIICPS1**
	01 – SDA on PTB6 and SCL on PTB7.	**bIICPS2**
	1x – SDA on PTB2 and SCL on PTB3.	**bIICPS3**
	On SG and SH devices: 0 – SDA on PTA2 and SCL on PTA3. 1 – SDA on PTB6 and SCL on PTB7.	**bIICPS_SH**
BKGDPE	BKGD pin enable bit: 0 – BKGD pin in I/O mode. 1 – BKGD pin in debug mode.	**bBKGDPE**
RSTPE	Reset pin enable bit: 0 – Reset pin in I/O mode. 1 – Reset pin in reset mode.	**bRSTPE**

 Writing into SOPT or SOPT1 is recommended, even if no changes are going to be made over its default state. After the first write into it, the register is locked and does not accept new changes until another reset takes place.

Notice that performing bit operations on the SOPT/SOPT1 registers must be avoided as they produce a RMW (read-modify-write) operation, causing all bits to be written. This way, a code sequence such as the following (supposing a QG device):

```
SOPT1_BKGDPE = 1;  // enable BKGD pin for debug mode
SOPT1_STOPE = 1;   // enable STOP instruction
```

The second write produces no effect !

Will produce the wrong result!

The first line will read SOPT1 (0xD2 following a reset), set bit 1 (BKGDPE) and write the result back on the SOPT1 register, causing the whole register to be written with 0xD2, thus enabling BKGD pin and disabling STOP instruction.

When the second line is run, it tries to write "1" on the STOPE bit, but it was previously written with "0" by the preceding instruction! Remember that SOPT/SOPT1 is a write-once register, meaning that once it is written by the application code, its content cannot be changed until another reset is issued!

The final result is that only BKGD pin was enabled and the STOP instruction is disabled (despite the explicit enabling instruction)!

The correct way for modifying multiple bits on the SOPT/SOPT1 register (also valid for any other write-once register) is to do all task within a single write operation, such as:

```
SOPT1 = bSTOPE | bBKGDPE;  // enable STOP instruction and BKGD pin
```

This sequence produces de desired effect!

5.5.2. SOPT2 Register

This register is only present on the following devices.

Name	Model		BIT 7	BIT 6	BIT 5	BIT 4	BIT 3	BIT 2	BIT 1	BIT 0
SOPT2	AC8, AC16, AW8A, AW16A, QD	Read	COPCLKS	0	0	0	0	0	0	0
		Write	COPCLKS	-	-	-	-	-	-	-
		Reset	0	0	0	0	0	0	0	0
	AC32, AC48, AC60, AC96, AC128	Read	COPCLKS	0	0	0	TPMCCFG	0	0	0
		Write	COPCLKS	-	-	-	TPMCCFG	-	-	-
		Reset	0	0	0	0	1	0	0	0
	DN, DV, DZ, EN	Read	COPCLKS	COPW	0	ADHTS	0	MCSEL		
		Write	COPCLKS	COPW	-	ADHTS	-	MCSEL		
		Reset	0	0	0	0	0	0	0	0
	EL, SL	Read	COPCLKS	COPW	0	ACIC	T2CH1PS	T2CH0PS	T1CH1PS	T1CH0PS
		Write	COPCLKS	COPW	-	ACIC	T2CH1PS	T2CH0PS	T1CH1PS	T1CH0PS
		Reset	0	0	0	0	0	0	0	0
	JM	Read	COPCLKS	COPW	0	0	0	SPI1FE	SPI2FE	ACIC
		Write	COPCLKS	COPW	-	-	-	SPI1FE	SPI2FE	ACIC
		Reset	0	0	0	0	0	1	1	0
	LC, QA	Read	COPCLKS	0	0	0	0	0	0	ACIC
		Write	COPCLKS	-	-	-	-	-	-	ACIC
		Reset	0	0	0	0	0	0	0	0
	LG	Read	COPCLKS	0	0	0	0	0	0	SPIFE
		Write	COPCLKS	-	-	-	-	-	-	SPIFE
		Reset	0	0	0	0	0	0	0	1
	LL	Read	COPCLKS	0	0	0	0	SPIPS	IICPS	ACIC
		Write	COPCLKS	-	-	-	-	SPIPS	IICPS	ACIC
		Reset	0	0	0	0	0	0	0	0
	QB	Read	COPCLKS	0	0	TPMCH0PS	0	0	0	ACIC
		Write	COPCLKS	-	-	TPMCH0PS	-	-	-	ACIC
		Reset	0	0	0	0	0	0	0	0
	QE4, QE8	Read	COPCLKS	0	TPM2CH2PS	TPM1CH2PS	0	ACIC2	IICPS	ACIC1
		Write	COPCLKS	-	TPM2CH2PS	TPM1CH2PS	-	ACIC2	IICPS	ACIC1
		Reset	0	0	0	0	0	0	0	0
	QE16/32/64/96/128	Read	COPCLKS	0	0	0	SPI1PS	ACIC2	IIC1PS	ACIC1
		Write	COPCLKS	-	-	-	SPI1PS	ACIC2	IIC1PS	ACIC1
		Reset	0	0	0	0	0	0	0	0
	QG	Read	COPCLKS	0	0	0	0	0	IICPS	ACIC
		Write	COPCLKS	-	-	-	-	-	IICPS	ACIC
		Reset	0	0	0	0	0	0	0	0
	SG, SH	Read	COPCLKS	COPW	0	ACIC	0	0	T1CH1PS	T1CH0PS
		Write	COPCLKS	COPW	-	ACIC	-	-	T1CH1PS	T1CH0PS
		Reset	0	0	0	0	0	0	0	0

COPCLKS is a write-once bit: after a reset this bit be can be modified only one time. Subsequent writes are ignored and produce no effect!

On EL and SL devices, the whole SOPT2 register is writing-once. Once it is modified, subsequent writes are ignored and produce no effect!

Bit Name	Description	C Symbol
COPCLKS	This bit selects the clock source for the COP watchdog timer: 0 – internal 1kHz clock.　　　1 – BUSCLK.	bCOPCLKS
COPW	This bit selects the window for watchdog operation: 0 – standard operation. Writing the correct sequence (0x55 and 0xAA) into SRS clears the COP count. 1 – windowed mode. The COP counter can only be cleared by writing the clear sequence (0x55 and 0xAA) into SRS on the last 25% portion of the COP timeout.	bCOPW

TPMCCFG	Clock source for the TPM modules: 0 – TPM1, TPM2 and TPM3 clocked by TPMCLK input. 1 – TPM1 clocked by TPM1CLK input, TPM2 clocked by TPM2CLK and TPM3 clocked by TPM3CLK.	**bTPMCCFG**
T2CH1PS	Alternate pin for TPM2 channel 1 input/output pin: 0 – TPM2CH1 connected to PTB4. 1 – TPM2CH1 connected to PTA7.	**bT2CH1PS**
T2CH0PS	Alternate pin for TPM2 channel 0 input/output pin: 0 – TPM2CH0 connected to PTA1. 1 – TPM2CH0 connected to PTA6.	**bT2CH0PS**
T1CH1PS	Alternate pin for TPM1 channel 1 input/output pin: 0 – TPM1CH1 connected to PTB5. 1 – TPM1CH1 connected to PTC1.	**bT1CH1PS**
T1CH0PS	Alternate pin for TPM1 channel 1 input/output pin: 0 – TPM1CH0 connected to PTA0. 1 – TPM1CH0 connected to PTC0.	**bT1CH0PS**
TPMCH0PS	Alternate pin for the TPM module: 0 – TPMCH0 connected to PTA0. 1 – TPMCH0 connected to PTB5.	**bTPMCH0PS**
ADHTS	ADC hardware trigger selection. This bit selects the trigger source for the ADC to start a new conversion when ADCSC2:ADCTRG = 1: 0 – ADC conversion initiated by an RTC overflow. 1 – ADC conversion initiated by IRQ pin.	**bADHTS**
MCSEL	These bits control the MCLK output on pin PTA0:	
	000 – MCLK disabled (PTA0 operating in I/O mode)	**MCLK_DISABLED**
	001 – PTA outputs a clock signal equal to BUSCLK/2	**MCLK_DIV2**
	010 – PTA outputs a clock signal equal to BUSCLK/4	**MCLK_DIV4**
	011 – PTA outputs a clock signal equal to BUSCLK/8	**MCLK_DIV8**
	100 – PTA outputs a clock signal equal to BUSCLK/16	**MCLK_DIV16**
	101 – PTA outputs a clock signal equal to BUSCLK/32	**MCLK_DIV32**
	110 – PTA outputs a clock signal equal to BUSCLK/64	**MCLK_DIV64**
	111 – PTA outputs a clock signal equal to BUSCLK/128	**MCLK_DIV128**
SPI1FE	SPI1 pins filtering: 0 – filter disabled. SPI can operate at maximum speed. 1 – filter enabled. The SPI speed is limited, but pins offer improved noise immunity.	**bSPI1FE**
SPI2FE	SPI2 pins filtering: 0 – filter disabled. SPI can operate at maximum speed. 1 – filter enabled. The SPI speed is limited, but pins offer improved noise immunity.	**bSPI1FE**
SPIPS	Alternate pins for SPI module: 0 – SPSCLK is connected to PTA1, MOSI to PTA3, MISO to PTA2 and SS to PTA0. 1 – SPSCLK is connected to PTB6, MOSI to PTB5, MISO to PTB4 and SS to PTB7.	**bSPIPS**
SPI1PS	Alternate pins for SPI 1 module: 0 – SPSCLK1 is connected to PTB2, MOSI1 to PTB3, MISO1 to PTB4 and SS1 to PTB5. 1 – SPSCLK1 is connected to PTE0, MOSI1 to PTE1, MISO1 to PTE2 and SS1 to PTE3.	**bSPI1PS**
IICPS	Alternate pins for I²C module: 0 – SDA line connected to PTA2, SCL line connected to PTA3. 1 – SDA line connected to PTB6, SCL line connected to PTB7.	**bIICPS**
IIC1PS	Alternate pins for I²C1 module: 0 – SDA line connected to PTA2, SCL line connected to PTA3. 1 – SDA line connected to PTB6, SCL line connected to PTB7.	**bIIC1PS**
ACIC	Analog comparator to input capture (TPM). On QG devices: 0 – ACMP output is not connected to TPM channel 0 input (TPMCH0). 1 – ACMP output is connected to TPM channel 0 input (TPMCH0).	
	Analog comparator to input capture (TPM). On LL devices: 0 – ACMP output is not connected to TPM2 channel 0 input (TPM2CH0). 1 – ACMP output is connected to TPM2 channel 0 input (TPMCH0).	**bACIC**
	On JM and LC devices: 0 – ACMP output is not connected to TPM1 channel 0 input (TPM1CH0). 1 – ACMP output is connected to TPM1 channel 0 input (TPM1CH0).	
	On EL, SG, SH and SL devices: 0 – ACMP output is not connected to TPM1 channel 0 input (TPM1CH0). 1 – ACMP output is connected to TPM1 channel 0 input (TPM1CH0).	**bACIC_SH**

ACIC1	Analog comparator 1 to input capture (TPM1): 0 – ACMP1 output is not connected to TPM1 input channel 0 (TPM1CH0). 1 – ACMP1 output is connected to TPM1 input channel 0 (TPM1CH0).	**bACIC1**
ACIC2	Analog comparator 2 to input capture (TPM2): 0 – ACMP2 output is not connected to TPM2 input channel 0 (TPM2CH0). 1 – ACMP2 output is connected to TPM2 input channel 0 (TPM2CH0).	**bACIC2**

For detailed information about capturing analog signals with the timer/PWM module (TPM), refer to topics 9.4.2.1 and 10.1.

Detailed information about the I^{2}C modules, SPI modules and their external connections can be found in chapter 11.

6

Input/Output Ports

I/O ports are the elements for inputting and outputting digital data to/from the microcontroller. The number of I/O pins varies from one device to another; the minimum number is 6 I/Os (on 8-pin devices) and the maximum is 71 I/Os (on some 80-pin devices).

These ports are grouped into 8-pin sets named PTA (port A), PTB, PTC, PTD, PTE, PTF, PTG, PTH and PTJ (port J).

When operating as an output, the pin can source or sink up to 2mA. Some pins can source/sink up to 10mA (all ports on AW, LC, LL, QA, QB, QD, QE and QG devices, ports C and F on GB/GT devices, port E on RC/RD/RE/RG devices). There is also a global limit of 60mA (when operating at 3V) or 100mA (when operating at 5V) for the sum of all currents sink/source by all ports.

 Pin IRO (Infrared Output on RC/RD/RE/RG devices) can sink up to 16mA!

All pins include internal clamping diodes to limit input voltage. These diodes can sink/source, each one, up to 200µA (with maximum 5mA global limit). Remember that positive current injection ($V_{PIN} > V_{DD}$) can cause the voltage regulator to go out of regulation. Power supply design must ensure proper operation in such cases.

 Pins with IRQ function (AW, LC, LL, QA, QB, QD, QE and QG devices) and PTA0 (RC, RD, RE and RG devices) do not include internal clamping diodes and should not be driven above V_{DD}!

The HCS08 devices also include internal pull up resistors on all port pins. These pull up devices can be independently enabled/disabled under software control. Some ports also include pull down resistors.

Earlier devices (RC, RD, RE and RG devices) include three registers for each port:

PTxD – the Port x Data register is used to read/write on port pins: each bit of this register accesses the corresponding external pin. Data written into PTxD register is externally available only if the corresponding pin is configured as an output. When the pin is configured as an input, data is only stored into the register but it does not change external pin state.

When the PTxD register is read, returned data depends on the pin configured direction: if the pin is an input, the corresponding PTxD bit reflects the external pin state. If the pin is an output, the returned data is the content of PTxD.

PTxDD – the Port x Data Direction register controls the direction of each port pin:
0 – the pin is an input.
1 – the pin is an output.

PTxPE –   the Port x Pull Enable register enables/disables the internal pull up device (pull up device is actually enabled only when the pin is configured as input):

0 – pull up device disabled.
1 – pull up device enabled.

Most devices (except the Rx ones) include two additional features: slew rate and drive strength control (not present on GB/GT devices). These features are controlled by two other registers:

PTxSE –   the Port x Slew rate Enable register is used to control the slew rate for each port pin:

0 – slew rate disabled (rise/fall time equals to 3ns for a 50pF load).
1 – slew rate enabled (rise/fall time equals to 30ns for a 50pF load).

PTxDS –   the Port x Drive Strength selection register is used to control the driving strength for each port pin:

0 – low output drive strength (up to 2mA).
1 – high output drive strength (up to 10mA).

> *Slew rate control is an important ally to reduce electromagnetic interference. By increasing rise and fall times, the harmonic components present on signal edges are substantially reduced, contributing to decrease electromagnetic interference.*
>
> *Drive strength control helps decrease overall device current consumption. The application can control which pins need to be configured to high-current mode, leaving the other pins in low-current mode.*

The new QE devices include special registers to speed up I/O operations: PTxSET (to set one or more pins), PTxCLR (to clear one or more pins) and PTxTOG (to toggle one or more pins). Using these registers is slightly different from using standard I/O registers; writes into PTxSET set all pins written with "1" (bits written with "0" are not changed) writes into PTxCLR clear all pins written with "1" (bits written with "0" are not changed) and finally, writes into PTxTOG toggle the state of all pins written with "1" (bits written with "0" are not changed). These features are present only on ports C and E.

6.1. I/O Operations

I/O operations are controlled by I/O registers. We divided I/O operations into two categories: software-controlled and peripheral-controlled.

6.1.1. Software-controlled

On software-controlled I/O operations, I/O pins are controlled directly by the I/O registers under software control. This is the default state after a reset.

As we saw earlier, there are two important registers for controlling an I/O port: PTxD and PTxDD registers. The first is used to read and write into port pins, while the last controls which pins operate as outputs and which ones operate as inputs.

6.1.1.1. Output Mode

Let us say we have a led connected to PTA0 pin (such as on SPYDER08 demonstration kit). To successfully control the led we need to configure PTA0 as an output.

```
PTADD = 1;    // PTA0 is an output, other pins of port A are configured as inputs
```

Although this should work perfectly, we have an important issue: unused pins are configured as inputs and this is not good. CMOS inputs left open act like excellent antennas picking up surrounding noise, drawing excessive current (due to the noise, the input circuitry can switch very fast, increasing current draw) and, in extreme cases, leading the circuit to failure or malfunctioning.

Unused pins should have one of the following configurations:

1. Configured as outputs and left disconnected. This is a common approach which leaves space for future use of those pins.

2. Configured as inputs and externally tied to V_{SS} or V_{DD} (preferably V_{SS}).

3. Configured as inputs with internal pull up/pull down enabled.

On Spyder08, we will configure all pins as outputs. To achieve this we simply write:

```
PTADD = 0x1F; // pins PTA0 to PTA4 are outputs, PTA6 and PTA7 are not implemented!
```

We can also use the symbols defined on the "hcs08.h" file:

```
PTADD = BIT_4 | BIT_3 | BIT_2 | BIT_1 | BIT_0;
```

On 8-pin devices, pins PTA6 and PTA7 are not implemented. 8 and 16-pin devices share pin PTA4 with BKGD function (used for communication with BDM) and PTA5 with reset function. On these devices, the PTA4 pin is output-only when not operating on BKGD function and the PTA5 pin is input-only when not operating on reset function! On SH devices, PTA5 can also operate in open-drain output mode with low drive strength.

Once the pin is correctly configured as an output, we can control its state by changing the respective bit in PTAD registers (or the PTxD register corresponding to the desired pin).

To turn on the led LD1 on the Spyder08 device, it is necessary to set bit 0 in PTAD register. Using the OR bitwise operation with a bitmask (all bits set in the mask are set in the port):

Bit:	7	6	5	4	3	2	1	0
Mask:	0	0	0	0	0	0	0	1
Hex:	0				1			

```
PTAD = PTAD | 0x01;        // set bit 0 of PTA (set pin PTA0)
```

Or:

```
PTAD = PTAD | BIT_0;       // set bit 0 of PTA (set pin PTA0)
```

Usually, we use the reduced form for these kind of attribution operations, so that:
```
PTAD = PTAD | BIT_0;
```
Becomes:
```
PTAD |= BIT_0;
```

Below we can see how this operation is really executed (supposing the previous content of PTAD was 0x30 or 00110000 binary):

Bit:	7	6	5	4	3	2	1	0
Port:	0	0	1	1	0	0	0	0
Mask:	0	0	0	0	0	0	0	1
Result:	0	0	1	1	0	0	0	1

Only bit 0 changes!

Notice that after the operation in PTAD, the previous state of all bits is maintained, except for bit 0, which is set by the operation.

We can also use the special symbols defined on the "derivative.h" file. This way, port A bit 0 can be directly accessed by writing:

```
PTAD_PTAD0 = 1;    // set bit 0 of PTA (set pin PTA0)
```

To clear a pin state, we can also use bitwise logical operations. In this case, it is an AND masking operation (all bits set in the mask remains set in the port):

Bit:	7	6	5	4	3	2	1	0
Mask:	1	1	1	1	1	1	1	0
Hex:	F				E			

```
PTAD = PTAD & 0xFE;          // clear bit 0 of PTA (clear pin PTA0)
```

An easier way to use AND masking is negating the same mask used to set the pin:

Bit:	7	6	5	4	3	2	1	0
OR Mask:	0	0	0	0	0	0	0	1
NOT:	1	1	1	1	1	1	1	0
Hex:	F				E			

Same Result!

```
PTAD = PTAD & ~0x01;         // clear bit 0 of PTA (clear pin PTA0)
```

Or:

```
PTAD = PTAD | ~BIT_0;        // clear bit 0 of PTA (clear pin PTA0)
```

 There is no timing impact on using ~0x01 or ~BIT_0 instead of 0xFE. These values are calculated by the compiler at compile time and generate the same code!

Below we can see how this operation is really executed (supposing the previous content of PTAD was 0x0F or 00001111 binary):

Bit:	7	6	5	4	3	2	1	0
Port:	0	0	0	0	1	1	1	1
Mask:	1	1	1	1	1	1	1	0
Result:	0	0	0	0	1	1	1	0

Only bit 0 changes!

Notice that after the operation in PTAD, the previous state of all bits is maintained, except for bit 0, which is cleared by the operation.

Again, we can also use the special symbols defined on the "derivative.h" file. This way, port A bit 0 can be directly accessed by writing:

```
PTAD_PTAD0 = 0;    // clear bit 0 of PTA (clear pin PTA0)
```

Another example of a led flasher, this time on the Spyder08 demonstration kit, is shown below.

```
// Spyder08 LED flasher
#include <hidef.h>           /* for EnableInterrupts macro */
#include "derivative.h"      /* include peripheral declarations */
#include "hcs08.h"           // This is our definition file!

void main(void)
{
  unsigned int temp;
  SOPT1 = bBKGDPE;  // configure SOPT1 register, enable pin BKGD for BDM
  PTADD = BIT_4 | BIT_3 | BIT_2 | BIT_1 | BIT_0; // bit 4 to bit 0 as outputs
  while (1)
  {
    PTAD |= BIT_0;                      // LED = on
    for (temp=10000; temp; temp--);    // wait for a while
    PTAD &= ~BIT_0;                    // LED = off
    for (temp=10000; temp; temp--);    // wait for a while
  }
}
```

Example 6.1 – LED flasher on the Spyder08

Of course, when using port C and port E on QE devices, we have the option to use the SET/RESET/TOGGLE registers. Notice, however, that setting and clearing single bits through SET and CLEAR registers is less efficient than performing the same operation on PTxD registers. This is because PTxD registers are usually located in the direct page memory and thus, accessed using BSET and BCLR instructions.

Nevertheless, when changing more than one bit or toggling single/multiple bits, using SET/RESET/TOGGLE registers is preferred.

Let us suppose we want to set pins PTC7 and PTC0 (on port C). Using standard I/O we can write:

```
PTCD |= 0x81;       // set PTC0 and PTC7, do not change other pins
```

Using our "hcs08.h" special symbols:

```
PTCD |= BIT_7 | BIT_0;      // set PTC0 and PTC7, do not change other pins
```

On QE devices, the same operation can be done using the PTCSET register:

```
PTCSET = BIT_7 | BIT_0;     // set PTC0 and PTC7, do not change other pins
```

Notice that the masking operation is done directly by the hardware: when using SET, CLEAR and TOGGLE registers, no software logical operation is needed!

To toggle pin states using standard I/O one can use the XOR bitwise operation. The next line demonstrates how to toggle pins PTC0 and PTC7 (in this case, we use the same mask for the OR operation):

```
PTCD = PTCD ^ 0x81;         // set PTC0 and PTC7, do not change other pins
```

This line compiles to the following code:

```
LDA       PTCD       // load PTCD into the accumulator (3 BUSCLK cycles)
EOR       #0x81      // set bits 7 and 0 of the accumulator (2 BUSCLK cycles)
STA       PTCD       // store the accumulator back into PTCD (3 BUSCLK cycles)
```

This is a typical read/modify/write operation: the port is read by LDA, its content is modified by EOR and the result is written by STA. The total time for the operation is 8 cycles or 400ns at 20MHz BUSCLK. Using the toggle register, the same operation can be done with:

```
PTCTOG = 0x81;
```

The compiled code is slightly faster:

```
LDA       #0x81      // load A with mask (2 BUSCLK cycles)
STA       PTCTOG     // Store into PTCTOG (4 BUSCLK cycles)
```

Notice the reduction in the cycle-count: from 8 to 6 (300ns at 20MHz BUSCLK). The reduction can be greater if the PTCTOG register is located on the direct page registers area (thus enabling the use of more optimized direct addressing mode).

6.1.1.2. Input Mode

To configure a pin as an input, all you have to do is clear the respective PTxDD register-bit. Pin state can be read by reading PTxD register: this returns the value for all pins of the port. For pins configured as outputs, the returned value is the last value written into the pin. For pins configured as inputs, the returned value is the current state read from the external pin.

To read the external pin state we need to configure it as an input (writing "0" into the respective PTxDD bit) and then read the pin state through the respective PTxD register.

On 5V devices, a logic level "1" is an input voltage higher than $0.65*V_{DD}$ (3.25V for a V_{DD} of 5V) and a "0" logic level is an input voltage lower than $0.35*V_{DD}$ (1.75V for a V_{DD} of 5V). On 1.8V devices, a logic level "1" is an input voltage higher than $0.7*V_{DD}$ (for $V_{DD} > 2.3V$) or $0.85*V_{DD}$ (for $V_{DD} \leq 2.3V$). A logic level "0" is an input voltage lower than $0.35*V_{DD}$ (for $V_{DD} > 2.3V$) or $0.3*V_{DD}$ (for $V_{DD} \leq 2.3V$).

Reading a single pin can be done by referring the respective bit in PTxD register. To read pin PTA2, all we need to do is to read PTAD bit 2:

```
pin_state = PTAD & 0x04;
```

Or:

```
pin_state = PTAD & BIT_2;
```

Or:

```
pin_state = PTAD_PTAD2;
```

A typical circuit to read a key or switch is shown in figure 6.1A. On this circuit the pull up resistor (R1) maintains a high level ("1") on input line (TO_MCU) while the key or switch (SW1) is open. When SW1 is closed, voltage on the input line TO_MCU drops to zero (logic "0"). While SW1 is being closed, a number of switching noise can (and will) be generated. A capacitor (C1) can be used to help reduce the switching noise (bouncing noise).

On devices such as the HCS08, we can use the internal pull up devices in place of external ones, thus eliminating R1. It is also possible to debounce keys and switches by software, eliminating the need for C1. This leads us to the simplified circuit shown in figure 6.1B.

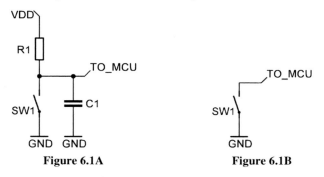

Figure 6.1A   Figure 6.1B

(i) *Internal pulling devices are in range of 17.5k to 52.5kΩ!*

Enabling an internal pull up device can be done by writing a "1" to the respective bit in PTxPE register. To enable the internal pull up on PTA2 we can write:

```
PTAPE = 0x04; // enable internal pull up only on PTA2
```

Or:

```
PTAPE = BIT_2;
```

To demonstrate the concepts learned until here, we will modify our led flasher program to turn it into a "key-controlled led blinker". Remember that to successfully configure pin directions it is necessary to know pin connections on the board. Figure 6.2 shows a simplified diagram for all MCU connections.

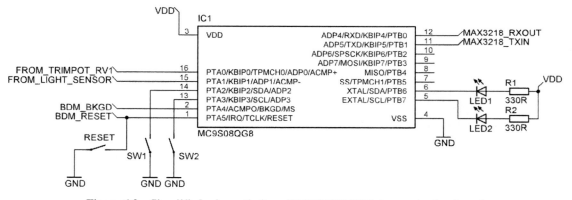

Figure 6.2 – Simplified schematic from DEMO9S08QG8 demonstration board

```
// DEMO9S08QG8 Key-controlled Led flasher
#include <hidef.h>              /* for EnableInterrupts macro */
#include "derivative.h"         /* include peripheral declarations */
#include "hcs08.h"              // This is our definition file!

void main(void)
{
  unsigned int temp;
  unsigned char blinking=0;
  SOPT1 = bBKGDPE;        // configure SOPT1 register, enable pin BKGD for BDM
  PTADD = 0;             // all PTA pins as inputs
  PTBDD = 0xFE;          // pin PTB0 as input, PTB1 through PTB7 as outputs
  PTBD = 0xC0;           // turn off LED1 and LED2, clear other PTB outputs
  PTAPE = BIT_2;         // PTA2 pull up enabled
  while (1)
  {
    if (blinking)
    {
      PTBD_PTBD6 = 0;                    // LED = on
      for (temp=3000; temp; temp--);    // wait for a while
      PTBD_PTBD6 = 1;                    // LED = off
      for (temp=30000; temp; temp--);   // wait for a while
    }
    if (!PTAD_PTAD2)                     // is SW1 pressed ?
    {
      // wait until SW1 is released:
      while (!PTAD_PTAD2) for (temp=100; temp; temp--);
      blinking = !blinking;             // change blinking state
    }
  }
}
```

Example 6.2 – Key-controlled LED flasher

Another approach for reading keys and switches is the use of interrupts. We will learn about this in topic 6.2.

6.1.2. Peripheral-controlled

A simple rule governs shared pins behavior: when no peripheral associated with it is using it, the pin is under I/O registers control.

When an internal peripheral is enabled, control of related pins is under peripheral control and depends on peripheral operating mode.

The priority order for shared pins is the following: the highest priority functions are the analog ones (the analog comparator has priority over the ADC, but the analog comparator and the ADC can use the same shared pins simultaneously), then we have digital peripherals (communication peripherals have priority over timing peripherals and both of them have priority over the keyboard interrupt modules). The lowest priority function is general-purpose I/O mode.

This information can be found on the "External Signal Description" chapter, on device datasheets.

 PTxPE, PTxDS and PTxSE registers continue performing their functions even when the pin is under peripheral control.

Table 6.1 shows the priority order for QG4 and QG8 devices (as shown in device datasheet).

Pin Number		Priority Lowest → Highest				
16-pin	8-pin	Port Pin	Alt 1	Alt 2	Alt 3	Alt 4
1	1	PTA5[1]	$\overline{IRQ}$	TCLK		RESET
2	2	PTA4		ACMPO	BKGD	$\overline{MS}$
3	3					V_{DD}
4	4					V_{SS}
5	-	PTB7		SCL[2]	EXTAL	
6	-	PTB6		SDA[2]	XTAL	
7	-	PTB5		TPMCH1	SS	
8	-	PTB4		MISO		
9	-	PTB3	KBIP7	MOSI	ADP7	
10	-	PTB2	KBIP6	SPSCK	ADP6	
11	-	PTB1	KBIP5	TxD	ADP5	
12	-	PTB0	KBIP4	RxD	ADP4	
13	5	PTA3	KBIP3	SCL[2]	ADP3	
14	6	PTA2	KBIP2	SDA[2]	ADP2	
15	7	PTA1	KBIP1		ADP1[3]	ACMP-[3]
16	8	PTA0	KBIP0	TPMCH0	ADP0[3]	ACMP+[3]

[1] Pin does not contain a clamp diode to V_{DD} and should not be driven above V_{DD}. On this pin, the pull up device is located on the internal gates inside the chip. The external measured level is lower than V_{DD}.

[2] I²C pins can be repositioned using IICPS in SOPT2, default reset locations are on PTA2 and PTA3.

[3] if ACMP and ADC are both enabled, both have access to the pin.

Table 6.1

6.1.3. Ganged Outputs

Some devices (such as SG and SH family) offer an option to internally tie multiple pins allowing a higher current sinking/sourcing (the outputs are ganged together).

This feature is available on pins PTB2, PTB3, PTB4, PTB5, PTC0, PTC1, PTC2 and PTC3 and it is controlled by the Ganged Output Drive Control Register (GNGC): each bit of this register controls whether the respective pin is ganged together with PTC0 or not.

The GNGC:GNGEN (ganged output drive enabled bit) controls the behavior of the selected pins: when GNGEN = 1, all selected pins (including PTC0) are configured as outputs (regardless of the state of their PTxDD bits) and their output state is controlled by the state of the PTC0 output. When PTC0 is cleared, all selected pins are cleared, when PTC0 is set, all selected pins are set.

Moreover, the slew rate and drive strength control for the selected pins is the same selected for the PTC0 pin.

Note that GNGC register is a write-once register. It can be written only once after reset. This means that once gang output selection is made, it cannot be changed or disabled until the next reset!

This feature is not present on 8-pin devices. On 16-pin devices, only PTB2 to PTB5 can be ganged, the control is still done through PTC0, despite the fact that port C is not present..

6.1.4. I/O Registers

I/O ports use up to eight registers for controlling I/O operations:

PTxD – for reading/writing into port pins.

PTxDD – for controlling pin directions.

PTxPE – for enabling/disabling internal pull up devices.

PTxSE – for enabling/disabling slew-rate control.

PTxDS – for enabling/disabling drive strength control.

PTxSET – for setting port pins.

PTxCLR – for clearing port pins.

PTxTOG – for toggling port pins.

GNGC – for controlling ganged outputs.

6.1.4.1. PTxD Register

Name	Model		BIT 7	BIT 6	BIT 5	BIT 4	BIT 3	BIT 2	BIT 1	BIT 0
PTAD PTBD PTCD	check data sheet	Read	PTxD7	PTxD6	PTxD5	PTxD4	PTxD3	PTxD2	PTxD1	PTxD0
PTDD PTED PTFD		Write								
PTGD PTHD PTID		Reset	0	0	0	0	0	0	0	0

6.1.4.2. PTxDD Register

Name	Model		BIT 7	BIT 6	BIT 5	BIT 4	BIT 3	BIT 2	BIT 1	BIT 0
PTADD PTBDD PTCDD	check data sheet	Read	PTxDD7	PTxDD6	PTxDD5	PTxDD4	PTxDD3	PTxDD2	PTxDD1	PTxDD0
PTDDD PTEDD PTFDD		Write								
PTGDD PTHDD PTIDD		Reset	0	0	0	0	0	0	0	0

PTxDDx – configure pin direction:
0 – input.
1 – output.

6.1.4.3. PTxPE Register

Name	Model		BIT 7	BIT 6	BIT 5	BIT 4	BIT 3	BIT 2	BIT 1	BIT 0
PTAPE PTBPE PTCPE	check data sheet	Read	PTxPE7	PTxPE6	PTxPE5	PTxPE4	PTxPE3	PTxPE2	PTxPE1	PTxPE0
PTDPE PTEPE PTFPE		Write								
PTGPE PTHPE PTIPE		Reset	0	0	0	0	0	0	0	0

PTxPEx – internal pull up:
0 – disabled.
1 – enabled (only when the pin is an input).

6.1.4.4. PTxSE Register

Name	Model		BIT 7	BIT 6	BIT 5	BIT 4	BIT 3	BIT 2	BIT 1	BIT 0
PTASE PTBSE PTCSE PTDSE PTESE PTFSE PTGSE PTHSE PTISE	check data sheet	Read	PTxSE7	PTxSE6	PTxSE5	PTxSE4	PTxSE3	PTxSE2	PTxSE1	PTxSE0
		Write								
		Reset	0	0	0	0	0	0	0	0

PTxSEx – slew rate control:
0 – disabled (rise/fall times in order of 3ns).
1 – enabled (rise/fall times in order of 30ns).

These registers are not available on RC, RD, RE and RG devices!

6.1.4.5. PTxDS Register

Name	Model		BIT 7	BIT 6	BIT 5	BIT 4	BIT 3	BIT 2	BIT 1	BIT 0
PTADS PTBDS PTCDS PTDDS PTEDS PTFDS PTGDS PTHDS PTIDS	check data sheet	Read	PTxDS7	PTxDS6	PTxDS5	PTxDS4	PTxDS3	PTxDS2	PTxDS1	PTxDS0
		Write								
		Reset	0	0	0	0	0	0	0	0

PTxDSx – output drive strength:
0 – standard mode (2mA maximum).
1 – high-power mode (10mA maximum).

These registers are not available on RC, RD, RE and RG devices!

6.1.4.6. PTxSET Register

Name	Model		BIT 7	BIT 6	BIT 5	BIT 4	BIT 3	BIT 2	BIT 1	BIT 0
PTCSET PTESET	check data sheet	Read	0	0	0	0	0	0	0	0
		Write	PTxSET7	PTxSET6	PTxSET5	PTxSET4	PTxSET3	PTxSET2	PTxSET1	PTxSET0
		Reset	0	0	0	0	0	0	0	0

PTxSETx – set the respective pin:
0 – no changes.
1 – set pin.

These registers are available only on QE devices!

6.1.4.7. PTxCLR Register

Name	Model		BIT 7	BIT 6	BIT 5	BIT 4	BIT 3	BIT 2	BIT 1	BIT 0
PTCCLR PTECLR	check data sheet	Read	0	0	0	0	0	0	0	0
		Write	PTxCLR7	PTxCLR6	PTxCLR5	PTxCLR4	PTxCLR3	PTxCLR2	PTxCLR1	PTxCLR0
		Reset	0	0	0	0	0	0	0	0

PTxCLRx – clear the respective pin:
0 – no changes.
1 – clear pin.

These registers are available only on QE devices!

6.1.4.8. PTxTOG Register

Name	Model		BIT 7	BIT 6	BIT 5	BIT 4	BIT 3	BIT 2	BIT 1	BIT 0
PTCTOG PTETOG	check data sheet	Read	0	0	0	0	0	0	0	0
		Write	PTxTOG7	PTxTOG6	PTxTOG5	PTxTOG4	PTxTOG3	PTxTOG2	PTxTOG1	PTxTOG0
		Reset	0	0	0	0	0	0	0	0

PTxTOGx –   toggle the respective pin:
                    0 – no changes.
                    1 – toggle pin.

These registers are available only on QE devices!

6.1.4.9. GNGC Register

Name	Model		BIT 7	BIT 6	BIT 5	BIT 4	BIT 3	BIT 2	BIT 1	BIT 0
GNGC	SG SH	Read	GNGPS7	GNGPS6	GNGPS5	GNGPS4	GNGPS3	GNGPS2	GNGPS1	GNGEN
		Write								
		Reset	0	0	0	0	0	0	0	0

Bit Name	Description	C Symbol
GNGPS7	PTB5 ganged output selection: 0 – PTB5 is not ganged. 1 – PTB5 is ganged with PTC0.	**bGNGPS7**
GNGPS6	PTB4 ganged output selection: 0 – PTB4 is not ganged. 1 – PTB4 is ganged with PTC0.	**bGNGPS6**
GNGPS5	PTB3 ganged output selection: 0 – PTB3 is not ganged. 1 – PTB3 is ganged with PTC0.	**bGNGPS5**
GNGPS4	PTB2 ganged output selection: 0 – PTB2 is not ganged. 1 – PTB2 is ganged with PTC0.	**bGNGPS4**
GNGPS3	PTC3 ganged output selection: 0 – PTC3 is not ganged. 1 – PTC3 is ganged with PTC0.	**bGNGPS3**
GNGPS2	PTC2 ganged output selection: 0 – PTC2 is not ganged. 1 – PTC2 is ganged with PTC0.	**bGNGPS2**
GNGPS1	PTC1 ganged output selection: 0 – PTC1 is not ganged. 1 – PTC1 is ganged with PTC0.	**bGNGPS1**
GNGEN	Ganged output drive enable bit: 0 – Ganged output drive disabled. 1 – Ganged output drive enabled. The selected pins are controlled by PTC0. PTC0 is always an output, regardless of the value of the PTCDD:0 bit.	**bGNGEN**

Note that GNGC register is a write-once register. It can be written only once after reset. This means that once gang output selection is made, it cannot be changed or disabled until the next reset!

6.2. Interrupts

There are two interrupt sources related to the I/O pins: the interrupt request pin ($\overline{\text{IRQ}}$) and the keyboard interrupt module (KBI). Some newer devices include a pin interrupt circuitry, instead of the KBI module (the operation of this module is virtually the same of KBI). While in stop or wait mode, these interrupts can also be used as a wake up source to bring the CPU back to run mode.

6.2.1. IRQ

The interrupt request pin ($\overline{\text{IRQ}}$) is part of the CPU core and allows interrupting the CPU on edge or edge/level detection.

The basic IRQ operation is very straightforward: once the pin is configured for IRQ operation (IRQSC:IRQPE = 1), it can be configured to generate interrupts on edges (IRQSC:IRQMOD = 0) or edges and levels (IRQSC:IRQMOD = 1). Most devices (except QA and QG ones) allow selecting the edge/level that triggers the interrupt. This is done through IRQSC:IRQEDG bit (1 = rising edge/high level, 0 = falling edge/low level).

When the selected edge/level is detected on $\overline{\text{IRQ}}$ pin, the IRQSC:IRQF flag is set and generates an interrupt if IRQSC:IRQIE is set. Once IRQF is set, it is cleared only by writing "1" to the acknowledge bit IRQSC:IRQACK.

 In order to be properly detected, the pulse width must be higher than 100ns when operating in asynchronous mode (internal clocks stopped), or 1.5 BUSCLK cycles when operating in synchronous mode (internal clocks running).

A pull up or pull down device is automatically enabled when the IRQ function is enabled (IRQPE = 1), depending on the edge/level selection (pull up for rising edge/high level, pull down for falling edge/low level).

IRQMOD	IRQEDG	IRQF is set on
0	0	falling edge on $\overline{\text{IRQ}}$
0	1	rising edge on $\overline{\text{IRQ}}$
1	0	falling edge / low level on $\overline{\text{IRQ}}$
1	1	rising edge / high level on $\overline{\text{IRQ}}$

Table 6.2 – IRQ pin modes

On all devices (except AW, GB, GT and Rx devices), the internal pulling device can be optionally disabled by IRQSC:IRQDD bit.

 On devices with a dedicated IRQ pin and devices with multiplexed functions on IRQ pin (when IRQPE = 1), reading the pin state can only be done using BIL and BIH instructions. When IRQPE=0, the pin state can be read using PTxD register.

The following example shows how to use IRQ interrupt. This is a modified version of example 6.2, but it uses the RESET key to change the led state. Press it once to start blinking and another time to stop blinking.

```
// DEMO9S08QG8 IRQ-controlled Led flasher
// LED1 cathode is connected to PTB6
// RESET key is connected to PTA5

#include <hidef.h>          /* for EnableInterrupts macro */
#include "derivative.h"     /* include peripheral declarations */
#include "hcs08.h"          // This is our definition file!
```

```
      unsigned char blinking=0;

      // This is the IRQ interrupt servicing routine
      void interrupt VectorNumber_Virq IRQ_isr()
      {
        IRQSC_IRQACK = 1;          // acknowledge IRQ interrupt (clear IRQF)
        blinking = !blinking;      // change blinking state
        // IRQ interrupt is disabled to avoid bouncing noise:
        IRQSC_IRQIE = 0;           // disable IRQ interrupt
      }

      void main(void)
      {
        unsigned int temp;
        SOPT1 = bBKGDPE;           // configure SOPT1 register, enable pin BKGD for BDM
        PTADD = 0;                 // all PTA pins as inputs
        PTBDD = 0xFE;              // pin PTB0 as input, PTB1 through PTB7 as outputs
        PTBD = 0xC0;               // turn off LED1 and LED2, clear other PTB outputs
        PTAPE = BIT_5;             // PTA5 pull up enabled
        IRQSC = bIRQPE | bIRQIE;   // IRQ pin enabled, IRQ interrupt enabled
        EnableInterrupts;          // enable interrupts (CCR:I = 0)
        while (1)
        {
          if (blinking)
          {
            PTBD_PTBD6 = 0;                  // LED = on
            for (temp=3000; temp; temp--);   // wait for a while
            PTBD_PTBD6 = 1;                  // LED = off
            for (temp=30000; temp; temp--);  // wait for a while
          }
          IRQSC_IRQIE = 1;         // enable IRQ interrupt
        }
      }
```

Example 6.3 – IRQ controlled LED flasher

Some comments about this example:

1. IRQ sensitivity is falling edge only as this is the only mode available on QG devices.

2. Variable "blinking" is now global due the necessity of changing its content outside the main() function.

3. The IRQ interrupt is disabled within the interrupt servicing routine IRQ_isr() and re-enabled on the end of the main loop. This helps debounce eventual noise generated by key pressing.

4. We could acknowledge IRQF and disable IRQ interrupt with a single statement:

```
      IRQSC = bIRQPE | bIRQACK;        // acknowledge IRQ interrupt and re-enable IRQ pin
```

In such case, our ISR code would be:

```
      void interrupt VectorNumber_Virq IRQ_isr()
      {
        IRQSC = bIRQPE | bIRQACK;      // acknowledge IRQ interrupt and re-enable IRQ pin
        blinking = !blinking;          // change blinking state
      }
```

The next example shows how to use the IRQ interrupt to create an event counter. In this application, we count the number of falling edges on the $\overline{\text{IRQ}}$ pin (there is no preview for debouncing key presses).

```
      // DEMO9S08QE128 Event counter using the IRQ pin
      // LEDS are connected to port C pins
      // RESET/IRQ key is connected to PTA5
      #include <hidef.h>              /* for EnableInterrupts macro */
      #include "derivative.h"         /* include peripheral declarations */
      #include "hcs08.h"              // This is our definition file!
```

```
unsigned char counter=0;

// This is the IRQ interrupt servicing routine
void interrupt VectorNumber_Virq IRQ_isr()
{
  IRQSC_IRQACK = 1;                    // acknowledge IRQ interrupt (clear IRQF)
  counter++;                           // increment counter
}

void main(void)
{
  SOPT1 = bBKGDPE;                     // configure SOPT1 register, enable pin BKGD for BDM
  PTCDD = 0xFF;                        // PTC0 through PTB7 as outputs
  PTEDD = BIT_7 | BIT_6;               // PTE7 and PTE6 as outputs
  PTAPE = BIT_5;                       // PTA5 pull up enabled
  IRQSC = bIRQPE | bIRQIE;             // IRQ pin enabled, IRQ interrupt enabled
  EnableInterrupts;                    // enable interrupts (CCR:I = 0)
  while (1)
  {
    PTCD = ~counter;                   // PTCD = negated counter (all bits inverted)
    PTED_PTED7 = PTCD_PTCD7;           // PTE7 state = PTC7 state
    PTED_PTED6 = PTCD_PTCD6;           // PTE6 state = PTE6 state
  }
}
```

Example 6.4 – Event counter

This example can be changed to count rising edges but we would be limited by the DEMO9S08QE128 hardware (the RESET key is connected to the GND and not to the V_{DD} as it should be for the example to work with rising edge detection).

6.2.1.1. IRQSC Register

Name	Model		BIT 7	BIT 6	BIT 5	BIT 4	BIT 3	BIT 2	BIT 1	BIT 0
IRQSC	All	Read	0	IRQDD	IRQEDG	IRQPE	IRQF	0	IRQIE	IRQMOD
		Write	-					IRQACK		
		Reset	0	0	0	0	0	0	0	0

Bit Name	Description	C Symbol
IRQDD	IRQ pulling device **disable** bit. This bit disables internal pull up/pull down device on IRQ pin: 0 – pulling device enabled.　　　1 – pulling device disabled. **This bit is not available on AW, GB, GT, and Rx devices!**	**bIRQDD**
IRQEDG	Edge selection for IRQ interrupt: 0 – IRQF set on falling edges or low level of IRQ. 1 – IRQF set on rising edges or high level of IRQ. **This bit is not available on QA and QG devices!**	**bIRQEDG**
IRQPE	IRQ pin enable bit. This bit enables the IRQ function on devices with multiplexed pins: 0 – IRQ function disabled.　　　1 – IRQ function enabled.	**bIRQPE**
IRQF	IRQ interrupt flag: 0 – no IRQ interrupt pending.　　　1 – IRQ interrupt pending. **This is a read-only bit!**	**bIRQF**
IRQACK	IRQ interrupt acknowledgment: 0 – no effect.　　　1 – acknowledge IRQ interrupt (clears IRQF)	**bIRQACK**
IRQIE	IRQ interrupt enable: 0 – IRQ interrupt disabled.　　　1 – IRQ interrupt enabled.	**bIRQIE**
IRQMOD	Mode selection for IRQ sensitivity: 0 – edges only.　　　1 – edges and levels.	**bIRQMOD**

6.2.2. Keyboard Interrupts (KBI)

The second source for external pin interrupts is the keyboard interrupt module (KBI). This module was initially designed to aid the development of keypads and keyboards, but it can be used as a simple way to detect external events and monitor pin states. The KBI module can also be used to wake up the CPU from low power modes.

Each KBI module comprises up to 8 inputs and a single control flip-flop which stores the KBI interrupt flag (KBF). The module operation is very similar to the IRQ module, only with more inputs.

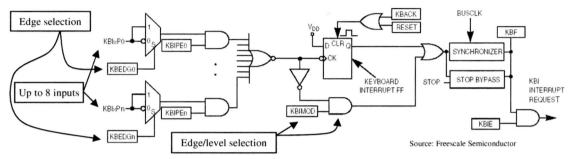

Figure 6.3 – KBI module block diagram

The operation principle is: the interrupt flag KBF (register KBIxSC) is set every time an edge is detected on one of the KBI inputs KBIxPx. Each input can be enabled by setting the corresponding KBIPEx bit in the KBI pin enable (KBIPE) register (disabled inputs do not generate any signal to the flip-flop).

 In order to be properly detected, the pulse width must be higher than 100ns when operating in asynchronous mode (internal clocks stopped) or 1.5 BUSCLK cycles, when operating in synchronous mode (internal clocks running).

Triggering KBF can occur on rising or falling edges, as well as high or low levels. Trigger polarity is selected independently for each input through the KBI edge selection (KBIxES) register, whereas edge/level detection is done through a single bit (KBIxSC:KBIMOD) which controls all inputs within a single module. On several devices, some KBI inputs work only on falling edge/low level sensing (refer to table 6.5 for further information).

KBIMOD	KBIEDGn	KBF is set on
0	0	falling edge on KBI input "n"
0	1	rising edge on KBI input "n"
1	0	falling edge / low level on KBI input "n"
1	1	rising edge / high level on KBI input "n"

Table 6.3 – KBI operating modes

Pins configured to KBI mode are also automatically configured as inputs, regardless of the PTxDD configuration. The PTxPE register can still control internal pulling devices: if the KBI input is configured to detect falling edges or low levels, a pull up device is enabled when the corresponding PTxPE bit is set. When a KBI input is configured to detect rising edges or high levels, a pull down device is enabled when the corresponding PTxPE bit is set.

PTxPEn	KBIEDGn	KBI input pulling device
0	0	no pulling device
0	1	no pulling device
1	0	pull up device
1	1	pull down device

Table 6.4 – KBI internal pulling devices

The KBF flag can optionally generate an interrupt if the keyboard interrupt enable (KBIxSC:KBIE) bit is set. Clearing KBF flag is done by writing "1" to its acknowledge bit (KBIxSC:KBACK).

6.2.2.1. KBI Module Connections

KBI inputs are connected to the pins shown in table 6.5.

Device Family	KBI Module(s)	KBI inputs							
		KBIP7	KBIP6	KBIP5	KBIP4	KBIP3	KBIP2	KBIP1	KBIP0
AC, AW	KBI	PTD7	PTD3	PTD2	PTG4	PTG3*	PTG2*	PTG1*	PTG0*
GB, GT	KBI	PTA7	PTA6	PTA5	PTA4	PTA3*	PTA2*	PTA1*	PTA0*
LC	KBI1	PTA7	PTA6	PTA5	PTA4	PTA3	PTA2	PTA1	PTA0
	KBI2	PTC7	PTC5	PTC4	PTB7	PTB6	PTB3	PTB1	PTB0
LL	KBI	PTA7	PTA6	PTA5	PTA4	PTA3	PTA2	PTA1	PTA0
JM	KBI	PTG3	PTG2	PTB5	PTB4	PTD3	PTD2	PTG1	PTG0
QA, QD	KBI	-	-	-	-	PTA3	PTA2	PTA1	PTA0
QB,QE4,QE8	KBI	PTB3	PTB2	PTB1	PTB0	PTA3	PTA2	PTA1	PTA0
QE16/32/64/96/128	KBI1	PTB3	PTB2	PTB1	PTB0	PTA3	PTA2	PTA1	PTA0
	KBI2	PTD7	PTD6	PTD5	PTD4	PTD3	PTD2	PTD1	PTD0
QG	KBI	PTB3	PTB2	PTB1	PTB0	PTA3	PTA2	PTA1	PTA0
Rx	KBI1	PTA7	PTA6	PTA5	PTA4	PTA3*	PTA2*	PTA1*	PTA0*
	KBI2	-	-	-	-	PTC3*	PTC2*	PTC1*	PTC0*

* Falling edge/low level only. These pins do not have KBI input polarity selection.

Table 6.5 – KBI external connections

Notice that AC, AW, GB, GT, RC, RD, RE and RG devices do not offer edge selection for all inputs. On these devices, only the four higher KBI inputs have selectable edges.

Eight-pin devices (such as the QA, QD and some QG devices) offer only four KBI inputs. The four higher KBI inputs (KBIP4 to KBIP7) are not implemented.

RC, RD, RE and RG devices implement two KBI modules, one with eight inputs (KBI1) and another with four inputs (KBI2). On both modules, KBIP0 to KBIP3 inputs are low level/falling edge sensible only. LC and QE32/64/96/128 devices include two eight-input KBI modules each. On these devices, the KBI modules are named KBI1 and KBI2.

The following example shows how to use KBI interrupts to read key presses. Leds LED1 and LED2 on DEMO9S08QG blink alternatingly. When SW1 is pressed, they blink faster and when SW2 is pressed they blink slower.

```
// DEMO9S08QG8 KBI-controlled led flasher
// LED1 cathode is connected to PTB6, LED2 cathode is connected to PTB7
// SW1 is connected to PTA2, SW2 is connected to PTA3

#include <hidef.h>          /* for EnableInterrupts macro */
#include "derivative.h"     /* include peripheral declarations */
#include "hcs08.h"          // This is our definition file!

unsigned int blinking_rate = 15000;

#define LED1    PTBD_PTBD6
#define LED2    PTBD_PTBD7
#define SW1     PTAD_PTAD2
#define SW2     PTAD_PTAD3

/*
  This is the KBI interrupt servicing routine
  SW1 and SW2 are falling edge sensitive (interrupt is triggered when key is pressed
*/
void interrupt VectorNumber_Vkeyboard KBI_isr()
{
  KBISC_KBACK = 1;  // acknowledge KBI interrupt (clear KBF)
  if (!SW1) blinking_rate = 15000; // if SW1 is pressed, change blinking rate
  if (!SW2) blinking_rate = 30000; // if SW2 is pressed, change blinking rate
}
```

```
void main(void)
{
  unsigned int temp;
  SOPT1 = bBKGDPE;              // configure SOPT1 register, enable pin BKGD for BDM
  PTADD = 0;                    // all PTA pins as inputs
  PTBDD = 0xFE;                 // pin PTB0 as input, PTB1 through PTB7 as outputs
  PTBD = 0xC0;                  // turn off LED1 and LED2, clear other PTB outputs
  PTAPE = BIT_5 | BIT_3 | BIT_2; // PTA5, PTA3 and PTA2 pull ups enabled
  KBIPE = BIT_3 | BIT_2;        // KBIP3 (PTA3) and KBIP2 (PTA2) inputs enabled
  KBISC = bKBIE;                // KBI interrupt enabled
  EnableInterrupts;             // enable interrupts (CCR:I = 0)
  while (1)
  {
    LED1 = 0; LED2 = 1;                              // LED1 = on, LED2 = off
    for (temp=blinking_rate; temp; temp--);  // wait for a while
    LED1 = 1; LED2 = 0;                              // LED1 = off, LED2 = on
    for (temp=blinking_rate; temp; temp--);  // wait for a while
  }
}
```

Example 6.5 – KBI-controlled LED flasher

Another example of KBI usage on the DEMO9S08AW60E board (using the MC9S08AW60):

```
// DEMO9S08AW60E KBI-controlled led sequencer
// SW1 is connected to PTC2
// SW2 is connected to PTC6
// SW3 is connected to PTD3
// SW4 is connected to PTD2
// the LEDs are connected to PTF

#include <hidef.h>            /* for EnableInterrupts macro */
#include "derivative.h"       /* include peripheral declarations */
#include "hcs08.h"            // This is our definition file!

unsigned int blinking_rate = 15000;
unsigned char shift_dir = 0;

#define SW4          PTDD_PTDD2
#define SW3          PTDD_PTDD3

// This is the KBI interrupt servicing routine
void interrupt VectorNumber_Vkeyboard1 KBI_isr()
{
  KBI1SC_KBACK = 1; // acknowledge KBI interrupt (clear KBF)
  if (!SW3)         // if PTD3 is equal to "0"
  {
    // SW3 is pressed, change blinking rate
    if (blinking_rate == 30000) blinking_rate = 15000; else blinking_rate = 30000;
  }
  if (!SW4)         // if PTD2 is equal to "0"
  {
    // SW4 is pressed, change shift direction
    shift_dir = !shift_dir;
  }
}

void main(void)
{
  unsigned int temp;
  unsigned char shift_register = 1;
  SOPT = 0;  // configure SOPT1 register, STOP instruction and COP disabled
  PTCDD = ~(BIT_2 | BIT_6);       // PTC2 and PTC6 as inputs, remaining pins as outputs
  PTCPE = BIT_2 | BIT_6;          // PTC2 and PTC6 internal pull up enabled
  PTDDD = ~(BIT_2 | BIT_3);       // PTD2 and PTD3 as inputs, remaining pins as outputs
  PTDPE = BIT_2 | BIT_3;          // PTD2 and PTD3 internal pull ups enabled
  PTFDD = 0xFF;// all PTF pins as outputs
  KBI1PE = BIT_5 | BIT_6;         // KBIP5 (PTD2) and KBIP6 (PTD3) inputs enabled
  KBI1SC = bKBIE;                 // KBI interrupt enabled
  EnableInterrupts;               // enable interrupts (CCR:I = 0)
  while (1)
  {
    PTFD = shift_register;                      // change led states
    for (temp=blinking_rate; temp; temp--);  // wait for a while
```

```
        if (shift_dir)
        {
          shift_register <<= 1;                    // shifting to right
          // did we reached the last led?
          if (!shift_register) shift_register=1;
        }
        else
        {
          shift_register >>= 1;                    // shifting to left
          // did we reached the last led?
          if (!shift_register) shift_register=128;
        }
      }
    }
```

Example 6.6 – KBI-controlled LED sequencer

6.2.2.2. KBI Module Versions

The following table shows the evolution of KBI modules since version 1:

KBI Version	Devices	Features added/modified
1	AC, AW, GB, GT, RC, RD, RE and RG	First release
2	JM, LC, QA, QB, QD, QE, QG	Rising/falling edge, high/low level detection

Table 6.6 – KBI module history

6.2.2.3. KBIxSC Register

Name	Model		BIT 7	BIT 6	BIT 5	BIT 4	BIT 3	BIT 2	BIT 1	BIT 0
KBI1SC*	AC, AW, GB, GT	Read	KBEDG7	KBEDG6	KBEDG5	KBEDG4	KBF	0	KBIE	KBMOD
		Write						KBACK		
		Reset	0	0	0	0	0	0	0	0
KBISC KBI1SC/KBI2SC	JM, LC, QA, QB, QD, QG, QE	Read	0	0	0	0	KBF	0	KBIE	KBMOD
		Write						KBACK		
		Reset	0	0	0	0	0	0	0	0
KBI1SC	RC, RD, RE, RG	Read	KBEDG7	KBEDG6	KBEDG5	KBEDG4	KBF	0	KBIE	KBMOD
		Write						KBACK		
		Reset	0	0	0	0	0	0	0	0
KBI2SC	RC, RD, RE, RG	Read	0	0	0	0	KBF	0	KBIE	KBMOD
		Write						KBACK		
		Reset	0	0	0	0	0	0	0	0

* On AC devices, this register is named KBISC.

Bit Name	Description		C Symbol
KBEDG7	Edge selection for KBI input KBIP7: 0 – falling edge/low level.	1 – rising edge/high level.	bKBEDG7
KBEDG6	Edge selection for KBI input KBIP6: 0 – falling edge/low level.	1 – rising edge/high level.	bKBEDG6
KBEDG5	Edge selection for KBI input KBIP5: 0 – falling edge/low level.	1 – rising edge/high level.	bKBEDG5
KBEDG4	Edge selection for KBI input KBIP4: 0 – falling edge/low level.	1 – rising edge/high level.	bKBEDG4
KBF	KBI interrupt flag: 0 – no interrupt.	1 – KBI interrupt pending.	bKBF
KBACK	KBI interrupt acknowledgement. This is a write-only bit. Writing "1" clears KBF, writing "0" has no effect.		bKBACK
KBIE	KBI interrupt enable: 0 – KBI interrupt disabled.	1 – KBI interrupt enabled.	bKBIE
KBMOD	Mode selection for triggering KBF: 0 – KBF is triggered on edges only.	1 – KBF is triggered on edges and levels.	bKBMOD

6.2.2.4. KBIxPE Register

Name	Model		BIT 7	BIT 6	BIT 5	BIT 4	BIT 3	BIT 2	BIT 1	BIT 0
KBIPE*	AC, AW, GB, GT, LC, QA, QB, QD, QE, QG, Rx	Read	KBIPE7	KBIPE6	KBIPE5	KBIPE4	KBIPE3	KBIPE2	KBIPE1	KBIPE0
		Write								
		Reset	0	0	0	0	0	0	0	0

* On AW, GB and GT devices, this register is named KBI1PE. On devices with two KBI modules, there are two registers: KBI1PE and KBI2PE.

* Not all bits are implemented on all devices; refer to table 6.5 for the available pins in each KBI module.

Bit Name	Description	C Symbol
KBIPE7	Enable KBIP7 input: 0 – KBI input disabled. 1 – KBI input enabled.	**bKBIPE7**
KBIPE6	Enable KBIP6 input: 0 – KBI input disabled. 1 – KBI input enabled.	**bKBIPE6**
KBIPE5	Enable KBIP5 input: 0 – KBI input disabled. 1 – KBI input enabled.	**bKBIPE5**
KBIPE4	Enable KBIP4 input: 0 – KBI input disabled. 1 – KBI input enabled.	**bKBIPE4**
KBIPE3	Enable KBIP3 input: 0 – KBI input disabled. 1 – KBI input enabled.	**bKBIPE3**
KBIPE2	Enable KBIP2 input: 0 – KBI input disabled. 1 – KBI input enabled.	**bKBIPE2**
KBIPE1	Enable KBIP1 input: 0 – KBI input disabled. 1 – KBI input enabled.	**bKBIPE1**
KBIPE0	Enable KBIP0 input: 0 – KBI input disabled. 1 – KBI input enabled.	**bKBIPE0**

6.2.2.5. KBIxES Register

Name	Model		BIT 7	BIT 6	BIT 5	BIT 4	BIT 3	BIT 2	BIT 1	BIT 0
KBIES	JM, LC, QA, QB, QD, QE, QG	Read	KBIEDG7	KBIEDG6	KBIEDG5	KBIEDG4	KBIEDG3	KBIEDG2	KBIEDG1	KBIEDG0
		Write								
		Reset	0	0	0	0	0	0	0	0

* On devices with two KBI modules, there are two registers: KBI1ES and KBI2ES.

* Not all bits are implemented on all devices; refer to table 6.5 for the available pins in each KBI module.

Bit Name	Description	C Symbol
KBIEDG7	Edge selection for KBIP7 input: 0 – falling edge/low level. 1 – rising edge/high level.	**bKBIEDG7**
KBIEDG6	Edge selection for KBIP6 input: 0 – falling edge/low level. 1 – rising edge/high level.	**bKBIEDG6**
KBIEDG5	Edge selection for KBIP5 input: 0 – falling edge/low level. 1 – rising edge/high level.	**bKBIEDG5**
KBIEDG4	Edge selection for KBIP4 input: 0 – falling edge/low level. 1 – rising edge/high level.	**bKBIEDG4**
KBIEDG3	Edge selection for KBIP3 input: 0 – falling edge/low level. 1 – rising edge/high level.	**bKBIEDG3**
KBIEDG2	Edge selection for KBIP2 input: 0 – falling edge/low level. 1 – rising edge/high level.	**bKBIEDG2**
KBIEDG1	Edge selection for KBIP1 input: 0 – falling edge/low level. 1 – rising edge/high level.	**bKBIEDG1**
KBIEDG0	Edge selection for KBIP0 input: 0 – falling edge/low level. 1 – rising edge/high level.	**bKBIEDG0**

6.3. Interfacing to External Signals and Loads

To close this chapter (but not the I/O subject) we will see some possible interfacing circuits and techniques to drive external loads and to read external signals.

6.3.1. Reading External Signals

Typical microcontroller circuits interface to external signals, which often are not of the same magnitude as microcontroller supply voltage.

A typical example is interfacing a 24 Volts signal (typically found in industrial environment) to a 5 or 3.3V microcontroller.

Figure 6.4 shows a possible solution using a voltage divider. To calculate values for R1 and R2 we will use a target input current of 1mA. Due the low maximum leakage pin input current (1μA) we will disregard the leakage current.

For a 3.3V target (input pin) voltage we need a 20.7kΩ resistor for R1 and a 3.3kΩ resistor for R2. We can approximate R1 to a commercial value of 22kΩ, resulting in a 3.13V input voltage for a 24V Vin.

For a 5V target (input pin) voltage, we need a 19kΩ resistor for R1 and a 5kΩ resistor for R2. We can approximate R1 to a commercial value of 18kΩ and R2 to 4.7kΩ, resulting in a 4.96V input voltage for a 24V Vin.

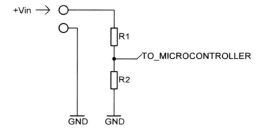

Figure 6.4

A simpler approach for interfacing slightly higher voltage circuits to low-voltage MCUs is based on the internal clamping diodes (figure 6.5). This approach relies on clamping diodes to limit voltage on the input pin. It is important to calculate the current limiting resistor so that the maximum sinking value is not exceeded (200μA maximum). To interface a 5V circuit to a 3.3V microcontroller we can use a 10kΩ resistor for R1 (figure 6.5).

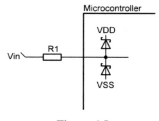

Figure 6.5

 Remember that sinking too much current into V_{DD} can cause the voltage regulator to work unreliably and going out of regulation!

It is always a good practice to include some protective circuits on critical pins. A varistor (R2) and/or a transient voltage suppressor (TVS)(D1) are good options to protect the MCU from voltage spikes that can otherwise cause damage to the chip. The current limiting resistor (R1) is necessary to limit peak current when R2/D1 starts conducting.

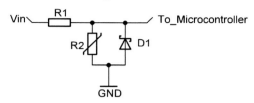

Figure 6.6

Another good option for interfacing to critical signals is an optocoupler. This is a good option when galvanic isolation is necessary. R1 resistor should be calculated according to the Vin maximum voltage. For high voltages it may be necessary to include additional circuitry to reduce voltage to the level of the LED voltage.

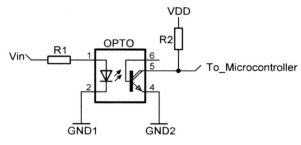

Figure 6.7

The pull up resistor (R2) can be omitted if we enable the internal pull up device on the HCS08 microcontroller. Remember that when using KBI interrupts, changing the sensitivity edge also changes the pulling device. When using this circuit to detect falling edges, the external pull up is still needed!

6.3.2. Driving External Loads

We will start by showing some possible ways to connect leds to a microcontroller pin. Figure 6.8 shows a led connected with its cathode to ground.

In such configuration, the led is activated by writing "1" into the microcontroller pin (previously configured as an output). This should be preferred when the pin sources more current than it sinks.

Using pin PTA0 we can write:

```
PTADD_PTADD0 = 1;          // pin PTA0 as output
PTADS_PTADS0 = 1;          // enable driving strength (10mA mode)
PTAD_PTAD0 = 1;            // turn LED1 on
```

Figure 6.8

Another possible configuration is shown in figure 6.9. The anode of the led is connected to V_{DD} and the led is turned on by writing "0" into the pin (configured as an output). This configuration should be preferred when the pin sinks more current than it sources.

Using pin PTA0 we can write:

```
PTADD_PTADD0 = 1;          // pin PTA0 as output
PTADS_PTADS0 = 1;          // enable driving strength (10mA mode)
PTAD_PTAD0 = 0;            // turn LED1 on
```

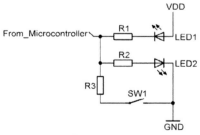

Figure 6.9

It is also possible to drive two leds through the same pin by using the circuit shown in figure 6.10. LED1 is turned on by writing "0" into the pin, while LED2 is turned on by writing "1" into the pin. To turn both leds off, all you have to do is change the pin direction to input. Of course, by using this configuration it is not possible to turn both leds on at the same time (but it is possible to alternate between the two leds so fast that it actually seems that they are both on).

Using pin PTA0 we can write:

```
PTADD_PTADD0 = 1;          // pin PTA0 as output
PTADS_PTADS0 = 1;          // enable driving strength (10mA mode)
PTAD_PTAD0 = 0;            // turn LED1 on / LED2 off
PTAD_PTAD0 = 1;            // turn LED2 on / LED1 off
PTADD_PTADD0 = 0;          // turn LED1 and LED2 off
```

Figure 6.10

 R1 and R2 are calculated according to the V_{DD} voltage level, the led junction voltage and the led current consumption. Typical values for red and green leds are in range of 330 to 470Ω.

Another interesting feature (common to almost any microcontroller) is the ability to change pin direction on the fly. To show how to do that, we will include a key to the circuit shown in figure 6.10. The resulting circuit (figure 6.11) provides two leds and a key, all controlled through a single pin.

Figure 6.11

Circuit operation is very simple: when the pin is an output, a high level turns LED2 on (and LED1 off) and a low level turns LED1 on (and LED2 off).

When the pin is configured as an input (with internal pull up enabled) the key (SW1) state can be read. Care should be taken when selecting R3: values too low cause LED1 to turn on when the key is pressed. Values too high prevent key press detection. Our experiment used 470Ω for R1 and R2 and 15kΩ for R3. This value might be slightly different from one chip to another due to changes on the internal pull up devices.

The following example shows how to control both leds and read key state through a single pin.

```
// Two LED flasher and key reading through a simple pin (tested on an MC9S08QG8)
#include <hidef.h>          /* for EnableInterrupts macro */
#include "derivative.h"     /* include peripheral declarations */
#include "hcs08.h"          // This is our definition file!

// delay function
void delay(unsigned int temp)
{
  for (;temp;temp--);
}

// Turn LED1 on
#pragma inline
void led1on(void)
{
  PTAD = 0;               // turn LED1 on
  PTADD_PTADD0 = 1;       // PTA0 as output
}

// Turn LED2 on
#pragma inline
void led2on(void)
{
  PTAD = 1;               // turn LED2 on
  PTADD_PTADD0 = 1;       // PTA0 as output
}

// Turn LED1 and LED2 off
#pragma inline
void ledsoff(void)
{
  PTADD_PTADD0 = 0;       // PTA0 as input (turn leds off)
}

// Return key state (0 = pressed, 1 = not pressed)
#pragma inline
char read_key(void)
{
  PTADD_PTADD0 = 0;       // PTA0 as input (turn leds off)
  if (PTAD & 0x01) return(1); else return(0);
}

void main(void)
{
  unsigned int aux;
  SOPT1 = bBKGDPE;        // Enable debug pin
  PTAPE_PTAPE0 = 1;       // PTA0 pull up enabled
  PTADS_PTADS0 = 1;       // Extra driving strength for PTA0
  aux = 40000;            // initial delay for leds
  while (1)
  {
    led1on();            // turn LED1 on
    delay(aux);          // wait for a while
    led2on();            // turn LED2 on
    delay(aux);          // wait for a while
    ledsoff();           // turn both leds off
    delay(aux);          // wait for a while
    // read key and set blinking rate accordingly (leds blink faster when
    // the key is pressed
    if (read_key()) aux = 40000; else aux = 10000;
  }
}
```

This directive instructs the compiler to inline the function (its code is inserted instead of function calls), speeding up the program.

Example 6.7 – Input/output multiplex

All these methods are good for driving low current leds. To drive higher current leds, an external component such as a transistor or MOSFET may be necessary. Figure 6.12 shows a typical circuit where a transistor acts as an electrical switch: a high level from the microcontroller saturates the transistor and thus, starts a current flow through the collector. This circuit can also drive a higher voltage over the led, because V_{CC} can be higher than MCU supply voltage. We can use a BC548 for Q1 (or the BC847 for SMT mounting). Typical values for R1 are in range of 1kΩ to 10kΩ (depending on the transistor h_{FE} parameter and the desired I_C current).

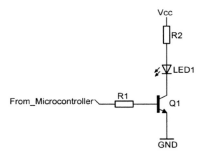

Figure 6.12

Another typical circuit is shown in figure 6.13. This circuit can be used to drive loads such as relays. V_{CC} can be above the microcontroller supply voltage. It is important to include D1 (flywheel diode) to protect Q1 and other circuits from the reverse voltage levels generated by the relay coil.

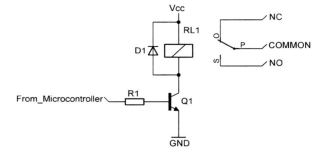

Figure 6.13

7

Clocking System

The clocking system is responsible for generating the clock signal used to synchronize operations within the CPU and peripherals. There are four different clocking systems available on HCS family devices:

- Oscillator module: the simplest module, available only on RC, RD, RE and RG devices. This module consists of a pierce oscillator capable of running quartz crystals or ceramic resonators in the range of 1 MHz to 16 MHz;

- Internal Clock Source (ICS): this module offers a good balance between performance and cost. It can run from an internal oscillator (with a center frequency of 31.25 kHz when untrimmed) or an external oscillator (capable of driving quartz crystals or ceramic resonators from 32,768 Hz up to 16 MHz). A frequency-locked loop (FLL) circuitry is available for multiplying the reference frequency by a predefined factor;

- Multi-purpose Clock Generator (MCG): this module offers the same flexibility of the ICS but also includes a phase-locked loop (PLL) circuitry for even more control over the output frequency and reduced jitter. This module is available only on EL, DN, DV, DZ and JM devices and is not studied in this book;

- Internal Clock Generator (ICG): this module is a little bit more complex than ICS. It is also based on an FLL circuitry and includes facilities to multiply and divide the reference clock signal, allowing finer control over the output frequency. Devices with ICG module offer two reference sources: the internal one (typically 243 kHz) and one external reference source (internal oscillator driving an external quartz crystal or ceramic resonator ranging from 32 kHz up to 16 MHz).

7.1. Oscillator

The oscillator module is a simple pierce oscillator and is found on RC, RD, RE and RG devices. This module can drive quartz crystals and ceramic resonators connected to the XTAL and EXTAL pins or can be driven by an external clock signal applied to the EXTAL pin. When using an external clock oscillator, the XTAL pin must be left disconnected.

The minimum frequency for external crystals and resonators is 1 MHz and the maximum is 16 MHz, there is no guarantee of operation outside this range.

Figure 7.1 shows the block diagram for the oscillator module.

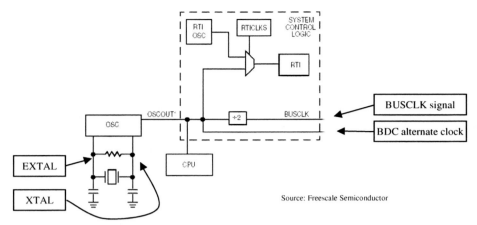

Source: Freescale Semiconductor

Figure 7.1 – Oscillator module on RC, RD, RE and RG devices

7.2. ICS

The ICS module is found on most HCS08 devices (EL, LL, QA, QD, QE, QG, SG and SL devices). It comprises an internal oscillator (with a center frequency around 32.7 kHz when untrimmed), an optional external oscillator and a frequency-locked loop circuitry.

The FLL circuitry comprises a digital oscillator controlled by a filter circuit. The DCO output signal (DCOOUT) is 512 times (1024 on ICSv2) the filter frequency, which is equal to the reference clock (internal or external, depending on the ICSC1:IREFS bit selection) divided by a programmable factor (1, 2, 4, 8, 16, 32, 64 or 128, depending on the ICSC1_RDIV bits selection). Figure 7.2 shows the ICS block diagram for the ICSv1 (QA, QD and QG devices) and ICSV2 (EN, SG, SH and SL devices). The ICSv3 has some differences that will be discussed in topic 7.2.4.

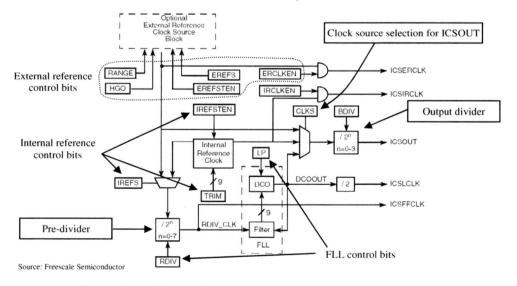

Source: Freescale Semiconductor

Figure 7.2 – ICS block diagram for EN, QD, QG, SG and SL devices

There are up to five output signals generated by ICS: the external reference clock (ICSERCLK), which is equal to the external reference frequency; the internal reference clock (ICSIRCLK), which is equal to the internal reference clock frequency; the local clock (ICSLCLK), which is equal to the DCO output divided by two; the fixed frequency clock (ICSFFCLK), which is equal to the reference clock divided by the factor programmed in RDIV and ICSOUT, which is the main output from the ICS module. The ICSOUT source is selected through ICSC1:CLKS bits, according to the table 7.1.

CLKS	Output signal
00	DCO
01	internal reference
10	external reference
11	reserved

Table 7.1 – ICS clock sources

When CLKS selects the DCO output, the ICSOUT is equal to the DCO output frequency, which can be calculated using the following formula:

$$DCOOUT = \frac{f_{REF} * FLL\_multiplier}{Pre\text{-}divider}$$

On devices with the ICSv1 module (QA, QD and QG), the FLL multiplier factor is 512. When operating with the internal reference clock ($f_{REF} \approx 31.25$ kHz) and using a pre-divider (RDIV) of 1, we have an output frequency of 16 MHz (8MHz BUSCLK).

On devices with the ICSv2 module (EL, SG, SH and SL), the FLL multiplier factor is 1024. When operating with the internal reference clock ($f_{REF} \approx 31.25$ kHz) and using a pre-divider (RDIV) of 1, we have an output frequency of 32 MHz (16MHz BUSCLK).

 Following a reset, CLKS=0, RDIV=0, IREFS=1 and BDIV=1, selecting the internal clock as the reference for the FLL. The DCO output is post-divided by 2, resulting in a 4 MHz BUSCLK (QA, QD and QG devices) or 8MHz BUSCLK (EL, SG, SH and SL devices)!

7.2.1. ICS Operating Modes

Based on the reference and output options, the manufacturer describes seven operating modes for the ICS module:

Mode	IREFS	CLKS	LP	ICSOUT	Description
FEI	1	00	0	$\dfrac{\frac{f_{INT}}{RDIV} * FLL}{BDIV}$	FLL engaged internal. The FLL operates referenced by the internal clock. The RDIV divider must be set to 000b (divide by one). **This is the default mode after a reset!**
FEE	0	00	0	$\dfrac{\frac{f_{ER}}{RDIV} * FLL}{BDIV}$	FLL engaged external. The FLL operates referenced by the external clock. Use RDIV bits to divide the reference signal so it stays in the range of 31.25 kHz to 39.0625 kHz.
FBI	1	01	0	$\dfrac{f_{INT}}{BDIV}$	FLL bypassed internal. The FLL operates referenced by the internal clock but the DCO output signal is not used.
FBILP	1	01	1	$\dfrac{f_{INT}}{BDIV}$	FLL bypassed internal low power. The FLL is disabled, reducing power consumption. When BDM debug is in use, FBI mode will be entered instead of FBILP.
FBE	0	10	0	$\dfrac{f_{ER}}{BDIV}$	FLL bypassed external. The FLL operates referenced by the external clock. Use RDIV bits to divide the reference signal so it stays in the range of 31.25 kHz to 39.0625 kHz. The output from DCO is not used.
FBELP	0	10	1	$\dfrac{f_{ER}}{BDIV}$	FLL bypassed external low power. The FLL is disabled, reducing power consumption. When BDM debug is in use, FBE mode will be entered instead of FBELP.
stop	x	x	x	-	All clock signals are stopped in stop modes (there are only two exceptions: ICSIRCLK and ICSERCLK can continue operating if enabled to operate in stop mode (IRCLKEN=1, IREFSTEN=1 / ERCLKEN=1, EREFSTEN=1)).

Table 7.2 – ICS operating modes

7.2.2. Internal Reference Modes

All internal reference modes (FEI, FBI and FBILP) are based on the internal reference clock. There are five controls for the internal reference:

ICSC1:IREFS – selects the internal (IREFS=0) or external (IREFS=1) clock as the reference for the FLL circuitry.

ICSC1:IRCLKEN – enables the IRCLK signal.

ICSC1:IREFSTEN – the internal reference stop enable bit (IREFSTEN) allows operation of the internal reference in stop modes.

ICSTRM – TRIM bits are stored into this register.

ICSSC:FTRIM – this bit allows the fine tuning of the internal reference oscillator.

The internal oscillator typically operates in the range of 25 to 41.66 kHz and can be trimmed to 31.25 kHz (for a 16 or 32 MHz DCO output), up to 39.0625 kHz (for a 20 or 40 MHz DCO output).

There are two registers to adjust the ICS output frequency: ICSTRM (where the TRIM bits are located) and ICSSC (where the Fine trim (FTRIM) bit is located). Writing 0xFF to ICSTRM results in the lowest output frequency whereas writing 0x00 results in highest possible output frequency (note that these values can be out of the allowed range for safe FLL operation).

The internal reference is automatically trimmed by the BDM tool to the default 31.25 kHz frequency. The BUSCLK frequency is measured using special BDM functions and the corresponding trim values are written into the appropriate memory addresses (0xFFAE for non-volatile FTRIM and 0xFFAF for non-volatile ICSTRM values). The internal reference can be trimmed up to 0.1% of the desired frequency (the manufacturer states a maximum deviation of 2% over the full voltage and temperature ranges).

To use these non-volatile trim values, the application needs to copy them to the ICS trim registers:

```
ICSSC = NV_FTRIM;          // configure FTRIM value (from the non-volatile value)
ICSTRM = NV_ICSTRM;        // configure TRIM value (from the non-volatile value)
```

It is also possible to change the DCO output frequency to another value within the allowed range of 16 to 20 MHz (QA, QD and QG devices). On P&E Multilink BDM and Cyclone Pro tools; this is done by selecting the "Advanced Debug/Programming Options" in the "MultilinkCyclonePro" main menu option. The desired DCO output frequency can be selected on the "Advanced Options" window (figure 7.3). This feature was not tested on ICSv2 devices.

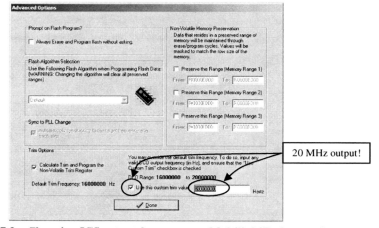

Figure 7.3 – Changing ICS output frequency on Multilink/Cyclone tools

Softec users can select the DCO output frequency on the "Target Connection" window (figure 3.28B). Pressing the "Communication Settings" button opens the window shown in figure 7.4.

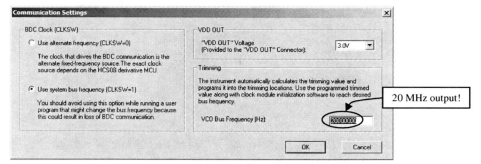

Figure 7.4 – Changing the ICS output on Softec tools

After changing the DCO output, the command window on the debugging application shows the new BUSCLK frequency as shown in figure 7.5 (considering a standard configuration for all other ICS registers).

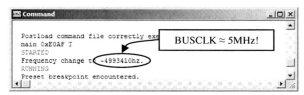

Figure 7.5

By changing BDIV to 00 (divide by 1) we have a BUSCLK frequency very close to 10 MHz:

```
ICSSC = NV_FTRIM;          // configure FTRIM value
ICSTRM = NV_ICSTRM;        // configure TRIM value
ICSC2 = 0;                 // force BDIV to 00 (BUSCLK=DCOOUT/2)(all other ICSC2 bits
                           // are also cleared)
```

Figure 7.6

Following a reset, the ICS is configured to work in FEI (FLL engaged internal) mode, with BDIV=01 (divide by two), producing a BUSCLK equal to 4 MHz. Changing from one ICS mode to another is done by simply changing CLKS, IREFS and LP bits (refer to table 7.2).

> *RDIV bits must be written on the same operation that changes CLKS (or before it) to ensure that the FLL clock input is within the allowed range (from 31.25 to 39.0625 kHz).*

On FBI (FLL bypassed internal) mode, the internal reference clock is used as ICSOUT (the CPU runs on a very low frequency). The FLL can be fully disabled (to help reduce power consumption) if it is not in use. This is done by setting the ICSC2:LP bit, which leads us to the FBILP (FLL bypassed internal low power) mode. FBI and FBILP modes are mostly used on low power applications that run on very low clock frequencies (thus spending very low power).

FBILP mode cannot be entered when debug is active. The FBI mode is entered instead. When debugging is not active, the FBILP mode operates as expected.

To select FBI mode use:

```
ICSSC = NV_FTRIM;              // configure FTRIM value
ICSTRM = NV_ICSTRM;            // configure TRIM value
ICSC1 = ICS_FBI | bIREFS;      // configure CLKS for FBI mode, IREFS = 1
```

To select FBILP mode use:

```
ICSSC = NV_FTRIM;              // configure FTRIM value
ICSTRM = NV_ICSTRM;            // configure TRIM value
ICSC1 = ICS_FBI | bIREFS;      // configure CLKS for FBI mode, IREFS = 1
ICSC2 = bICS_LP;               // set LP (FLL disabled when not used)
```

 When stopped, the internal oscillator needs at least 60µs (100µs maximum) to restart operation properly!
The FLL, when disabled or following a power-up, needs up to 1.5ms to lock to the desired frequency!

The next example shows how to configure an MC9S08QG8 ICS module to operate in FEI mode at 16 MHz frequency. Figure 7.7 shows the command window after running the application. The BDM tool measures a BUSCLK frequency of 8,002,027 Hz (DCO output frequency is equal to 16,004,054 Hz, a deviation of 0.025% from the nominal 16 MHz frequency).

```
// DEMO9S08QG8 ICS demo (16 MHz)
#include <hidef.h>            /* for EnableInterrupts macro */
#include "derivative.h"       /* include peripheral declarations */
#include "hcs08.h"            // This is our definition file!

void main(void)
{
  unsigned int temp;
  SOPT1 = bBKGDPE;            // configure SOPT1 register, enable pin BKGD for BDM
  ICSSC = NV_FTRIM;           // configure FTRIM value
  ICSTRM = NV_ICSTRM;         // configure TRIM value
  ICSC1 = ICS_FLL | bIREFS;   // select FEI mode (ICSOUT = DCOOUT = 512 * IRCLK)
  ICSC2 = BDIV_1;             // ICSOUT = DCOOUT / 1
  PTBDD = 0xFF;               // PTB0 through PTB7 as outputs
  PTBD = 0xC0;                // turn off LED1 and LED2, clear other PTB outputs
  while (1)
  {
    PTBD_PTBD6 = 0;                    // LED = on
    for (temp=3000; temp; temp--);     // wait for a while
    PTBD_PTBD6 = 1;                    // LED = off
    for (temp=30000; temp; temp--);    // wait for a while
  }
}
```

Example 7.1 – ICS example (FEI mode)

Figure 7.7

The following example shows how to configure an MC9S08QG8 ICS module to operate in FBILP mode (internal reference clock with FLL disabled). Notice the reduction in the number of iterations in the **for** loops: with a lower clock (512 times less than the last example), the loops need to

---

be 512 times faster. Figure 7.8 shows the BUSCLK frequency measured by the BDM tool. According to the tool, the BUSCLK is 15,555 Hz, which in turn results in a frequency of 31.11 kHz for the internal reference oscillator (a 0,448% deviation from the nominal 31.25 kHz).

```
// DEMO9S08QG8 ICS example (FBI mode, BUSCLK = 15.625 kHz)
#include <hidef.h>          /* for EnableInterrupts macro */
#include "derivative.h"     /* include peripheral declarations */
#include "hcs08.h"          // This is our definition file!


void main(void)
{
  unsigned int temp;
  SOPT1 = bBKGDPE;          // configure SOPT1 register, enable pin BKGD for BDM
  ICSSC = NV_FTRIM;         // configure FTRIM value
  ICSTRM = NV_ICSTRM;       // configure TRIM value
  ICSC1 = ICS_FBI;          // select FBI mode (ICSOUT = IRCLK)
  ICSC2 = BDIV_1 | bICS_LP; // ICSOUT = IRCLK / 1, FLL is disabled
  PTBDD = 0xFF;             // PTB0 through PTB7 as outputs
  PTBD = 0xC0;              // turn off LED1 and LED2, clear other PTB outputs
  while (1)
  {
    PTBD_PTBD6 = 0;            // LED = on
    for (temp=6; temp; temp--);  // wait for a while
    PTBD_PTBD6 = 1;            // LED = off
    for (temp=59; temp; temp--); // wait for a while
  }
}
```

Example 7.2 – ICS demo (FBI mode)

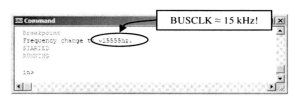

Figure 7.8

7.2.3. External Reference Modes

Most ICS enabled devices offer the option to run ICS from an external clock reference (a quartz crystal, ceramic resonator or external oscillator).

The external reference is controlled by six bits:

ICSC1:IREFS – selects the internal (IREFS=0) or external (IREFS=1) clock as the reference for the FLL circuitry.

ICSC2:EREFS – selects the source for the external reference clock (0 = crystal/resonator, 1 = external oscillator).

ICSC2:ERCLKEN – enables the ERCLK signal.

ICSC2:EREFSTEN – the external reference stop enable (EREFSTEN) bit allows enabling the external reference operation in stop modes (EREFSTEN = 1). When EREFSTEN = 0, the external reference is disabled in stop mode entry.

ICSC2:RANGE – selects the operation range for the external oscillator.

ICSC2:HGO – selects the low/high operation gain for the oscillator.

ICSSC:OSCINIT – indicates when the external crystal/resonator has completed its initialization and is fully operational.

There are three external reference based modes: FLL engaged external (FEE), FLL bypassed external (FBE) and FLL bypassed external low power (FBELP). The FEE and FBE modes use the external oscillator as the reference for the FLL circuitry while FBELP does not.

All these modes require the use of the external oscillator (XOSC) module (not present on 8-pin devices, such as QA, QD and some QG). The XOSC module can be configured to drive a quartz crystal or ceramic resonator (ICSC2:EREFS = 0) or to be driven by an external oscillator (ICSC2:EREFS = 1).

When configured to drive a quartz crystal or ceramic resonator it is possible to use ICSC2:RANGE and ICSC2:HGO bits to configure the oscillator according to the crystal/resonator:

RANGE	HGO (gain)	Crystal/Resonator
0	0	Quartz crystals up to 40 kHz
0	1	Quartz crystals up to 40 kHz
1	0	Quartz crystals/ceramic resonators up to 8 MHz
1	1	Quartz crystals/ceramic resonators up to 16 MHz

Table 7.3 – External oscillator frequency ranges

When using FEE and FBE modes, the external reference must use a crystal/resonator in range of 1 MHz up to 5 MHz or an external oscillator running in the same range.

When using FBELP mode, it is possible to use crystals/resonators up to 16 MHz or an external oscillator running up to 20 MHz (40 MHz on EL, SG, SH and SL devices).

Figure 7.9 shows the typical connections to the XOSC module. C_1 and C_2 are in range of 5 to 25 pF and should be selected according to the crystal manufacturer specification (do not forget to consider EXTAL and XTAL pins capacitance (typically 7pF per pin) as well as the printed circuit board (PCB) capacitance). R_F should be selected according to the selected range (typically 10 MΩ for 32 kHz crystals and 1 MΩ for crystals ranging from 1 up to 16 MHz). R_S must be selected according to the range and gain configured on the ICSC2 register, but typically it is not necessary.

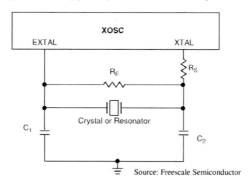

Figure 7.9 – External oscillator connections

When using an external oscillator, the clock signal must be supplied on EXTAL pin and XTAL must be left unconnected (it can be used in I/O mode).

 The XOSC module and related external reference circuitry are not implemented on 8-pin devices such as QA, QD and some QG devices! On all other devices the EXTAL pin is multiplexed with PTB7 and the XTAL pin is multiplexed with PTB6.

The application can check at any time the current clock mode by reading the clock status bits (ICSSC:CLKST).

CLKST	Current Clock Mode
00	FLL output (FEI or FEE)
01	FLL bypassed, internal clock mode (FBI or FBILP)
10	FLL bypassed, external clock mode (FBE or FBELP)
11	Reserved

Table 7.4 – Clock status bits (CLKST)

What happens if the external oscillator stops working?

On FEE mode, the FLL continues operating without reference, but the output frequency cannot be guaranteed. As soon as the reference returns to full operation, the FLL re-engages and locks to it.

On FBE and FBELP mode, an oscillator failure completely stops clocks for all internal subsystems (including the CPU). There is no option to automatically switch to the internal reference. As soon as the reference returns to full operation, the internal clocks are again supplied by it.

7.2.4. ICS Version 3

The ICS module found on the LL, QB and QE devices presents some differences compared to the previous versions:

* Extended DCO range (from 16 MHz and up to 60 MHz). There are three selectable ranges: 16 to 20 MHz, 32 to 40 MHz and 48 to 60 MHz (the range is actually selected through ICSSC:DRS bits). Refer to table 7.6. **Note that DRS bits can only be changed when FLL is not in low power mode (LP = 0). Changes made to DRS when LP = 1 produce no effect!**

* Extended pre-divider for external reference clock (10-bit divider capable of divide by up to 1024).

RDIV	RANGE = 0	RANGE = 1
0	1	32
1	2	64
2	4	128
3	8	256
4	16	512
5	32	1024
6	64	Reserved
7	128	Reserved

Table 7.5 – ICSv3 pre-dividers

* New FLL factors including special values for maximum performance when running the external reference with 32,768 Hz crystals (bit ICSSC:DMX32). Refer to table 7.6 below.

DRS	DMX32	Reference Range	FLL Factor	DCO Output Range
00 (low range)	0	31.25 to 39.0625 kHz	512	16 to 20 MHz
	1	32.768 kHz	608	19.92 MHz
01 (mid range)	0	31.25 to 39.0625 kHz	1024	32 to 40 MHz
	1	32.768 kHz	1216	39.85 MHz
10 (high range)	0	31.25 to 39.0625 kHz	1536	48 to 60 MHz
	1	32.768 kHz	1824	59.77 MHz
11		Reserved (do not use)		

Table 7.6 – ICSv3 special FLL configurations

Figure 7.10 shows the block diagram for the ICSv3 module.

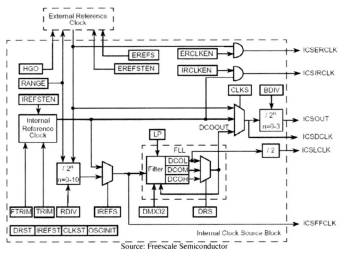

Figure 7.10 – ICSv3 module

The next example shows how to configure the ICS module to operate in FEI mode (DCO output of 48 MHz and BUSCLK = 24 MHz). Example 8.4 in chapter 8 will show how to configure the ICS module to run in FEI and FBE modes.

```
// DEMO9S08QE128 ICS example (FEI mode, BUSCLK = 24 MHz)

#include <hidef.h>          /* for EnableInterrupts macro */
#include "derivative.h"     /* include peripheral declarations */
#include "hcs08.h"          // This is our definition file!

#define LED1 BIT_0          // PTC0
#define LED2 BIT_1          // PTC1

// A simple software loop delay
void delay(unsigned int value)
{
  for (;value; value--);
}

void main(void)
{
  SOPT1 = bBKGDPE;                    // configure SOPT1 register, enable pin BKGD for BDM
  ICSSC = DCO_HIGH | NVFTRIM;         // configure FTRIM value, select DCO high range
  ICSTRM = NVICSTRM;                  // configure TRIM value
  ICSC1 = ICS_FLL | bIREFS;           // select FEI mode (ICSOUT = DCOOUT = 1536 * IRCLK)
  ICSC2 = BDIV_1;    // ICSOUT = DCOOUT / 1
  PTCDD = 0xFF;      // PTC pins as outputs
  PTCD = 0xFE;       // set all PTC pins, except PTC0 (LED1 is on, LED2 is off)
  while (1)
  {
    PTCTOG = LED1;   // toggle LED1 (bit 0 of PTC)
    PTCTOG = LED2;   // toggle LED2 (bit 1 of PTC)
    delay(60000);    // a big delay
    PTCTOG = LED1;   // toggle LED1 (bit 0 of PTC)
    PTCTOG = LED2;   // toggle LED2 (bit 1 of PTC)
    delay(60000);    // a big delay
  }
}
```

> This example also shows how to use the special I/O toggling registers!

Example 7.3 – ICSv3 example (FEI mode)

7.2.5. ICS Module Versions

The following table shows the evolution of ICS modules since version 1:

ICS Version	Devices	Features added/modified
1	QA, QD, QG	First release.
2	EL, SG, SL, SH	IREFST bit for internal reference status. FLL multiplier = 1024.
3	QE, LL	Extended DCO range and FLL multipliers. Extended pre-divider for external reference. 32,768 Hz specific FLL mode.

Table 7.7 – ICS module history

On QA, QD and QG devices (ICSv1), the manufacturer recommends not to use the FEI mode with BDIV = 00b (divide by 1), as the ICS module can sometimes generate short clock pulses that can cause the device to malfunction. This is a documented peripheral bug which was corrected on the newer versions of the ICS module.

On QB, QE4, QE8 and LL devices, ICS only supports the low range DCO (the multiplier can be set only to 512 or 608)!

7.2.6. ICS Registers

There are four registers associated with the ICS module:

* Internal Clock Source Control 1 (ICSC1) register.
* Internal Clock Source Control 2 (ICSC2) register.
* Internal Clock Source Status and Control (ICSSC) register.
* Internal Clock Source Trim (ICSTRM) register.

7.2.6.1. ICSC1 Register

Name	Model		BIT 7	BIT 6	BIT 5	BIT 4	BIT 3	BIT 2	BIT 1	BIT 0
ICSC1	QA, QD	Read	CLKS		0	0	0	1	1	IREFSTEN
		Write								
		Reset	0	0	0	0	0	1	1	0
ICSC1	EL, QB, QE, QG, SG, SL	Read	CLKS		RDIV			IREFS	IRCLKEN	IREFSTEN
		Write								
		Reset	0	0	0	0	0	1	0	0

Bit Name	Description	C Symbol
CLKS	Clock source selection:	
	00 – FLL output.	**ICS_FLL**
	01 – internal reference.	**ICS_FBI**
	10 – external reference.	**ICS_FBE**
	11 – reserved.	-
RDIV	Reference divider selection:	
	000 – reference divided by 1.	**RDIV_1**
	001 – reference divided by 2.	**RDIV_2**
	010 – reference divided by 4.	**RDIV_4**
	011 – reference divided by 8.	**RDIV_8**
	100 – reference divided by 16.	**RDIV_16**
	101 – reference divided by 32.	**RDIV_32**
	110 – reference divided by 64.	**RDIV_64**
	111 – reference divided by 128.	**RDIV_128**

Bit Name	Description	C Symbol
IREFS	Internal reference select: 0 – external reference selected. 1 – internal reference selected.	bIREFS
IRCLKEN	Internal reference clock (IRCLK) enable: 0 – ICSIRCLK disabled. 1 – ICSIRCLK enabled.	bIRCLKEN
IREFSTEN	Internal reference stop enable: 0 – internal reference clock is disabled in stop mode. 1 – the internal reference remains enabled if IRCLKEN is set before entering stop mode.	bIREFSTEN

 On QE32/64/96/128 devices the RDIV factor is multiplied by 32 when RANGE=1. The maximum RDIV factor is 1024 (RANGE = 1 and RDIV = 101)!

7.2.6.2. ICSC2 Register

Name	Model		BIT 7	BIT 6	BIT 5	BIT 4	BIT 3	BIT 2	BIT 1	BIT 0
ICSC2	QA, QD	Read	BDIV		0	0	LP	0	0	0
		Write								
		Reset	0	1	0	0	0	0	0	0
ICSC2	EL, QB, QE, QG, SG, SL	Read	BDIV		RANGE	HGO	LP	EREFS	ERCLKEN	EREFSTEN
		Write								
		Reset	0	1	0	0	0	0	0	0

Bit Name	Description	C Symbol
BDIV	Bus frequency divider: 00 – selected clock source is divided by 1. 01 – selected clock source is divided by 2. 10 – selected clock source is divided by 4. 11 – selected clock source is divided by 8.	BDIV_1 BDIV_2 BDIV_4 BDIV_8
RANGE	Selects the operating range of the external reference: 0 – low frequency (up to 40 kHz). 1 – high frequency (1 to 16 MHz).	bICS_RANGE
HGO	High gain oscillator. Selects the gain for the external oscillator: 0 – low gain (lower power). 1 – high gain (higher power).	bICS_HGO
LP	Low power mode (disables the FLL circuitry): 0 – FLL remains active when not in use. 1 – FLL disabled when not in use.	bICS_LP
EREFS	Selects the external reference oscillator: 0 – external oscillator (external clock signal applied into EXTAL pin). 1 – internal oscillator driving a crystal or resonator connected to EXTAL and XTAL pins.	bEREFS
ERCLKEN	External reference clock (ICSERCLK) signal enable: 0 – ICSERCLK disabled. 1 – ICSERCLK enabled.	bERCLKEN
EREFSTEN	External reference stop enable: 0 – external reference is disabled in stop modes. 1 – external reference remains enabled if ERCLKEN = 1 or ICS mode is FEE.	bEREFSTEN

 External reference controls are not available on 8-pin devices (such as QA, QD and some QG devices)!

7.2.6.3. ICSSC Register

Name	Model		BIT 7	BIT 6	BIT 5	BIT 4	BIT 3	BIT 2	BIT 1	BIT 0
ICSSC	QA,QD	Read	0	0	0	0	0	CLKST	0	FTRIM
		Write								
		Reset	0	0	0	0	0	0	0	0
ICSSC	QG	Read	0	0	0	0	CLKST		OSCINIT	FTRIM
		Write								
		Reset	0	0	0	0	0	0	0	0
ICSSC	EL, SG, SL	Read	0	0	0	IREFST	CLKST		OSCINIT	FTRIM
		Write								
		Reset	0	0	0	1	0	0	0	0
ICSSC	QB, QE, LL	Read	DRST		DMX32	IREFST	CLKST		OSCINIT	FTRIM
		Write	DRS							
		Reset	0	0	0	1	0	0	0	0

Bit Name	Description	C Symbol
DRST	DCO range status bits (the status bits are not updated immediately after writing into ICSSC register, some clock cycles are needed due the different clock domains): 00 – low range. 01 – mid range. 10 – high range. 11 – reserved.	- - - -
DRS	DCO range selection (these bits cannot be changed while LP = 1): 00 – low range. 01 – mid range. 10 – high range. 11 – reserved.	 DCO_LOW DCO_MID DCO_HIGH -
DMX32	Set the DCO to maximize output frequency when working with a 32,768 Hz reference: 0 – default frequency range for the DCO. 1 – DCO tuned for maximum frequency with a 32,768 Hz reference.	bDMX32
IREFST	Internal reference status bits: 0 – reference clock is sourced by external clock. 1 – reference clock is sourced by internal clock.	bIREFST
CLKST	Clock source status (these bits are not updated immediately after writing into ICSC1 register, some clock cycles are needed due the different clock domains): 00 – FLL output. 01 – internal reference. 10 – external reference. 11 – reserved.	 - - - -
OSCINIT	External reference oscillator (XOSC) initialized: 0 – XOSC not initialized. 1 – XOSC initialized.	bOSCINIT
FTRIM	Fine tuning bit for internal reference oscillator: 0 – internal reference frequency not changed. 1 – internal reference frequency increased by 0.1%.	bFTRIM

 DRST, DRS and DMX32 are only available on QE32/64/96/128 devices!

7.2.6.4. ICSTRM Register

Name	Model		BIT 7	BIT 6	BIT 5	BIT 4	BIT 3	BIT 2	BIT 1	BIT 0
ICSTRM	EL, QA, QB, QD, QE, QG, SG, SL	Read				TRIM				
		Write								
		Reset	1	0	0	0	0	0	0	0

The internal reference oscillator frequency is modified by writing into this register. Increasing ICSTRM decreases the reference frequency, while decreasing ICSTRM increases the reference frequency.

7.3. ICG

The internal clock generator (ICG) is another clocking module found on some HCS08 devices (AC, AW, GB, GT and LC devices).

The ICG is a more complex version of the ICS, with two separate internal references (243 kHz and 8 MHz); one external reference oscillator (XOSC), a digital oscillator (DCO), whose operating frequency can be digitally programmed and a frequency locked loop (FLL) circuitry, which interacts with the DCO and automatically adjusts its frequency according to a reference signal. The ICG output signal (ICGOUT) feeds the CPU and its peripherals.

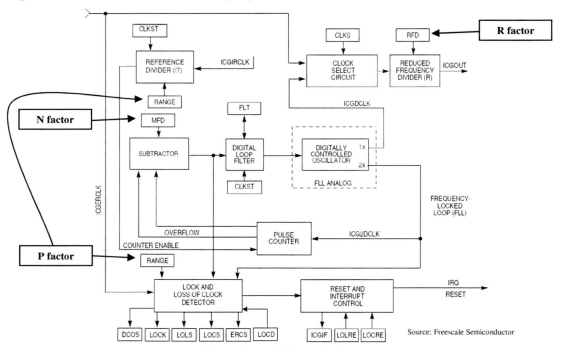

Figure 7.11 – ICG block diagram

Some of the advanced features of the internal clock generator are:

+ FLL lock detection.

+ Interrupt or reset on loss of lock.

+ Interrupt or reset on loss of clock.

To fully understand the ICG operation principles it is necessary to know that there are three numeric factors: two multiplicative factors (P and N) and one dividing factor (R).

The N factor multiplies the reference frequency by a value of 4, 6, 8, 10, 12, 14, 16 or 18. This factor is programmed on the MFD bits (register ICGC2).

Factor P multiplies the reference frequency by a value of 1 or 64 and it depends on the selected reference source. When using the internal reference clock (ICGIRCLK) the P factor is always equal to 64. When using the external reference clock (ICGERCLK), the P factor depends on the oscillator range (selected by ICGC1:RANGE bit). When RANGE=0, P=64 and when RANGE=1, P=1.

The last factor (R) is used to divide the DCO output by a fixed value between 1 and 128. This factor is programmed through RFD bits (register ICGC2).

It is important to note that the FLL circuitry needs some time to effectively lock the DCO to the selected reference. We say that the FLL is unlocked when it is not synchronized with the reference clock and that it is locked when fully synchronized.

There are four ICG operating modes (one more if we consider the stop mode). They are selected by CLKS bits (register ICGC1) and may also be entered due to some especial conditions we will see later:

CLKS	Mode	Output signal
00	Self-clocked mode (**SCM**)	ICGOUT is derived from DCO but it is not controlled by the FLL. The DCO output can be manually controlled by the digital loop filter (FLT).
01	FLL engaged internal (**FEI**)	ICGOUT is derived from DCO and it is corrected by the FLL. The 243 kHz internal reference (ICGIRCLK) is the FLL reference.
10	FLL bypassed external (**FBE**)	ICGOUT is derived from the external reference signal (ICGERCLK) divided by R. The FLL is disabled.
11	FLL engaged external (**FEE**)	ICGOUT is derived from DCO and it is corrected by the FLL. The external reference (ICGERCLK) is the FLL reference.

Table 7.8 – ICG operating modes

When in self-clocked mode (**SCM**), the FLL does not correct the DCO output frequency. The DCO output frequency can be manually adjusted by using the digital loop filter (registers ICGFLTU and ICGFLTL). The minimum frequency is around 8 MHz and the maximum frequency is around 40 MHz.

Following a reset, ICGC1:CLKS = "00" and ICGDCLK is approximately equal to 8 MHz (BUSCLK = 4 MHz). The internal reference must be properly trimmed (register ICGTRM) to achieve maximum precision.

The FLL engaged internal (**FEI**) mode allows using the 243 kHz internal reference generator (IRG) as the FLL reference. ICGOUT will be higher or lower than the reference, depending on the N and R factors. When operating in this mode and if the **FLL** is locked, the ICGOUT frequency is equal to:

$$f_{ICGOUT} = \frac{243 \text{kHz} * 64 * N}{7 * R}$$

The internal reference must be properly trimmed to achieve maximum precision.

FEI and SCM modes are the least expensive ones as they do not need any external components!

In FLL bypassed external (**FBE**) mode, the ICG does not use the FLL (the FLL is disabled). ICGOUT is equal to:

$$f_{ICGOUT} = \frac{ICGERCLK}{R}$$

In FLL engaged external (**FEE**) mode, the ICG uses the FLL to multiply the external reference (ICGERCLK) by a specific value programmed by the user. In the FLL synchronization phase (FLL unlocked, ICGS1:LOCK=0), ICGOUT is equal to:

$$f_{ICGOUT} = \frac{ICGERCLK * P * N}{2 * R}$$

As soon as the FLL is locked (ICGS1:LOCK=1), ICGOUT is equal to:

$$f_{ICGOUT} = \frac{ICGERCLK * P * N}{R}$$

The ICG module also has some status bits that can be used to detect its operating mode and states:

- **REFST** – indicates the current FLL reference clock source (1 – crystal/resonator, 0 – external oscillator).

- **LOLS** – indicates a loss of lock condition. This bit is only modified by the FLL when the loss of lock condition is caused by external events. It is not changed on mode changes.

- **LOCK** – indicates the FLL is locked to the selected reference.

- **LOCS** – indicates the loss of ICG reference clock (on some devices this feature can be disabled by writing "1" into the ICGC1:LOCD bit).

- **ERCS** – indicates the loss of external reference clock.

- **DCOS** – indicates the digital controlled oscillator stability. The DCO stability is measured by the subtractor circuitry. It is considered stable when DCO output does not vary more than 4*N counts for two consecutive samples of the DCO clock. The ICG uses this bit to determinate the time to switch to the user selected mode. This bit is always cleared upon entering stop mode.

When the clock mode is changed and if the new mode uses the FLL to multiply the reference clock, there is a delay between clock mode selection (ICGC1:CLKS) and the actual change in the clock status bits (ICGS1:CLKST). Due to FLL synchronization time (up to 5 ms), the ICS clock output is different from the selected mode until the FLL locks (ICGS1:LOCK=1) and the DCO is stable (ICGS2:DCOS=1).

7.3.1. Internal References

As we said earlier, there are two internal clock references for the ICG module: one operating at 243 kHz and another operating at 8 MHz.

The 243 kHz internal reference generator (IRG) is used on SCM and FEI modes. This clock source can vary up to +/- 25% due to manufacturing processes, but it can be fine-tuned up to 0.5% of its regular frequency.

To tune the IRG oscillator all you have to do is write the appropriate value into the ICG trim register (ICGTRM).

The BDM tool can automatically trim the oscillator. This is done by reading the actual BUSCLK signal and calculating the trim value to tune the IRG. After calculating this value, it is written into the FLASH memory at the 0xFFBE address. The application needs to copy this value to the ICGTRM register to trim the oscillator (this is automatically done upon entering a debug section) as shown below:

```
ICGTRM = NVICGTRM;      // trim IRG oscillator
```

If your device does not have a defined NVICGTRM symbol, you can define it at the beginning of your program:

```
// this may be necessary for AW devices
#define NVICGTRM (*(const char * __far)0xFFBE)
```

The 8 MHz oscillator is dedicated exclusively to the background debug controller (BDC) and is only used as an alternate clock source for debugging purposes.

7.3.2. External References

It is possible to use an external reference clock source to drive the ICG in FBE and FEE modes.

The external reference clock (ICGERCLK) can work in two different modes: with a quartz crystal or ceramic resonator, or with an external clock oscillator. This selection is made through the ICGC1: REFS bit.

When a crystal/resonator is selected (REFS=1), ICGC1:RANGE and ICGC1:HGO bits allow selecting the frequency range and oscillator gain.

When RANGE = 0, the low frequency range is selected (from 32 to 100 kHz). This mode presents lower power consumption and a higher startup time (in order of 430 ms). The P factor is automatically set to 64.

When RANGE = 1, the high frequency range is selected (from 2 up to 10 MHz). This mode presents higher power consumption and a lower startup time (in order of 4 ms). The P factor is automatically set to 1.

The HGO bit can be used to reduce (HGO = 0) or to increase oscillator gain (HGO = 1). Its configuration depends on the crystal properties and affects power consumption and noise immunity.

When HGO = 1 and RANGE = 1, the oscillator can run with crystals in range of 1 MHz up to 16 MHz (FBE mode) or 2 MHz up to 10 MHz (FEE mode). When HGO = 0 and RANGE = 1, the oscillator can run crystals in range of 1 MHz up to 8 MHz (FBE mode) or 2 MHz up to 8 MHz (FEE mode).

The oscillator can continue to run in stop modes if ICGC1:OSCSTEN bit is set to "1" before entering stop mode (this bit has no effect if HGO = 1 and RANGE = 1).

Figure 7.12A shows the necessary connections for operation with an external quartz crystal. C_1 and C_2 are in range of 8.2 to 15 pF and should be selected according to the crystal manufacturer specification (do not forget to consider EXTAL and XTAL pins capacitance (typically 7pF per pin) as well as the printed circuit board (PCB) capacitance). R_F should be selected according to the selected range (typically 10 MΩ for 32 kHz crystals and 1 MΩ for crystals ranging from 1 up to 16 MHz). R_S must be selected according to the range and gain configured on ICGC1 register, but typically it is not necessary.

When an external oscillator is connected to the EXTAL pin (figure 7.12B), it is necessary to configure ICGC1:REFS bit to "0". The clock signal can be anywhere between 0 and 40 MHz (FBE mode), 32 up to 100 kHz (FEE mode with RANGE = 0) or 2 up to 10 MHz (FEE mode with RANGE = 1).

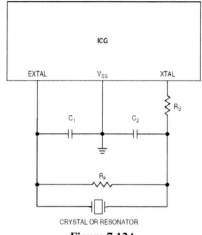

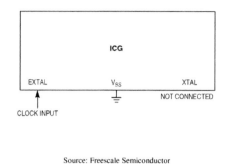

Source: Freescale Semiconductor

Figure 7.12A   **Figure 7.12B**

 If a crystal or resonator is used as the external reference for the ICG, this selection must be done on the first write into the ICSC1 register (CLKS and REFS can be written only once after reset). Failure to do so locks the EXTAL and XTAL pins into their I/O function, disabling the external oscillator option until the next reset!

7.3.3. ICG Events

The ICG module can generate up to two different events: loss of lock and loss of clock. These events can generate an interrupt or a reset, depending on the LOLRE and LOCRE configuration (register ICGC2).

LOLRE – loss of lock reset enable. When LOLRE = 1, a reset is generated when ICGS1:LOLS is set.

LOCRE – loss of clock reset enable. When LOCRE = 1, a reset is generated when ICGS1:LOCS is set. A loss of clock event also changes ICG operating mode. If the ICG is in FEE, a loss of reference clock causes the ICG to enter SCM mode, and a loss of DCO clock causes the ICG to enter FBE mode. If the ICG is in FBE mode, a loss of clock causes the ICG to enter SCM mode. In all cases, the CLKS and CLKST bits are automatically changed according to the new ICG settings.

LOLS and LOCS, when set, also set ICGIF (register ICGS1) and it generates an interrupt request to the CPU.

Clearing LOLS and LOCS is done by any reset or by writing "1" to ICGIF (which also clears ICGIF).

 Notice that interrupt mode is the default operating mode for LOLS and LOCS. Any application running on an ICG enabled device must provide an ISR to deal with such interrupts (unless LOLRE and LOCRE are both set (reset mode), or LOLRE and LOCD are both set)!

7.3.4. ICG External Connections

The ICG module uses only two pins when running from an external reference. Table 7.9 shows the ICG connections for current HCS08 devices.

Device	ICG Pins	
	EXTAL	XTAL
AW	PTG6	PTG5
GB/GT	PTG2	PTG1
LC	PTB0	PTB1

Table 7.9 – ICG connections

AC and AW devices offer an additional feature: the ability to output the BUSCLK signal through an I/O pin (PTC2). The SMCLK register controls this functionality:

Name	Model		BIT 7	BIT 6	BIT 5	BIT 4	BIT 3	BIT 2	BIT 1	BIT 0
		Read	0	0	0	MPE	0		MCSEL	
MCLK	AC AW	Write								
		Reset	0	0	0	0	0	0	0	0

The BUSCLK signal is output through PTC2 (MCLK) when MPE = 1. This output signal is called MCLK and it is divided by the factor specified into MCSEL. The MCLK frequency can be calculated with the following equation (when MCSEL = 0, MCLK = 0 and PTC2 stays low):

$$f_{MCLK} = \frac{BUSCLK}{2*MCSEL}$$

7.3.5. ICG Module Versions

The following table shows the evolution of ICG modules since version 1:

ICG Version	Devices	Features added/modified
1	GB, GT	First release.
2	GB(A), GT(A)	High gain oscillator added. Loss of clock detection disable-bit (LOCD) added.
4	AC, AW, LC	Only internal updates and bug fixes.

Table 7.10 – ICG module history

7.3.6. ICG Registers

There are seven registers associated with the ICG module:

- Internal Clock Generator Control 1 (ICGC1) register.

- Internal Clock Generator Control 2 (ICGC2) register.

- Internal Clock Generator Status (ICGS1) register 1.

- Internal Clock Generator Status (ICGS2) register 2.

- Internal Clock Generator Upper Filter (ICGFLTU) register.

- Internal Clock Generator Lower Filter (ICGFLTL) register.

- Internal Clock Generator Trim (ICGTRM) register.

7.3.6.1. ICGC1 Register

Name	Model		BIT 7	BIT 6	BIT 5	BIT 4	BIT 3	BIT 2	BIT 1	BIT 0
ICGC1	GB, GT	Read	0	RANGE	REFS	CLKS		OSCSTEN	-	0
		Write	-	RANGE	REFS	CLKS		OSCSTEN	0	-
		Reset	0	1	0	0	0	1	0	0
ICGC1	AC, AW, GB(A), GT(A) LC	Read	HGO	RANGE	REFS	CLKS		OSCSTEN	LOCD	0
		Write	HGO	RANGE	REFS	CLKS		OSCSTEN	LOCD	
		Reset	0	1	0	0	0	1	0	0

Bit Name	Description	C Symbol
HGO	High gain oscillator (not available on original GB/GT devices): 0 – low power operation (standard oscillator). 1 – high gain oscillator. **This is a write-once bit and can be programmed only once after a reset!**	bICG_HGO
RANGE	This bit controls the oscillator range, reference divider and FLL factor P. It only has an effect when ICG is configured to FBE or FEE modes: 0 – low frequency range (up to 100 kHz) (P = 64). 1 – high frequency range (up to 16 MHz) (P = 1). **This is a write-once bit and can be programmed only once after a reset!**	bICG_RANGE
REFS	External reference selection: 0 – external oscillator connected to EXTAL. 1 – quartz crystal or ceramic resonator.	bREFS
CLKS	ICG operating mode selection (ICGOUT signal). These bits can be changed at any time, except if the first write after a reset set the clock source to the internal reference (CLKS=00 or CLKS=01). In this case, it is not possible to change CLKS to an external reference mode until the next reset.	
	00 – self-clocked mode (SCM): ICGOUT is supplied by DCO without FLL correction.	ICG_SCM
	01 – FLL engaged internal (FEI): DCO output multiplied by the internal reference.	ICG_FEI
	10 – FLL bypassed external (FBE): ICGOUT is supplied by the external reference, FLL is disabled.	ICG_FBE
	11 – FLL engaged external (FEE): DCO output multiplied by the external reference.	ICG_FEE
OSCSTEN	External reference oscillator stop enable bit: 0 – External oscillator disabled in stop mode. 1 – External oscillator enabled in stop mode.	bOSCSTEN
LOCD	Loss of clock disable bit: 0 – loss of clock detection is enabled (LOCS is set when reference clock stops). 1 – loss of clock detection is disabled (LOCS is always zero).	bLOCD

7.3.6.2. ICGC2 Register

Name	Model		BIT 7	BIT 6	BIT 5	BIT 4	BIT 3	BIT 2	BIT 1	BIT 0
ICGC2	AC, AW, GB, GT, LC	Read	LOLRE	MFD			LOCRE	RFD		
		Write								
		Reset	0	0	0	0	0	0	0	0

Bit Name	Description	C Symbol
LOLRE	Loss of lock reset enable bit: 0 – an interrupt is generated when LOLS is set. 1 – a reset is generated when LOLS is set.	bLOLRE
MFD	FLL N multiplying factor: 000 – N = 4.	MFDx4
	001 – N = 6.	MFDx6
	010 – N = 8.	MFDx8
	011 – N = 10.	MFDx10
	100 – N = 12.	MFDx12
	101 – N = 14.	MFDx14
	110 – N = 16.	MFDx16
	111 – N = 18.	MFDx18
LOCRE	Loss of clock reset enable bit: 0 – an interrupt is generated when LOCS is set. 1 – a reset is generated when LOCS is set.	bLOCRE
RFD	ICG dividing factor R: 000 – R = 1.	RFD_DIV1
	001 – R = 2.	RFD_DIV2
	010 – R = 4.	RFD_DIV4
	011 – R = 8.	RFD_DIV8
	100 – R = 16.	RFD_DIV16
	101 – R = 32.	RFD_DIV32
	110 – R = 64.	RFD_DIV64
	111 – R = 128.	RFD_DIV128

7.3.6.3. ICGS2 Register

Name	Model		BIT 7	BIT 6	BIT 5	BIT 4	BIT 3	BIT 2	BIT 1	BIT 0
ICGS2	AC, AW, GB, GT, LC	Read	0	0	0	0	0	0	0	DCOS
		Write	-	-	-	-	-	-	-	-
		Reset	0	0	0	0	0	0	0	0

Bit Name	Description	C Symbol
DCOS	DCO clock stable status bit: 0 – DCO output is unstable. 1 – DCO output is stable.	bDCOS

7.3.6.4. ICGS1 Register

Name	Model		BIT 7	BIT 6	BIT 5	BIT 4	BIT 3	BIT 2	BIT 1	BIT 0
ICGS1	AC, AW, GB, GT, LC	Read	CLKST		REFST	LOLS	LOCK	LOCS	ERCS	ICGIF
		Write	-		-	-	-	-	-	1
		Reset	0	0	0	0	0	0	0	0

Bit Name	Description	C Symbol
CLKST	ICG clock status bits: 00 – self-clocked mode (SCM). 01 – FLL engaged internal (FEI). 10 – FLL bypassed external (FBE). 11 – FLL engaged external (FEE).	ICG_SCM_MODE ICG_FEI_MODE ICG_FBE_MODE ICG_FEE_MODE
REFST	External reference status bit: 0 – external oscillator connected to the external reference input EXTAL. 1 – crystal/resonator connected to the external reference oscillator inputs.	bREFST
LOLS	Loss of lock status bit: 0 – the FLL is locked to the reference source. 1 – the FLL unexpectedly lost lock since LOLS was last cleared. ICGIF is also set and generates an interrupt if LOLRE=0. If LOLRE=1 a reset is generated.	bLOLS
LOCK	FLL lock status bit: 0 – FLL is unlocked. 1 – FLL is locked.	bLOCK
LOCS	Loss of clock status bit: 0 – the ICG reference clock is operating regularly. 1 – the ICG unexpectedly lost the clock since LOCS was last cleared. ICGIF is also set and generates an interrupt if LOLRE=0. If LOLRE=1 a reset is generated.	bLOCS
ERCS	External reference clock status: 0 – external reference clock is not stable (outside specs). 1 – external reference clock is stable (within specs).	bERCS
ICGIF	ICG interrupt flag (this bit is set along with LOLS and LOCS): 0 – no ICG interrupt pending. 1 – an ICG interrupt is pending (write "1" to clear it).	bICGIF

7.3.6.5. ICGFLTU and ICGFLTL Register

Name	Model		BIT 7	BIT 6	BIT 5	BIT 4	BIT 3	BIT 2	BIT 1	BIT 0
ICGFLTU	AC, AW, GB, GT, LC	Read	0	0	0	0	FLT			
		Write	-	-	-	-				
		Reset	0	0	0	0	0	0	0	0
ICGFLTL	AC, AW, GB, GT, LC	Read	FLT							
		Write								
		Reset	1	1	0	0	0	0	0	0

In FEI, FBE and FEE modes, these registers are read-only. When ICG is operating in SCM mode, any writes to these registers change the DCO operating frequency.

This register has a latching mechanism that prevents modification until both registers are changed. To change the filter value, first write to ICGFLTU (upper 4 bits) and then to ICGFLTL (lower 8 bits). The filter value is only updated on the ICGFLTL write operation.

7.3.6.6. ICGTRM Register

Name	Model		BIT 7	BIT 6	BIT 5	BIT 4	BIT 3	BIT 2	BIT 1	BIT 0
ICGTRM	AC, AW, GB, GT, LC	Read				TRIM				
		Write								
		Reset	1	0	0	0	0	0	0	0

The internal reference generator (IRG) frequency is modified by writing into this register. Increasing ICGTRM decreases the reference frequency, while decreasing ICGTRM increases the reference frequency.

7.3.7. ICG Examples

The first example demonstrates ICG FEI mode. This application is intended to run on the DEMO9S08LC60 demonstration board (MC9S08LC60 MCU).

Leds LD401 to LD405 blink sequentially at an initial BUSCLK = 20 MHz (the ICG is initially configured to run at 40 MHz with an internal reference of 243 kHz, N = 18, P = 64 and R = 1.

Pressing SW304 decreases N factor thus decreasing the BUSCLK speed. Pressing SW305 increases N factor and increase BUSCLK speed.

```
// DEMO9S08LC60 ICG example (FEI mode, BUSCLK = 20 MHz)
// LD401 is connected to PTB3, LD402 is connected to PTB4, LD403 is connected to PTB5
// LD404 is connected to PTB6, LD405 is connected to PTB7
// SW305 is connected to PTC7, SW304 is connected to PTC5
// SW303 is connected to PTC4, SW302 is connected to PTC3

#include <hidef.h>           /* for EnableInterrupts macro */
#include "derivative.h"      /* include peripheral declarations */
#include "hcs08.h"           // This is our definition file!

#define LD401 BIT_3          // PTB3
#define LD402 BIT_4          // PTB4
#define LD403 BIT_5          // PTB5
#define LD404 BIT_6          // PTB6
#define LD405 BIT_7          // PTB7
#define LD406 PTAD_PTAD5     // PTA5

unsigned char my_mfd;

// ICG interrupt (process LOLS and LOCS interrupts)
void interrupt VectorNumber_Vicg icg_isr(void)
{
  ICGS1_ICGIF = 1;           // clear ICG interrupt
  // do something ...
  LD406 = 1;                 // turn LD406 on
}

// KBI ISR (process SW305 and SW304 key presses)
void interrupt VectorNumber_Vkeyboard2 kbi2_isr(void)
{
  KBI2SC_KBACK = 1; // clear KBI interrupt flag
  if (!PTCD_PTCD7)
  {
    // if SW305 is pressed, increase MFD (speed up BUSCLK)
    my_mfd++;
    if (my_mfd>7) my_mfd = 7;        // limit my_mfd to 7 (N = 18)
  }
  if (!PTCD_PTCD5)
  {
    // if SW304 is pressed, decrease MFD (slow down BUSCLK)
    if (my_mfd) my_mfd--;    // do not decrease below 0
  }
  ICGC2 = my_mfd << 4;       // change MFD factor
}

// A simple software loop delay
void delay(unsigned int value)
{
  for (;value; value--);
}
```

```
void main(void)
{
  SOPT1 = bBKGDPE;               // configure SOPT1 register, enable pin BKGD for BDM
  // ICG configuration:
  ICGTRM = NVICGTRM;             // trim IRG oscillator
  ICGC2 = MFDx18;                // MFD = 111b (N = 18)
  ICGC1 = ICG_FEI;               // ICG in FEI mode (CLKS = 01b)
  // I/O ports configuration:
  PTCPE = BIT_7 | BIT_5;         // enable pull ups on PTC7 and PTC5
  PTBDD = 0xFF;                  // PTB0 through PTB7 as outputs
  PTBD = 0;                      // turn off LD401 through LD405
  PTADD = BIT_5;                 // PTA5 as an output
  // KBI2 configuration:
  KBI2SC = bKBIE;                // turn on KBI2 interrupt
  KBI2PE = BIT_7 | BIT_6;        // enable KBI inputs corresponding to PTC7 and PTC5
  my_mfd = 7;                    // initialize "my_mfd" factor
  EnableInterrupts;              // clear CCR:I
  while (1)
  {
    PTBD = LD401;                // LD401 on
    delay(30000);                // delay
    PTBD = LD402;                // LD402 on
    delay(30000);                // delay
    PTBD = LD403;                // LD403 on
    delay(30000);                // delay
    PTBD = LD404;                // LD404 on
    delay(30000);                // delay
    PTBD = LD405;                // LD405 on
    delay(30000);                // delay
  }
}
```

Example 7.4 – ICG example using an MC9S08LC60

The application starts running at 40 MHz (20 MHz BUSCLK). Each press on SW304 decreases MFD by one (and N by 2). The actual BUSCLK speed is shown in the command window (figure 7.13) as it is constantly measured by the BDM tool.

The eight possible speeds are shown in table 7.11.

my_mfd	MFD	ICGOUT	BUSCLK
0	4	8,886,857 Hz	4,443,429 Hz
1	6	13,330,286 Hz	6,665,143 Hz
2	8	17,773,714 Hz	8,886,857 Hz
3	10	22,217,143 Hz	11,108,571 Hz
4	12	26,660,571 Hz	13,330,286 Hz
5	14	31,104,000 Hz	15,552,000 Hz
6	16	35,547,429 Hz	17,773,714 Hz
7	18	39,990,857 Hz	19,995,429 Hz

Table 7.11

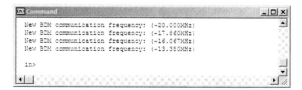

Figure 7.13

We can also run the same example using the 32,768 Hz quartz crystal available on the board. All you have to do is change the two lines that configure the ICG module at the beginning of the **main**() function (the **while** loop was added to wait for the DCO to stabilize).

```
ICGC1 = ICG_FEE | bREFS;       // ICG in FEE mode (CLKS = 11), external crystal
ICGC2 = MFDx18;                // MFD = 111 (N = 18)
while (!ICGS2_DCOS);           // wait for the DCO to stabilize
```

With the new example running, it is possible to notice the new speeds on the debugger command window (reflecting the use of the external 32,768 Hz instead of the IRG).

The next example shows the same implementation on the DEMO9S08AW60E demonstration board. We needed to change the led sequencing mechanism due to the need to show the LOLS bit status on LED7 (which is in the same port of the leds used in the sequencing mechanism). Therefore, an algorithm similar to the one shown in example 6.6 is used here.

The application starts operating in FEE mode with an external 4 MHz quartz crystal reference. We choose N = 10 and R = 1, resulting in a nominal 20 MHz BUSCLK (ICGOUT = 4 MHz * 10/1 = 40 MHz).

Each press on the SW key decreases the MFD factor (from the maximum of 3 down to 0). The minimum ICSOUT frequency is equal to: 4 MHz * 4/1 = 16 MHz (BUSCLK = 8 MHz).

```
// DEMO9S08AW60E ICG example (FEE mode, 4 MHz quartz crystal, BUSCLK = 20 MHz)
// SW1 is connected to PTC2, SW2 is connected to PTC6
// SW3 is connected to PTD3, SW4 is connected to PTD2
// LEDs are connected to PTF

#include <hidef.h>          /* for EnableInterrupts macro */
#include "derivative.h"     /* include peripheral declarations */
#include "hcs08.h"          // This is our definition file !

#define LED7 PTFD_PTFD7
#define NVICGTRM (*(const char * __far)0x0000FFBE)

unsigned char my_mfd, icg_lols;

// ICG interrupt (process LOLS and LOCS interrupts)
void interrupt VectorNumber_Vicg icg_isr(void)
{
  // if it was a loss of clock, we need to restore FEE mode
  if (ICGS1_LOCS)
  {
    // ICG in FEE mode (CLKS = 11), external crystal:
    ICGC1 = ICG_FEE | bREFS | bICG_RANGE;
    ICGC2 = MFDx10;             // MFD = 111 (N = 10), RFD = 000 ( R = 1)
    while (!ICGS2_DCOS);        // wait for the DCO to stabilize
    my_mfd = 3;                 // restore my_mfd factor
  }
  // if it was a loss of lock, flag LOLS condition
  if (ICGS1_LOLS) icg_lols = 1;
  ICGS1_ICGIF = 1;             // clear ICG interrupt
}

// KBI ISR (process SW3 and SW4 key presses)
void interrupt VectorNumber_Vkeyboard1 kbi1_isr(void)
{
  KBI1SC_KBACK = 1;           // clear KBI interrupt flag
  if (!PTDD_PTDD3)            // if SW3 is pressed, increase MFD (speed up BUSCLK)
  {
    my_mfd++;
    if (my_mfd>3) my_mfd = 3;  // limit my_mfd to 3 (N = 10)
  }
```

> This line may be necessary if your version of Codewarrior does not include the NVICGTRM definition for the AW devices. Remove it if you have any compiling error.

```
    if (!PTDD_PTDD2)                    // if SW4 is pressed, decrease MFD (slow down BUSCLK)
    {
      if (my_mfd) my_mfd--;             // do not decrease below 0
    }
    ICGC2 = (my_mfd << 4);              // change MFD factor
}

// A simple software loop delay
void delay(unsigned int value)
{
  for (;value; value--);
}

void main(void)
{
  unsigned char shift_register = 1;
  SOPT = 0;   // configure SOPT1 register, STOP instruction and COP disabled
  // ICG configuration:
  // ICG in FEE mode (CLKS = 11), external crystal:
  ICGC1 = ICG_FEE | bREFS | bICG_RANGE;
  ICGC2 = MFDx10;                       // MFD = 111 (N = 10), RFD = 000 ( R = 1)
  ICGTRM = NVICGTRM;                    // trim IRG oscillator
  while (!ICGS2_DCOS);                  // wait for the DCO to stabilize
  // I/O ports configuration:
  PTDPE = BIT_3 | BIT_2;                // enable pull ups on PTC7 and PTC5
  PTFDD = 0xFF;                         // PTB0 through PTB7 as outputs
  PTFD = 0;                             // turn off LD401 through LD405
  PTADD = BIT_5;                        // PTA5 as an output
  KBI1SC = bKBIE;                       // turn on KBI2 interrupt
  KBI1PE = BIT_6 | BIT_5;               // enable KBI inputs corresponding to PTD3 and PTD2
  my_mfd = 3;                           // initialize "my_mfd" factor
  icg_lols = 0;                         // initialize "icg_lols"
  EnableInterrupts;
  while (1)
  {
    PTFD = shift_register;              // change led states
    if (icg_lols) PTFD_PTFD7 = 1; // if icg_lols = 1, turn LED7 on
    delay(30000);                      // delay
    shift_register <<= 1;              // shift one bit to right
    // did we reached the last led?
    if (shift_register==32) shift_register=1; // restart from the first led
  }
}
```

Example 7.5 – ICG example using an MC9S08AW60

8

Low Power Modes

We live in a planet with limited natural resources and a major concern nowadays concerns power consumption. The more the population increases, the more the economy enhances, the bigger the need for energy is (in our present case, electrical energy).

That is why semiconductor manufacturers have been investing more and more on the research of new devices which can work more efficiently and draw less power.

Other big players on the balance are the battery powered applications: handheld devices, portable gadgets, intelligent sensors and many other applications are pushing the power consumption race to µA and even nA levels.

In response to this new level of competition, Freescale answered with a number of new features and operating modes that make the HCS08 very power-efficient devices. In fact, there are five operating modes available on the HCS08 devices: run, wait, stop3, stop2 and stop1.

Run mode is the standard mode when the CPU is running code at a speed defined by the bus clock (BUSCLK). In wait mode, the CPU is halted but all peripherals remain active. In stop mode, some peripherals are turned off (depending on selected stop mode). The lowest possible power consumption is achieved with stop1 mode: almost all peripherals are turned off (including the internal power regulator, RAM memory and internal registers).

The new Flexis devices include two new low power modes: low power run and low power wait. The mode selection on these devices is slightly different from that of other devices. We will further study about the low power modes on these devices in topic 8.3.

Figure 8.1 shows the relative power consumption for all HCS08 operating modes.

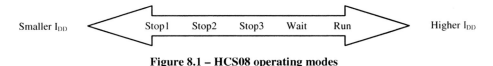

Figure 8.1 – HCS08 operating modes

Switching between run and low power modes is done by using two assembly instructions: WAIT and STOP. Selection of the different stop modes is done through SPMSC2 register, as we will see later in this chapter.

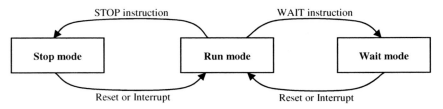

Figure 8.2 – Low power modes

8.1. Wait Mode

The wait mode is a low power mode entered by running WAIT instruction. While in wait mode, the chip state is defined as follows:

1. CPU clock is stopped, halting program execution. This happens immediately after the execution of the assembly instruction WAIT.

2. The memory and registers contents are preserved as long as the chip remains powered.

3. All peripherals continue operating normally.

Exiting wait mode can be done by either an interrupt or a reset event. An interrupt in this mode makes the CPU return to the run mode and initiate the interrupt servicing routine (ISR).

8.2. Stop Modes

On Ex, Dx, Gx, LC, QA, QG, Sx and Rx devices, selection among the three stop modes is made through PDC and PPDC bits, located in the SPMSC2 register:

PDC	PPDC	Mode
0	0	stop3
0	1	stop3
1	1	stop2
1	0	stop1

Table 8.1 – Selection of the stop modes

On AC, AW and QD devices, selection between the two available stop modes is made through PPDC bit, located in the SPMSC2 register:

PPDC	Mode
0	stop3
1	stop2

Table 8.2 – Selection of the stop modes

After selecting the desired stop mode, executing the STOP instruction makes the CPU enter the selected stop mode.

Warning: using STOP and WAIT instructions within an ISR should be avoided as they automatically enable interrupts (CCR:I=0). Enabling interrupts within an ISR results in the so called "nested interrupts", which can easily lead to stack overflows!

On the referred devices the PDC and PPDC bits are write-once and can be modified only once after device reset. Once modified, any additional writes to them have no effect until the next reset. User is advised to carefully choose the desired stop mode, as using different stop modes in the same application is not possible (except on QE devices as we will see in topic 8.3).

While in active debug mode (by using the BDM tool), it is not possible to enter stop2 or stop1 mode! Entering stop3 mode while in debug mode causes the internal voltage regulator and the oscillator circuitry to remain operating.

8.2.1. Stop3 Mode

The stop3 mode works in the same way as the stop mode found on the HC08 devices. In this mode, the chip spends less power than in wait mode thus conserving power (good for battery-powered systems). To select this mode, the PDC bit (register SPMSC2) must be cleared and the STOP instruction must be executed.

Table 8.3 shows the stop3 mode current consumption for some HCS08 devices (typical values specified by the manufacturer at $25°C$) and the additional current spent by RTI, LVD and oscillator modules when active in stop3 mode.

Device	Supply Voltage (V)	Stop3 Current Consumption (nA)	RTI/RTC Adder (nA)	LVD Adder (μA)	Internal Clock Reference Adder (μA)	External Clock Reference Adder (μA)
AW	5	975	300	110	N/A	5
	3	825	300	90	N/A	5
GB/GTxx	3	675	300	70	N/A	5
A	2	500	300	60	N/A	5
LC	3	840	350	75	N/A	4
	2	660	350	70	N/A	3.5
QD	5	900	400	110	75	-
	3	900	350	90	65	-
QE	3	450	200	100	70	0.5
	2	350	???	???	???	???
QG	3	750	300	70	???	5
	2	680	300	60	???	4
Rx	3	600	300	70	N/A	N/A
	2	500	300	60	N/A	N/A

Table 8.3 – Stop3 mode currents and adders

Once in stop3 mode, the chip state is defined as follows:

1. BUSCLK signal is stopped, halting the CPU and all digital peripherals as soon as the STOP instruction gets executed.

2. Memory and registers contents are preserved as long as the chip remains powered.

3. The clocking system, analog-to-digital converter (ADC), real-time interrupt (RTI)/real-time clock (RTC) and internal voltage regulator can optionally operate while in stop3 mode:

   a. The internal voltage regulator continues operating in stop3 mode if bits SPMSC1:LVDE and SPMSC1:LVDSE are set before entering stop mode.

   b. RTI/RTC continues operating in stop3 mode if its clock source is the 1 kHz internal oscillator or the external reference.

   c. The ADC operates in stop3 mode if its clock source is the ADC internal oscillator and if the LVD module is enabled.

Exiting from a stop3 mode can be done by a reset or one of the following interrupts:

- External (IRQ) interrupt pin;

- Keyboard interrupt pin (KBI);

- Low-Voltage Detect (LVD) module;

- Real-time interrupt (RTI) timer or the real-time clock (RTC) timer on some devices;

- Analog-to-digital converter (ADC) (only on devices with ADC10 or ADC12 converter);

 The interrupt source must be enabled prior to executing the STOP instruction!

When the stop3 mode is exited by a reset signal, the chip restarts program execution exactly in the same way as a cold start (the reset vector is taken and all peripheral registers assume their reset state). In case the exit was originated by an interrupt, the program flow restarts and the CPU takes the appropriate interrupt vector. After processing the ISR, the program flow returns to the instruction following the STOP instruction.

The next example shows the usage of the stop3 mode. This small application blinks led1 (connected to PTB6 on the DEMO9S08QG8 board). By pressing SW2 we can toggle LED2 state and by pressing SW1 we can make the CPU enter stop3 mode or leave this mode. Actually any key (SW1 or SW2), when pressed, wakes up the CPU.

```
// DEMO9S08QG8 stop3 mode example
// LED1 cathode is connected to PTB6
// LED2 cathode is connected to PTB7
// SW1 is connected to PTA2
// SW2 is connected to PTA3

#include <hidef.h>          /* for EnableInterrupts macro */
#include "derivative.h"     /* include peripheral declarations */
#include "hcs08.h"          // This is our definition file !

#define LED1    PTBD_PTBD6
#define LED2    PTBD_PTBD7
#define SW1     PTAD_PTAD2
#define SW2     PTAD_PTAD3

unsigned char mode;         // current mode

// A simple software loop delay
void delay(unsigned int value)
{
  for (;value; value--);
}

// This is the KBI interrupt servicing routine
void interrupt VectorNumber_Vkeyboard KBI_isr()
{
  KBISC_KBACK = 1;          // acknowledge KBI interrupt (clear KBF)
  if (!SW2) LED2 = !LED2;   // if SW2 is pressed then toggle LED2 state
  if (!SW1) mode = !mode;   // if SW1 is low (pressed) then toggle mode
}

void main(void)
{
  // configure SOPT1 register, enable STOP instruction and BKGD pin
  SOPT1 = bSTOPE | bBKGDPE;
  SPMSC2 = 0;               // stop3 mode selected
  PTBDD = 0xFF;             // all PTB pins as outputs
  PTAPE = BIT_2 | BIT_3;    // turn on internal pull ups on PTA2, PTA3
  KBISC = bKBIE;            // enable KBI interrupt
  KBIPE = BIT_2 | BIT_3;    // enable KBI inputs 2 and 3 (PTA2 and PTA3)
  mode = 0;
  EnableInterrupts;         // enable interrupts (CCR:I = 0)
  while (1)
  {
    LED1 = 0;               // LED1 is on
    delay(30000);           // delay
    LED1 = 1;               // LED1 is off
    delay(30000);           // delay
    // if mode is true (different from zero) then enter stop3 mode
    if (mode)
    {
      STOP;                 // enter stop3 mode
      mode = 0;             // after wakeup, return to mode = 0 (stop mode disabled)
    }
  }
}
```

> Do not forget to enable the STOP instruction (STOPE=1), otherwise the execution of STOP generates an illegal opcode (ILOP) reset!

Example 8.1 – Stop3 example

We can change example 8.1 so that only SW1 can be used to enter/exit stop3 mode: just modify the last **if** statement in the KBI ISR, as shown in listing 8.1.

```
// This is the KBI interrupt servicing routine
void interrupt VectorNumber_Vkeyboard KBI_isr()
{
  KBISC_KBACK = 1;          // acknowledge KBI interrupt (clear KBF)
  if (!SW2)     LED2 = !LED2; // if SW3 is pressed then toggle LED2 state
  if (!SW1) mode = !mode;   // if SW is low (pressed) then toggle mode
  // if mode is true (different from zero) then enter stop3 mode
  if (mode)
  {
    KBIPE = BIT_2;           // only PTA2 is enabled as a KBI input
    STOP;                    // enter stop3 mode
    mode = 0;                // after wakeup, return to mode = 0 (stop mode disabled)
    KBIPE = BIT_2 | BIT_3;   // enable KBI inputs 2 and 3 (PTA2 and PTA3)
  }
}
```

Listing 8.1

8.2.2. Stop2 Mode

The stop2 mode is a new low power mode available on the HCS08 family. This stop mode enables a smaller current consumption (when compared to the stop3 mode) by disabling all internal peripherals and registers. On GB, GT, LC, QG, RC, RD, RE and RG devices, this mode is selected by setting both PDC and PPDC bits to "1". On AC, AW and QD devices, this mode is selected by setting PPDC to "1".

In fact, everything powered by the internal voltage regulator is turned off. There are only four modules that remain powered and can continue operating normally: the RAM memory, the I/O ports, the IRQ circuitry and the real-time interrupt (RTI) timer or the real-time clock (RTC) timer.

It is important to note that all peripheral registers content is lost upon entering stop2 mode. However, there is a special circuit to latch the I/O pin state while in stop2 mode. This circuit retains the state of the I/O pins in stop2 mode and even after leaving stop2 mode. The control of the I/O pins remains latched until the PPDACK (SPMSC2 register) bit is set, releasing the control to the I/O port registers.

A good programming practice is to save the state of any important peripheral registers to RAM before entering stop2 mode (including the state of the I/O port registers) and restore them after exiting stop2 mode.

Table 8.4 shows the stop2 mode current consumption for some HCS08 devices (typical values specified by the manufacturer at $25°C$) and the additional current spent by RTI module when active in the stop2 mode:

Device	Supply Voltage (V)	Stop2 Current Consumption (nA)	RTI/RTC Adder (nA)
AW	5	900	300
	3	720	300
GB/GTxxA	3	550	300
	2	400	300
LC	3	650	350
	2	600	350
QD	5	800	400
	3	800	350
QE	3	350	200
	2	250	200
QG	3	600	300
	2	550	300
Rx	3	500	300
	2	500	300

Table 8.4 – Stop2 mode currents and adders

If the low-voltage detection (LVD) circuit is enabled before entering stop mode, the execution of the STOP instruction enters stop3 mode regardless of the state of PDC and PPDC bits!

While in stop2 mode, the LVD circuit is automatically enabled in the reset mode!

Exiting stop2 mode can be done by one of the following events:

1. Reset (low level on reset pin).

2. External interrupt (IRQ).

3. Real-time interrupt (RTI) or real-time clock (RTC) when it was previously enabled before entering the stop2 mode.

*The IRQ pin is automatically enabled upon entering any stop mode! On devices with multiplexed IRQ/reset functions, it is not possible to use the reset pin as the source for wake up (IRQ is used instead). **Additionally, the internal pull up on this pin is automatically disabled when the STOP instruction is executed. This pin must be externally driven or pulled up while in stop2 mode.***

When the wake up event is a reset, the chip restarts program execution (the reset vector is taken and all peripheral registers assume their reset state). Any other wake up events listed above initiates stop2 mode recovery sequence, which is almost identical to a power-on reset (POR) sequence, except for two differences:

1. PPDF bit (register SPMSC2) is set. This bit enables the application to differentiate between a reset event (PPDF = 0) and a wake up from stop2 event (PPDF = 1).

2. The state of the I/O pins remains latched into the same state as before entering stop2 mode. It is important to note that the I/O port registers (as well as all other peripheral registers) are reset to their default state.

The device initialization function must be written to differentiate between a POR and a wake up event:

*   On a POR (PPDF = 0), the initialization function initializes the peripheral registers to the application default state. The contents of the global variables must be initialized to their default state.

*   On a wake up from stop2 (PPDF = 1), the initialization function should restore the peripheral registers to their saved state (if their state was previously saved to RAM). It is also important to initialize the I/O port registers to their previous state, before setting the PPDACK bit (which restores the I/O pin control back to the I/O port registers). On a wake up from stop2, the contents of the global variables must not be changed, and neither must any other RAM location. **The I/O pins state remains latched until the application writes "1" into PPDACK!**

The minimal startup code option should be preferred for applications using stop2 mode. This option generates a simpler startup code that does not initialize the global and static variables. Further information on the project options is presented in topic 3.2.

Another option is the use of NO_INIT segments in the linker configuration file (.PRM). Listing 9.1 and example 9.2 in chapter 9 shows how to do this.

The following example demonstrates how to use stop2 mode. When the MCU is in run mode, LED1 blinks. Pressing SW2 toggles LED2 and pressing SW1 makes the MCU enter stop2 mode.

Once in stop2 mode, all registers are cleared as well as the RAM contents (the state of the I/O pins is maintained). Pressing RESET wakes up the MCU and initiates a POR (power-on reset).

The SPMSC2:PPDF bit is tested at the beginning of the program. If it is set, the stop2 mode is acknowledged (by writing "1" into SPMSC2:PPDACK). If this operation is not done, the I/O pin states can not be changed after recovering from stop2 mode (I/O pins would remain locked on the state they were before the STOP instruction was executed). LED2 is also lit to indicate a recovery from stop2 mode.

This example can only be tested when the debugging software is closed. If a debug session is in progress, the chip enters stop3 instead of stop2.

```
// DEMO9S08QG8 stop2 mode example
// LED1 cathode is connected to PTB6, LED2 cathode is connected to PTB7
// SW1 is connected to PTA2, SW2 is connected to PTA3
// RESET is connected to PTA5/IRQ/RESET

#include <hidef.h>          /* for EnableInterrupts macro */
#include "derivative.h"     /* include peripheral declarations */
#include "hcs08.h"          // This is our definition file!

#define LED1   PTBD_PTBD6
#define LED2   PTBD_PTBD7
#define SW1    PTAD_PTAD2
#define SW2    PTAD_PTAD3

// A simple software loop delay
void delay(unsigned int value)
{
  for (;value; value--);
}

// This is the KBI interrupt servicing routine
void interrupt VectorNumber_Vkeyboard KBI_isr()
{
  KBISC_KBACK = 1;            // acknowledge KBI interrupt (clear KBF)
  if (!SW2) LED2 = !LED2;     // if SW2 is pressed then toggle LED2 state
  if (!SW1) STOP;            // if SW1 is pressed then enter stop2 mode
}

void main(void)
{
  // configure SOPT1 register, enable STOP instruction and BKGD pin
  SOPT1 = bSTOPE | bBKGDPE;
  SPMSC1 = 0;                // disable LVD
  SPMSC2 = bPDC | bPPDC;     // stop2 mode selected
  PTBDD = 0xFF;              // PTB pins as outputs
  PTAPE = BIT_2 | BIT_3;     // Turn on pull ups on PTA2 and PTA3
  KBISC = bKBIE;             // KBI interrupt enabled
  KBIPE = BIT_2 | BIT_3;     // PTA2 and PTA3 as KBI inputs
  if (SPMSC2_PPDF)           // if PPDF is set
  {
    SPMSC2_PPDACK = 1;       // acknowledge PPDF flag
    LED2 = 0;                // turn LED2 on
  }
  EnableInterrupts;          // enable interrupts (CCR:I = 0)
  while (1)
  {
    LED1 = 0;                // LED1 is on
    delay(30000);            // delay
    LED1 = 1;                // LED1 is off
    delay(30000);            // delay
  }
}
```

> In stop2 mode we can use the STOP instruction within the ISR, because the wakeup from this low power mode actually restarts the stack pointer to its default value.

> Do not forget to disable LVDE and LVDSE, or the CPU will enter stop3 instead of stop2!

> Do not forget to acknowledge the power down flag (writing "1" into PPDACK)!
> Failing to do so prevents any external I/O operation.

Example 8.2 – Stop2 example

 The DEMO9S08QG8 board has a pull up device attached to the PTA5/IRQ/RESET pin (actually this pull up device is part of the BDM circuitry). This external pull up must be included when designing an application that uses this pin as a wake up source from stop2 mode!

8.2.3. Stop1 Mode

The stop1 mode is the most efficient low power mode as it presents the lowest possible power consumption, but it has the drawback of erasing all RAM and peripheral registers content. This mode is available only on some low-voltage devices (GB, GT, LC, QG, RC, RD, RE and RG models).

To select the stop1 mode it is necessary to set the PDC and clear the PPDC bits on the SPMSC2 register (this operation can be done only one time after a reset!).

In stop1 mode, all internal modules (CPU, clock generation, peripheral registers and RAM memory) are turned off. The IRQ circuitry and optionally the real-time interrupt (only on QG and LC devices) can still operate while in this mode.

Table 8.5 shows some stop1 mode current consumption for the HCS08 devices (typical values specified by the manufacturer at 25°C) and the additional current spent by the RTI module when active in the stop1 mode (the RTI can operate in stop1 mode only on QG and LC devices):

Device	Supply Voltage (V)	Stop1 Current Consumption (nA)	RTI Adder (nA)
GB/GTxxA	3	25	*
	2	20	*
LC	3	770	350
	2	600	350
QG	3	475	300
	2	470	300
RC/RD/RE/RG	3	100	*
	2	100	*

* RTI is not available in stop1 mode on these devices

Table 8.5 – Stop1 mode currents and adders

If the low-voltage detection (LVD) circuit is enabled before entering stop mode, the execution of the STOP instruction enters stop3 mode regardless of the state of PDC and PPDC bits!

While in stop1 mode, the LVD circuit is automatically enabled in the reset mode!

One of the following events can cause a wake up from stop1 mode:

1. Reset (low level on the reset pin).

2. External interrupt (IRQ).

3. Real-time interrupt (RTI) when the RTI module was previously enabled before entering the stop1 mode (this option is only possible on QG and LC devices).

The IRQ pin is automatically enabled while entering any stop mode! On devices with multiplexed IRQ/reset functions, it is not possible to use the reset pin as the source for wake up (IRQ is used instead).

Any of the wake up events listed above initiates stop1 mode recovery sequence, which is almost identical to a power-on reset (POR) sequence, with only one difference: the PDF bit (register SPMSC2) is set. This bit allows the application to differentiate between a reset event (PDF = 0) and a wake up from stop1 event (PDF = 1).

Remember that upon entering stop1 mode all RAM content is lost!

The next example shows how to use the stop1 mode on the MC9S08QG8. When the MCU is in run mode, LED1 blinks. Pressing SW2 toggles LED2 and pressing SW1 makes the MCU enter stop1 mode.

Once in stop1 mode, all registers are cleared as well as the RAM contents (all I/O pins default to input mode). Pressing RESET wakes up the MCU and initiates a POR (power-on reset).

The SPMSC2:PDF bit is tested at the beginning of the program and if set, the stop1 mode is acknowledged (by writing "1" into SPMSC2:PPDACK) and LED2 is lit to signal this condition.

This example can only be tested when the debugging software is closed. If a debug session is in progress, the chip enters stop3 instead of stop1.

```
// DEMO9S08QG8 stop1 mode example
// LED1 cathode is connected to PTB6, LED2 cathode is connected to PTB7
// SW1 is connected to PTA2, SW2 is connected to PTA3
// RESET is connected to PTA5/IRQ/RESET

#include <hidef.h>          /* for EnableInterrupts macro */
#include "derivative.h"     /* include peripheral declarations */
#include "hcs08.h"          // This is our definition file!

#define LED1    PTBD_PTBD6
#define LED2    PTBD_PTBD7
#define SW1     PTAD_PTAD2
#define SW2     PTAD_PTAD3

// A simple software loop delay
void delay(unsigned int value)
{
  for (;value; value--);
}

// This is the KBI interrupt servicing routine
void interrupt VectorNumber_Vkeyboard KBI_isr()
{
  KBISC_KBACK = 1;              // acknowledge KBI interrupt (clear KBF)
  if (!SW2) LED2 = !LED2;       // if SW2 is pressed then toggle LED2 state
  if (!SW1) STOP;              // if SW1 is pressed then enter stop1 mode
}

void main(void)
{
  // configure SOPT1 register, enable STOP instruction and BKGD pin
  SOPT1 = bSTOPE | bBKGDPE;
  SPMSC1 = 0;                   // disable LVD
  SPMSC2 = bPDC;               // stop1 mode selected
  PTBDD = 0xFF;                // PTB pins as outputs
  PTBD = 0xFF;                 // Set all PTB outputs
  PTAPE = BIT_2 | BIT_3;       // Turn on pull ups on PTA2 and PTA3
  KBISC = bKBIE;               // KBI interrupt enabled
  KBIPE = BIT_2 | BIT_3;       // PTA2 and PTA3 as KBI inputs
  if (SPMSC2_PDF)              // if PDF is set
  {
    SPMSC2_PPDACK = 1;         // acknowledge PDF flag
    LED2 = 0;                  // turn LED2 on
  }
  EnableInterrupts;            // enable interrupts (CCR:I = 0)
  while (1)
  {
    LED1 = 0;                  // LED1 is on
    delay(30000);             // delay
    LED1 = 1;                  // LED1 is off
    delay(30000);             // delay
  }
}
```

> In stop1 mode we can use the STOP instruction within the ISR, because the wakeup from this low power mode restarts all registers within the MCU.

> Do not forget to disable LVDE and LVDSE, the CPU will enter stop3 instead of stop1!

> This clears the PDF flag.

Example 8.3 – Stop1 example

 The DEMO9S08QG8 board has a pull up device attached to the PTA5/IRQ/RESET pin (actually this pull up device is part of the BDM circuitry). This external pull up must be included when designing an application that uses this pin as a wake up source from stop1 mode!

8.3. Advanced Power Management Modes

Newer devices (such as LL, QB and QE devices) offer two new low power modes: low power run and low power wait. These modes are designed to allow intermediate power consumption (a middle term between run and stop modes). While on these modes, the internal voltage regulator is placed in stand-by, limiting the maximum bus clock speed to 125 kHz for the CPU and peripherals.

Stop3 and stop2 modes are also present and work almost in the same way of the other HCS08 devices. Stop1 mode is not present on these devices.

One major difference from other HCS08 devices is the ability to switch between low power modes on-the-fly (PPDC is a read/write bit and can be written any number of times after a reset).

We will now take a closer look on the operation of each low power mode on these devices (the wait mode is not seen as it works exactly as shown for the other devices).

8.3.1. Low Power Run Mode

This mode was designed to allow CPU and peripherals operation with minimal power consumption.

The internal voltage regulator is placed in stand-by mode (under software control), thus limiting maximum clock speed to 250 kHz (125 kHz BUSCLK).

To enter low power run mode, all you have to do is set the low power regulator control bit (SPMSC2:LPR). Notice that before doing so, some conditions must be met:

1. ICS must be operating in FBELP mode and the external oscillator must be configured to low gain (ICSC2:HGO = 0). Refer to topic 7.2.

2. LVD module must be disabled (SPMSC1:LVDE = 0 and SPMSC1:LVDSE = 0). Refer to topic 5.2. Note that these bits are always set after a reset (LVE module is enabled following a reset).

3. The internal bandgap reference is disabled and cannot be used as a reference for the analog comparator or as an input channel for the analog-to-digital converter.

4. The analog-to-digital converter must be configured to run from its own internal clock (ADACK). The ADC clock dividers must be configured in order to reduce ADC clock to the lowest frequency allowed.

5. It is not allowed to program or erase FLASH memory while in low power run mode.

6. It is not possible to enter low power run mode while a debug session is in progress. Trying to do so automatically clears LPR bit and restores full power run mode.

While in low power run mode the low power regulator status (SPMSC2:LPRS) bit is set to indicate that the internal voltage regulator is in stand-by mode. When LPRS = 0, the regulator is operating in full power mode.

There is a special circuitry to automatically exit low power run mode upon an interrupt event. This is controlled by the low power wake up on interrupt (SPCSC2:LPWUI) bit. When LPWUI = 0, the internal voltage regulator remains in standby mode on an interrupt. If LPWUI = 1, the internal voltage regulator is automatically configured to full power mode upon an interrupt event. This enables the ISR to configure ICS to operate at a higher frequency during the ISR processing. Before returning from the interrupt, the application would configure ICS back to FBELP mode and set LPR bit to return to low power run mode.

The application can restore full power mode at any time by clearing the LPR bit.

Example 8.4 demonstrates how to use the low power run mode on the DEMO9S08QE128 board. After a reset, the two leds connected to PTC0 and PTC1 blink very fast, as the delay is very small and the ICS module is configured to operate in FEI mode at 40 MHz (20 MHz BUSCLK) (actually both leds appear to be steady). Pressing PTA2 key switches ICS to FBELP mode. That makes the BUSCLK be reduced to 16,384 Hz (a half of the nominal crystal frequency of 32,768 Hz). The leds now blink at a perceptible rate. LED3 (connected to PTC2) lights when the MCU operates in low power run mode (SPMSC2:LPRS bit is set).

Notice that when this program is run with the board connected to the host computer and a debug session is in progress, the internal regulator never enters stand-by mode and LPRS bit is not set. **To successfully test it, it is necessary to stop the debugger, turn the board off and then on again.** Now, when PTA2 key is pressed, LED3 lights, indicating that the low power run mode is operational.

Before attempting to run it, it is necessary to solder a 32,768 Hz quartz crystal on Y1. The other components (C8, C9, Rf and Rs) should not be mounted. It is also necessary to solder two jumpers on the J17 position (CLOCK EN jumper). To run this example, these jumpers must be mounted.

```
// DEMO9S08QE128 low power run example (FBELP mode, BUSCLK = 16384 Hz)
// LED1 is connected to PTC0, LED2 is connected to PTC1, LED3 is connected to PTC2
// SW1 is connected to PTA2, SW2 is connected to PTA3

#include <hidef.h>          /* for EnableInterrupts macro */
#include "derivative.h"     /* include peripheral declarations */
#include "hcs08.h"          // This is our definition file !

#define LED1 BIT_0          // PTC0
#define LED2 BIT_1          // PTC1
#define LED3 BIT_2          // PTC2
#define SW1  PTAD_PTAD2
#define SW2  PTAD_PTAD3

// A simple software loop delay
void delay(unsigned int value)
{
  for (;value; value--);
}

void main(void)
{
  SOPT1 = bBKGDPE;              // configure SOPT1 register, enable pin BKGD for BDM
  ICSC2 = bEREFS;              // disable FLL and select the external oscillator
  ICSSC = DCO_MID | bDMX32;    // select DCO mid-range and ajusts it to 32768 Hz
  ICSC1 = ICS_FLL;             // ICS in FEE mode (FLL engaged external because IREFS = 0)
  while (ICSSC_CLKST != ICSC1_CLKS); // wait for the clock change to take effect
  SPMSC1 = 0;                  // disable LVD module
  PTCDD = 0xFF;                // PTC pins as outputs
  PTCD = 0xFE;                 // set all PTC pins, except PTC0 (LED1 is on, LED2 is off)
  PTAPE = BIT_2 | BIT_3;       // PTA2 and PTA3 internal pull ups enabled
  while (1)
  {
    PTCTOG = LED1;             // toggle LED1 (bit 0 of PTC)
    PTCTOG = LED2;             // toggle LED2 (bit 1 of PTC)
    delay(60);                 // a little delay
    PTCTOG = LED1;             // toggle LED1 (bit 0 of PTC)
    PTCTOG = LED2;             // toggle LED2 (bit 1 of PTC)
    delay(60);                 // a little delay
    // now we test LPRS and turn LED3 on if it is set (low power run mode is enabled)
    if (SPMSC2_LPRS) PTCCLR = LED3; else PTCSET = LED3;
    if (!SW1)
    {
      ICSC2 = bICS_LP | bEREFS; // FLL low power mode, crystal oscillator
      ICSC1 = ICS_FBE;          // ICS in FBE mode (BUSCLK = 16384 Hz)
      SPMSC2_LPR = 1;           // enter low power run mode
    }
}
```

```
   if (!SW2)
   {
     ICSC2 = bEREFS;              // External crystal oscillator
     ICSC1 = ICS_FLL;             // ICS in FEE mode (BUSCLK = 20MHz)
     SPMSC2_LPR = 0;              // exit low power run mode
   }
 }
}
```

Example 8.4 – Low power run example

8.3.2. Low Power Wait Mode

This mode is entered by executing the WAIT instruction while the MCU is operating in the low power run mode. The same restrictions stated for the low power run mode are valid for the low power wait mode.

While in low power wait mode, the CPU is halted and the active peripherals continue operating (with a maximum BUSCLK of 125 kHz as stated for low power run mode).

To reduce the power consumption even more, the application can disable clock for unused peripherals by clearing the corresponding SCGC1 and SCGC2 bits (refer to topic 8.5.1 for further information on clock gating registers).

It is also possible to automatically exit low power wait mode upon an interrupt event. This is controlled by LPWUI bit and its operation is the same described for low power run mode.

8.3.3. Stop Modes

Devices with advanced power management modes allow two stop modes: stop3 and stop2. Their operation is very similar to that described for the other HCS08 devices, with some differences:

1. The desired stop mode is selected by SPMSC2:PPDE and SPMSC2:PPDC bits as shown in table 8.6:

PPDE	PPDC	Mode
0	0	stop3
0	1	stop3
1	0	stop3
1	1	stop2

Table 8.6 – Selection of the stop mode

2. The application can select, at any time, the desired stop mode. The stop mode selection bit (PPDC) can be written any number of times. Be careful on the first write into SPMSC2: PPDE is a write-once bit and cannot be modified after written (until the next reset). If using stop2 mode is planned, set PPDE on the first write into SPMSC2.

3. LPR and PPDC bits (both located in SPMSC2 register) are mutually exclusive: they cannot be both set. When one is set, it is not possible to set the other. **This way, it is not possible to enter stop2 mode from a low power run mode (LPR = 1).** Stop2 mode can only be entered from run mode (LPR = 0).

8.4. Low Power Control Registers

Selection and control of low power modes is done through a single register:

♦   System Power Management Status and Control 2 (SPMSC2) Register.

8.4.1. SPMSC2 Register

Name	Model		BIT 7	BIT 6	BIT 5	BIT 4	BIT 3	BIT 2	BIT 1	BIT 0
SPMSC2	AC, AW, QD	Read	LVWF	0	LVDV	LVWV	PPDF	0	-	PPDC*
		Write	-	LVWACK	LVDV	LVWV	-	PPDACK		PPDC*
		Reset	0	0	0	0	0	0	0	0
	QB, QE, LL	Read	LPR	LPRS	LPWUI	0	PPDF		PPDE*	PPDC
		Write	LPR	-	LPWUI	-	-	PPDACK	PPDE*	PPDC
		Reset	0	0	0	0	0	0	0	0
	LC, QA, QG	Read	0	0	0	PDF	PPDF	0	PDC*	PPDC*
		Write	-	-	-	-	-	PPDACK	PDC*	PPDC*
		Reset	0	0	0	0	0	0	0	0
	GB, GT	Read	LVWF	0	LVDV	LVWV	PPDF	0	PDC*	PPDC*
		Write	-	LVWACK	LVDV	LVWV	-	PPDACK	PDC*	PPDC*
		Reset	0	0	0	0	0	0	0	0
	RC, RD, RE, RG	Read	LVWF	0	0	0	PPDF	0	PDC*	PPDC*
		Write	-	LVWACK	-	-	-	PPDACK	PDC*	PPDC*
		Reset	0	0	0	0	0	0	0	0

 These are write-once bits and can be modified only once after a reset. All other writes into them are ignored and produce no effect!

Bit Name	Description	C Symbol
LPR	Low power regulator control bit. This bit controls the internal voltage regulator standby mode: 0 – standby mode disabled (low power run and low power wait modes disabled). 1 – standby mode enabled (low power run and low power wait modes enabled).	bLPR
LPRS	Low power regulator status bit. This bit indicates if the internal voltage regulator is operating in standby mode or full power mode: 0 – the internal regulator is in full power mode. 1 – the internal regulator is in standby mode.	bLPRS
LPWUI	Low power wake up on interrupt bit. This bit controls the behavior of the internal voltage regulator on an interrupt event: 0 – the regulator remains in standby mode. 1 – the regulator exits standby mode on an interrupt event.	bLPWUI
PDF	Power down flag. This status bit indicates whether the MCU has recovered from stop1 mode: 0 – the chip has not recovered from stop1 mode. 1 – the chip has recovered from stop1 mode.	bPDF
PPDF	Partial power down flag. This status bit indicates whether the MCU has recovered from stop2 mode: 0 – the chip has not recovered from stop2 mode. 1 – the chip has recovered from stop2 mode.	bPPDF
PPDACK	Partial power down acknowledge bit. Writing "1" into this bit clears PDF and PPDF flags. Writing "0" produces no effect.	bPPDACK
PPDE	Partial power down enable bit. This write-once bit enable or disable the partial power down feature. 0 – partial power down disabled. 1 – partial power down enabled.	bPPDE
PDC	Power down control bit. This bit selects power down modes stop2 or stop1 (depending on PPDC configuration): 0 – stop3 mode selected. 1 – stop2 or stop1 mode selected (depending on PPDC configuration).	bPDC
PPDC	Partial power down control bit. This bit selects stop mode 1 or 2, depending on PDC configuration. If PDC = 0, this bit has no effect. If PDC = 1, the selected stop mode is: 0 – stop1 mode selected. 1 – stop2 mode selected.	bPPDC

8.5. Clock Gating

QB, QE and LL devices include a clock gating circuitry, which allows controlling clock supply for each internal peripheral, enabling the application to dynamically control which peripherals are supplied with the BUSCLK signal and which ones are not.

 Digital circuits draw less power when they are not being clocked!

After a reset, all peripheral clocks are enabled. The application software must write a "1" to the related clock gating register bit to enable the related peripheral BUSCLK supply and "0" to disable clock supply. Changes take effect immediately after the write operation.

Notice that once the peripheral clock is disabled, writing into peripheral registers produces no effect until the peripheral clock is re-enabled.

 To reactivate a disabled peripheral, first re-enable its clock signal and then reconfigure its registers to the desired mode. This is a mandatory operation, failing it can lead to unexpected peripheral behavior!

8.5.1. Clock Gating Registers

There are two registers dedicated to controlling the clock supply for internal peripherals:

- System Clock Gating Control 1 (SCGC1) register.
- System Clock Gating Control 2 (SCGC2) register.

8.5.1.1. SCGC1 Register

Name	Model		BIT 7	BIT 6	BIT 5	BIT 4	BIT 3	BIT 2	BIT 1	BIT 0
SCGC1	QE, LL	Read	TPM3*	TPM2	TPM1	ADC	IIC2*	IIC1	SCI2*	SCI1
		Write								
		Reset	1	1	1	1	1	1	1	1
	QB	Read	MTIM	1	TPM	ADC	1	1	1	SCI
		Write		-			-	-	-	
		Reset	1	1	1	1	1	1	1	1

* These bits are not present on QE4, QE8 and LL devices and always read "1".

Bit Name	Description	C Symbol
MTIM	MTIM BUSCLK signal supply: 0 – clock disabled 1 – clock enabled	**bCKENMTIM**
TPM3	TPM3 BUSCLK signal supply: 0 – clock disabled 1 – clock enabled	**bCKENTPM3**
TPM2	TPM2 BUSCLK signal supply: 0 – clock disabled 1 – clock enabled	**bCKENTPM2**
TPM1/TPM	TPM1 or TPM BUSCLK signal supply: 0 – clock disabled 1 – clock enabled	**bCKENTPM1**
ADC	ADC BUSCLK signal supply: 0 – clock disabled 1 – clock enabled	**bCKENADC**
IIC2	I²C 2 BUSCLK signal supply: 0 – clock disabled 1 – clock enabled	**bCKENIIC2**
IIC1	I²C 1 BUSCLK signal supply: 0 – clock disabled 1 – clock enabled	**bCKENIIC1**

SCI2	SCI 2 BUSCLK signal supply: 0 – clock disabled 1 – clock enabled	**bCKENSCI2**
SCI1/SCI	SCI 1 or SCI BUSCLK signal supply: 0 – clock disabled 1 – clock enabled	**bCKENSCI1**

8.5.1.2. SCGC2 Register

Name	Model		BIT 7	BIT 6	BIT 5	BIT 4	BIT 3	BIT 2	BIT 1	BIT 0
SCGC2	LL	Read	DBG	FLS	IRQ	KBI	ACMP	TOD	LCD	SPI
		Write								
		Reset	1	1	1	1	1	1	1	1
	QB	Read	1	FLS	IRQ	KBI	ACMP	RTC	1	1
		Write	-						-	-
		Reset	1	1	1	1	1	1	1	1
	QE	Read	DBG	FLS	IRQ	KBI	ACMP	RTC	SPI2*	SPI1
		Write								
		Reset	1	1	1	1	1	1	1	1

* This bit are not present on QE4 and QE8 devices and always read "1".

Bit Name	Description	C Symbol
DBG	DBG module BUSCLK signal supply: 0 – clock disabled 1 – clock enabled	**bCKENDBG**
FLS	Flash controller BUSCLK signal supply: 0 – clock disabled 1 – clock enabled **Program execution is not affected by this bit, but erasing and writing into the flash array is not allowed when FLS=0.**	**bCKENFLS**
IRQ	IRQ BUSCLK signal supply: 0 – clock disabled 1 – clock enabled	**bCKENIRQ**
KBI	KBI BUSCLK signal supply: 0 – clock disabled 1 – clock enabled	**bCKENKBI**
ACMP	ACMP (either ACMP1 and ACMP2) BUSCLK signal supply: 0 – clock disabled 1 – clock enabled	**bCKENACMP**
RTC	RTC BUSCLK signal supply (this bit only affects the BUSCLK signal, the RTC can still operate from ICSERCLK or LPOCLK signals): 0 – clock disabled 1 – clock enabled	**bCKENRTC**
TOD	Time of Day BUSCLK signal supply (this bit only affects the BUSCLK signal, the TOD can still operate from OSCOUT, ICSIRCLK or LPOCLK signals): 0 – clock disabled 1 – clock enabled	**bCKENTOD**
SPI2	SPI 2 BUSCLK signal supply: 0 – clock disabled 1 – clock enabled	**bCKENSPI2**
SPI1/SPI	SPI 1 or SPI BUSCLK signal supply: 0 – clock disabled 1 – clock enabled	**bCKENSPI1**

8.6. Some Words on Low Power Techniques

Designing low power devices implies in spending as little power as possible. This is done by using lower voltage supplies (such as 1.8 or 3V), lower clock rates and caring about every extra current flowing within the circuit.

One way to reduce power consumption is to reduce the voltage supply to the circuit. Taking a brief look into the MC9S08QG8 data sheet (figure 8.3) we can see that running at 8 MHz (BUSCLK) with a 3V supply, this MCU typically draws 3.5 mA (RI_{DD} parameter). By reducing the voltage supply to 2V, the supply current falls down to 2.6mA (that is a power saving of about 26%).

A.6 Supply Current Characteristics

This section includes information about power supply current in various operating modes.

Table A-6. Supply Current Characteristics

Parameter	Symbol	V_{DD} (V)[1]	Typical[2]	Max	T (°C)
Run supply current [3] measured in FBE mode at f_{Bus} = 8 MHz	RI_{DD}	3	3.5 mA	5 mA	85
		2	2.6 mA	—	85
Run supply current [3] measured in FBE mode at f_{Bus} = 1 MHz	RI_{DD}	3	490 µA	1 mA	85
		2	370 µA	—	85
Wait mode supply current [4] measured in FBE at 8 MHz	WI_{DD}	3	1 mA	1.5 mA	85
Stop1 mode supply current	$S1I_{DD}$	3	475 nA	1.2 µA	85
		2	470 nA	—	85
Stop2 mode supply current	$S2I_{DD}$	3	600 nA	2 µA	85
		2	550 nA	—	85
Stop3 mode supply current	$S3I_{DD}$	3	750 nA	6 µA	85
		2	680 nA	—	85
RTI adder to stop1, stop2 or stop3 [4]	—	3	300 nA	—	85
		2	300 nA	—	85
LVD adder to stop3 (LVDE = LVDSE = 1)	—	3	70 µA	—	85
		2	60 µA	—	85
Adder to stop3 for oscillator enabled [5] (EREFSTEN =1)	—	3	5 µA	—	85
		2	4 µA	—	85

[1] 3-V values are 100% tested; 2-V values are characterized but not tested.
[2] Typicals are measured at 25°C.
[3] Does not include any DC loads on port pins.
[4] Most customers are expected to find that auto-wakeup from a stop mode can be used instead of the higher current wait mode.
[5] Values given under the following conditions: low range operation (RANGE = 0), Loss-of-clock disabled (LOCD = 1), low-power oscillator (HGO = 0).

Figure 8.3 – MC9S08QG8 supply current characteristics

Another efficient way to reduce power consumption is to reduce the clock rate. Figure 8.3 shows that an MC9S08QG8 can draw only 490µA when running at 1 MHz clock speed with the same 3V supply voltage. By reducing the voltage supply to 2V we have a current draw of as little as 370µA! That is a power saving of about 90% over our initial assumption (MCU running at 8 MHz with a 3V supply voltage).

Still on the voltage supply topic, it is also important to pay extra attention on the voltage regulator: these little chips are among the villains of power consumption. It does not matter if your microcontroller draws only 10µA when the voltage regulator draws ten or a hundred times that. Prefer low dropout regulators with a very low quiescent current (there are some regulators on the market with quiescent currents as low as 5µA!). If your application is running from higher voltage supplies (let us say, above 12V), prefer using switching regulators, as they are more efficient and spend less power than the linear ones.

Some other procedures to reduce power consumption are:

1. Turn off all unnecessary peripherals.

2. Turn off all unnecessary external loads (such as leds and everything that is not necessary).

3. Whenever possible, configure the output pins to operate in low power mode (drive strength control disabled).

4. Certify that all inputs are stable and with definite logic levels (remember that an analog voltage on a digital input can make the pin draw more power). If possible, switch unused inputs to output mode and place them on an acceptable logic level (of course this also depends on the external circuitry connected to the pin).

5. Enter stop mode (stop1 whenever possible and available).

9

Timing Peripherals

The timing peripherals are among the most important single hardwares of any microcontroller. They can be used to perform many tasks such as: periodic interval interrupts, precise timing delays, signal generation (including, but not limited to PWM), external signal measurements (period and active/inactive cycles), internal (within the application) timing measurements and so on.

The HCS08 family has two basic timer modules: one dedicated to periodic interruption generation (the RTI or RTC) and another for general-purpose usage (TPM and/or MTIM).

The RTI and the RTC are similar in functionality and their main function is to generate periodical interrupts for application synchronization or interval timing. When using an external quartz reference, they can also be used as a real-time clock system to keep precise track of the current time. The RTI module is a simpler and more limited version of the RTC. We will learn about them on the next topics.

The timer/PWM module (TPM) comprises a 16-bit counter, a 16-bit modulo register and up to eight independent channels dedicated to measuring time, generating periodic interrupts or generating PWM.

The MTIM module is a simple 8-bit timer used for general-purpose timing. It comprises an 8-bit counter and an 8-bit modulo register. The MTIM module is not capable of generating PWM nor measuring external signals (at least not directly by the hardware).

 The MTIM and TPM modules do not operate while in stop mode. Only the RTI and RTC modules are capable of operating in stop mode!

Table 9.1 shows a summary of the timing peripherals found on most HCS08 devices.

Device Family	RTI	RTC	TPM	MTIM
AC/AW	1	-	2	-
DN/DV/DZ	-	1	2	-
EL	-	1	2	-
EN	-	1	1	-
GB/GT	1	-	2	-
JM	-	1	2	-
LC	1	-	2	-
LL	*	*	2	-
QA	1	-	1	1
QB	-	1	1	1
QD	1	-	2	-
QE	-	1	3	-
QG	1	-	1	1
Rx	1	-	1	-
SG	-	1	1	1
SH	-	1	2	1
SL	-	1	2	-

* LL devices include a time of day (TOD) module instead of an RTI or RTC

Table 9.1 – Timing peripherals summary

9.1. RTI

The real-time interrupt (RTI) module is a very simple timer dedicated to periodic interrupt generation. It can be used as a time base generator and can also periodically wake up the CPU from a low power mode. This module is found on Ax, Gx, LC, QA, QD, QG and Rx devices.

The RTI module block diagram is shown in figure 9.1. The heart of the module is a binary counter with eight different dividing outputs: 8, 32, 64, 128, 256, 512 and 1024. One of these outputs can be selected to generate an interrupt signal to the CPU. This selection is actually done through SRTISC:RTIS bits.

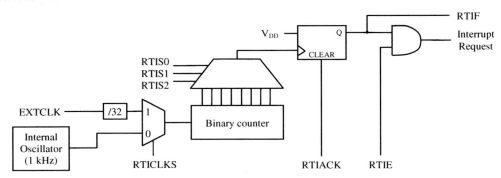

Figure 9.1 – RTI module block diagram

It is also possible to select between two different clock sources for the binary counter: an internal 1 kHz oscillator (shared with the COP module) and an external clock input (from the external reference on ICS or ICG modules).

Table 9.2 shows the possible timeouts for all RTI operating modes (t_{EXT} refers to the external clock source period). After a reset, the RTI module is disabled (RTIS = 000).

| RTIS | Clock source | |
	Internal (RTICLKS=0)	External (RTICLKS=1)
000	RTI disabled	RTI disabled
001	8 ms	$t_{EXT} * 256$
010	32 ms	$t_{EXT} * 1024$
011	64 ms	$t_{EXT} * 2048$
100	128 ms	$t_{EXT} * 4096$
101	256 ms	$t_{EXT} * 8192$
110	512 ms	$t_{EXT} * 16384$
111	1024 ms	$t_{EXT} * 32768$

Table 9.2 – RTI timeouts

The external clock option is used for precise timing generation and allows embedding a real-time clock into any application.

Despite its poor precision (+/- 30% around the 1 kHz center frequency), the internal oscillator is a good option when precision timing is not critical (most applications can tolerate such precision).

The RTI module is also an excellent option for periodically waking up the device from a low power mode, as it continues operating in all stop modes. This enables easy implementation of clocks, smart sensors, loggers and many other battery-operated applications.

A typical example can be a smart RF temperature sensor: it stays in low power mode and it is awaken periodically by the RTI module. Once woken up, the sensor is sampled and the current temperature is sent over the RF link. After the transmission, the whole system gets back to the low power mode.

The RTI module can be configured to generate an interrupt every time it overflows. This is controlled by the SRTISC:RTIE bit. When RTIE = 1, an interrupt is generated when the RTI flag (RTIF) is set. When RTIE = 0, no interrupt is generated.

Notice that it is not necessary to enable RTIE for the module to be able to wake up the CPU from a low power stop1 or stop2 mode. When the CPU is operating in wait or stop3 mode, it is necessary to enable RTIE (as well as CCR:I), in order for the RTI interrupt to be able to wake up the device.

The following example shows how to use the RTI to wake up the MCU from a stop3 mode. The application periodically flashes LED1 and enters stop3 mode. The RTI is configured to operate from the internal 1 kHz clock source with a timeout of 512ms.

```
// DEMO9S08QG8 RTI in stop3 mode example
// LED1 cathode is connected to PTB6

#include <hidef.h>          /* for EnableInterrupts macro */
#include "derivative.h"     /* include peripheral declarations */
#include "hcs08.h"          // This is our definition file!

#define LED1        PTBD_PTBD6

// A simple software loop delay
void delay(unsigned int value)
{
  for (;value; value--);
}

// ISR for RTI interrupt (only clears RTIF flag)
void interrupt VectorNumber_Vrti rti_isr(void)
{
  SRTISC_RTIACK = 1;         // acknowledge RTI interrupt (clear RTIF)
}

void main(void)
{
  // configure SOPT1 register, enable STOP instruction and BKGD pin
  SOPT1 = bSTOPE | bBKGDPE;
  SPMSC1 = 0;                // disable LVD
  SPMSC2 = 0;                // stop3 mode selected
  PTBDD = 0xFF;              // PTB pins as outputs
  PTBD = 0xFF;               // Set all PTB outputs
  // enable RTI (internal clock source, 512ms timeout, interrupt enabled)
  SRTISC = bRTIE | RTI_512ms;
  EnableInterrupts;          // enable interrupts
  while (1)
  {
    LED1 = 0;                // LED1 is on
    delay(20000);            // delay
    LED1 = 1;                // LED1 is off
    STOP;                    // enter stop1 mode
  }
}
```

This line can be omitted as the STOP instruction always enables interrupts!

Example 9.1 – RTI example (stop3)

The next example shows how to use the RTI to wake up the CPU from a stop2 mode on the MC9S08QG8 device. The application lights each led for a time equal to eight times the RTI configured timeout (512ms), that is, each led lights for about 4096ms (4 seconds).

Notice that it is necessary to change the linker parameter file (.PRM). This is necessary to include a new segment for non-initialized variables.

Remember that, unless you use the minimal startup code option (refer to topic 3.2), all variables are automatically initialized by the C startup code (the "led_state" and "counter" variables are always initialized to zero), preventing our application to work as expected.

By allocating those variables into a NO_INIT segment, we prevent the startup code from initializing them. The new PRM file should look like the one shown in listing 9.1 below.

```
/* This is a linker parameter file for the mc9s08qg8 */

NAMES END /* CodeWarrior will pass all the needed files to th
you may add your own files too. */

SEGMENTS /* Here all RAM/ROM areas of the device are listed.
    Z_RAM                = READ_WRITE   0x0060 TO 0x00FF;
    RAM                  = READ_WRITE   0x0110 TO 0x025F;
    NO_INIT_RAM          = NO_INIT      0x0100 TO 0x010F;
    ROM                  = READ_ONLY    0xE000 TO 0xFFAD;
    ROM1                 = READ_ONLY    0xFFC0 TO 0xFFCF;
END

PLACEMENT /* Here all predefined and user segments are placed into the SEGMENTS defined above.
*/
    DEFAULT_RAM                          /* non-zero page variables */
                                         INTO  RAM;

    _PRESTART,                           /* startup code */
    STARTUP,                             /* startup data structures */
    ROM_VAR,                             /* constant variables */
    STRINGS,                             /* string literals */
    VIRTUAL_TABLE_SEGMENT,               /* C++ virtual table segment */
    DEFAULT_ROM,
    COPY                                 INTO  ROM;
    _DATA_ZEROPAGE,                      /* zero page variables */
    MY_ZEROPAGE                          INTO  Z_RAM;
    MY_NO_INIT_SEG                       INTO  NO_INIT_RAM;
END

STACKSIZE 0x50

    VECTOR 0 _Startup // Reset vector: this is the default entry point for an application.
```

> This line creates a region called "NO_INIT_RAM" with 16 bytes in range of 0x100 to 0x010F. The NO_INIT qualifier excludes this area from initialization.

> Add these lines!

> This line places a segment named "MY_NO_INIT_SEG" into the "NO_INIT_RAM" area created above.

Listing 9.1

The application code is shown in example 9.2 below.

```
// DEMO9S08QG8 RTI in stop2 mode example
// LED1 cathode is connected to PTB6
// LED2 cathode is connected to PTB7

#include <hidef.h>              /* for EnableInterrupts macro */
#include "derivative.h"         /* include peripheral declarations */
#include "hcs08.h"              // This is our definition file!

#define LED1        PTBD_PTBD6
#define LED2        PTBD_PTBD7
#define TIMEOUT     8

#pragma DATA_SEG MY_NO_INIT_SEG
unsigned char led_state, counter;

#pragma DATA_SEG DEFAULT
volatile char aux, temp;

void main(void)
{
    // configure SOPT1 register, enable STOP instruction and BKGD pin
    SOPT1 = bSTOPE | bBKGDPE;
    SPMSC1 = 0;                 // disable LVD
    SPMSC2 = bPDC | bPPDC;      // stop2 mode selected
    PTBDD = 0xFF;               // PTB pins as outputs
    PTBD = 0xFF;                // Set all PTB outputs
```

> Non-initialized variables!

> Initialized variables!

```
        // enable RTI (internal clock source, 512ms timeout, no interrupt)
        SRTISC = RTI_512ms;
        if (SPMSC2_PPDF)                    // if PPDF is set
        {
          SPMSC2_PPDACK = 1;                // acknowledge PPDF flag
          if (!--counter)                   // decrement counter and if it is zero...
          {
            if (led_state)                  // if led_state is true
            {
              LED1 = 0;  // turn LED1 on
              LED2 = 1;  // turn LED2 off
            } else                          // if led_state is false
            {
              LED1 = 1;  // turn LED1 off
              LED2 = 0;                     // turn LED2 on
            }
            aux++;                          // this is just a dummy variable
            led_state = !led_state;         // invert led_state
            counter = TIMEOUT;              // restart counter
          }
        } else counter = TIMEOUT;           // if PPDF=0, restart counter
        STOP;                               // enter stop1 mode
        // the program does not continue past the STOP instruction above!
      }
```

Example 9.2 – RTI example (stop2)

The MAP file reflects the placement of the new segment and variables associated with it:

```
MODULE:                    -- main.c.o --
- PROCEDURES:
    main                    E092      54      84      1   .text
- VARIABLES:
    led_state               100       1       1       3   MY_NO_INIT_SEG
    counter                 101       1       1       3   MY_NO_INIT_SEG
    OSC_TRIM                FFAF      1       1       0   .abs_section_ffaf
    OSC_FTRIM               FFAE      1       1       0   .abs_section_ffae
    aux                     110       1       1       1   .common
```

Listing 9.2

 As it is, example 9.2 will not work as expected. Instead of each LED being turned on for about 4 seconds, each one will quickly blink each 4 seconds. Can you explain why this happens? What changes must be made in the program so that each LED stays on for 4 seconds as expected?

The next example shows how to use the RTI to wake up the MCU from a stop1 mode. The application periodically flashes LED1 and enters stop1 mode. The RTI is configured to operate from the internal 1 kHz clock source with a timeout of 512ms. Notice that it does not enable the RTI interrupt as it is not necessary for waking up from stop1 mode.

```
// DEMO9S08QG8 RTI in stop1 mode example
// LED1 cathode is connected to PTB6

#include <hidef.h>            /* for EnableInterrupts macro */
#include "derivative.h"       /* include peripheral declarations */
#include "hcs08.h"            // This is our definition file!

#define LED1    PTBD_PTBD6

// A simple software loop delay
void delay(unsigned int value)
{
  for (;value; value--);
}
```

```
void main(void)
{
  // configure SOPT1 register, enable STOP instruction and BKGD pin
  SOPT1 = bSTOPE | bBKGDPE;
  SPMSC1 = 0;                 // disable LVD
  SPMSC2 = bPDC;              // stop1 mode selected
  PTBDD = 0xFF;               // PTB pins as outputs
  PTBD = 0xFF;                // Set all PTB outputs (turn leds off)
  SRTISC = RTI_512ms;         // enable RTI (internal clock source, 512ms timeout)
  if (SPMSC2_PDF)             // if PDF is set
  {
    SPMSC2_PPDACK = 1;        // acknowledge PDF
  }
  while (1)
  {
    LED1 = 0;                 // LED1 is on
    delay(20000);             // delay
    LED1 = 1;                 // LED1 is off
    STOP;                     // enter stop1 mode
    // the program does not continue past the STOP instruction above!
  }
}
```

These lines can be omitted if the reset origin is not important!

Example 9.3 – RTI example (stop1)

Before continuing, let us see what needs to be changed in example 9.2 so that it works as expected.

```
...
// enable RTI (internal clock source, 512ms timeout, no interrupt)
SRTISC = RTI_512ms;
if (SPMSC2_PPDF)                 // if PPDF is set
{
  SPMSC2_PPDACK = 1;             // acknowledge PPDF flag
  if (led_state)
  {
    LED1 = 0;
    LED2 = 1;
  } else
  {
    LED1 = 1;
    LED2 = 0;
  }
  aux++;
  if (!--counter)
  {
    led_state = !led_state;
    counter = TIMEOUT;
  }
} else counter = TIMEOUT;        // if PPDF=0, restart counter
STOP;                            // enter stop1 mode
// the program does not continue past the STOP instruction above!
}
```

led_state testing must be done outside the counter testing!!!
Testing led_state within counter testing causes leds state to be changed only when counter is zero (and it lasts for just 512ms)!

Listing 9.3

9.1.1. RTI Registers

There is only one register to control the RTI module:

♦   System Real-Time Interrupt Status and Control (SRTISC) register.

9.1.1.1. SRTISC Register

Name	Model		BIT 7	BIT 6	BIT 5	BIT 4	BIT 3	BIT 2	BIT 1	BIT 0
SRTISC	AC, AW, GB, GT, LC, QA, QD, QG, Rx	Read	RTIF	0	RTICLKS	RTIE	0	RTIS		
		Write	-	RTIACK			-			
		Reset	0	0	0	0	0	0	0	0

Bit Name	Description	C Symbol
RTIF	Real-time interrupt flag. This read-only bit indicates an overflow of the RTI counter: 0 – no timeout. 1 – RTI has timed out.	bRTIF
RTIACK	Real-time interrupt acknowledge bit. This write-only bit is used to acknowledge RTI interrupt. Writing "1" into it erases the RTIF flag. Writing "0" has no effect.	bRTIACK
RTICLKS	Real-time interrupt clock source selection bit. This bit selects the clock source for the RTI counter: 0 – internal 1 kHz oscillator. 1 – external oscillator.	bRTICLKS
RTIE	Real-time interrupt enable bit. This bit controls whether the RTIF (if set) generates an interrupt request: 0 – RTI interrupt disabled. 1 – RTI interrupt enabled.	bRTIE
RTIS	Real-time interrupt timeout selection bits. These bits select the timeout for the RTI module. Refer to table 9.3 and 9.4 for possible values.	See tables 9.3 and 9.4

RTIS	Timeout	C Symbol
000	RTI disabled	RTI_OFF
001	8 ms	RTI_8ms
010	32 ms	RTI_32ms
011	64 ms	RTI_64ms
100	128 ms	RTI_128ms
101	256 ms	RTI_256ms
110	512 ms	RTI_512ms
111	1024 ms	RTI_1024ms

Table 9.3 – RTI timeout C symbols (internal clock mode)

RTIS	Timeout	C Symbol
000	RTI disabled	RTI_OFF
001	$t_{EXT} * 256$	RTI_DIV256
010	$t_{EXT} * 1024$	RTI_DIV1024
011	$t_{EXT} * 2048$	RTI_DIV2048
100	$t_{EXT} * 4096$	RTI_DIV4096
101	$t_{EXT} * 8192$	RTI_DIV8192
110	$t_{EXT} * 16384$	RTI_DIV16384
111	$t_{EXT} * 32768$	RTI_DIV32768

Table 9.4 – RTI timeout C symbols (external clock mode)

9.1.2. A Real-Time Clock with the RTI Module

The next example shows how to set up the RTI module to build a real-time clock application. The example was designed to run on an MC9S08QG8 device but can be easily adapted to run on other devices.

A 32,768Hz quartz crystal must be installed at Y1 (do not forget to mount the associated components: R8, R9, C23, C24 and the OSC_EN jumper). LED1 and LED2 must be disabled by opening the corresponding USER_EN jumpers.

```
// DEMO9S08QG8 Real-time clock example using the RTI module

#include <hidef.h>          /* for EnableInterrupts macro */
#include "derivative.h"     /* include peripheral declarations */
#include "hcs08.h"          // This is our definition file!

unsigned char seconds, minutes, hours;

// RTI ISR (implements a real-time clock)
void interrupt VectorNumber_Vrti rti_isr(void)
{
  SRTISC_RTIACK = 1;        // acknowledge RTI interrupt (clear RTIF)
  seconds++;                // increment seconds
  if (seconds>59)           // do we have more than 59 seconds ?
  {
    seconds = 0;            // yes, so we force seconds to zero ...
    minutes++;              // and increment minutes
    if (minutes>59)         // do we have more than 59 minutes ?
    {
      minutes = 0;          // yes, so we force minutes to zero ...
      hours++;              // and increment hours
      if (hours>23) hours = 0;
    }
  }
}

void main(void)
{
  // configure SOPT1 register, enable STOP instruction and BKGD pin
  SOPT1 = bSTOPE | bBKGDPE;
  // Configure ICS: external crystal, ICSERCLK enabled in stop mode
  // ICSOUT = 32768 * 512 = 16.7 MHz, BUSCLK = 8.4 MHz
  // IRCLK = disabled, ERCLK = 32768 Hz
  ICSC2 = bEREFS | bERCLKEN | bEREFSTEN;
  // Configure ICS: FEE mode, external reference, RDIV = divide by 1
  ICSC1 = ICS_FLL;          // ICS in FEE mode (FLL engaged external because IREFS = 0)
  while (ICSSC_CLKST != ICSC1_CLKS); // wait for the clock change to take effect
  SPMSC1 = 0;               // disable LVD
  SPMSC2 = 0;               // stop3 mode selected
  PTBDD = 0xFF;             // PTB pins as outputs
  PTBD = 0xFF;              // Set all PTB outputs
  // Configure RTI: interrupt enabled, external clock, divide by 32768
  SRTISC = bRTIE | RTI_DIV32768;
  EnableInterrupts;         // enable interrupts (CCR:I = 0)
  while (1)
  {
    //STOP;
  }
}
```

> Uncomment this line to enable stop3 mode. The RTI module continues operating in stop3 mode provided the external reference clock (ICSERCLK) is enabled (ERCLKEN = 1) and enabled to operate in stop mode (EREFSTEN = 1).

Example 9.4 – Real-time clock using the RTI

 For better accuracy it is important to use a good quartz crystal and to carefully select the load capacitors C1 and C2 and the series resistor R_S (refer to figure 7.9).

9.2. RTC

The RTC is a more flexible version of the RTI module. It is found on some newer HCS08 devices (such as the DN, DV, DZ, EN, EL, JM, QB, QE, SG, SH and SL devices).

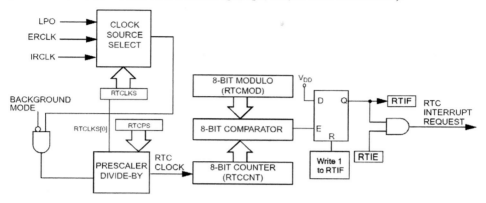

Figure 9.2 – RTC block diagram

The heart of the RTC module is an 8-bit counter (RTCCNT). This counter can operate with one of the three possible clock sources: the low power oscillator (LPO), the external reference clock (ERCLK) or the internal reference clock (IRCLK). On QE devices, the ERCLK is supplied by the external reference oscillator and the IRCLK is supplied by the internal oscillator (typically 31.25 kHz).

The desired clock source is selected through RTCSC:RTCLKS bits, and it is then divided by a prescale factor (the actual dividing factor is selected through RTCSC:RTCPS bits; refer to tables 9.5 and 9.6). When the RTCPS bits are all zeroes, the RTC module is disabled (default condition after a reset).

After divided by the prescaler, the clock signal (RTCCLOCK) feeds the clock input of the RTC counter. The RTC counter counts up to the value stored into the modulo register (RTCMOD). When their values are equal, an overflow signal is triggered (RTIF) on the next clock event and can generate an interrupt if RTIE is set.

This way, it is possible to configure the modulo register (RTCMOD) so that the RTC module generates an interrupt request (RTIF) at a programmable rate. When RTCMOD = 0, RTIF is set on each rising edge of RTCCLOCK, an RTCMOD value greater than zero produces an interrupt rate of:

$$F_{RTIF} = \frac{F_{Source}}{Prescaler * (RTCMOD + 1)}$$

Where F_{RTIF} is the frequency of the RTC interrupt, F_{Source} is the frequency of the selected clock source (according to the RTCSC:RTCLKS bits); prescaler is the dividing factor programmed into RTCSC:RTCPS bits and RTCMOD is the value programmed into the RTCMOD register.

It is possible to use the RTC module to wake up the CPU from a low power mode. For this to be possible, it is necessary that the clock source for the RTC remains active while in low power mode and that the module is properly configured before entering low power mode.

The same rules stated for the RTI module also apply to the RTC module when concerning to waking up from low power modes.

9.2.1. RTC Registers

There are three registers associated with the RTC module:

- ◆ RTC Status and Control (RTCSC) register.
- ◆ RTC Counter (RTCCNT) register.
- ◆ RTC Modulo (RTCMOD) register.

9.2.1.1. RTCSC Register

Name	Model		BIT 7	BIT 6	BIT 5	BIT 4	BIT 3	BIT 2	BIT 1	BIT 0
RTCSC	Dx, Ex, JM, QB, QE, Sx	Read	RTIF	RTCLKS		RTIE	RTCPS			
		Write								
		Reset	0	0	0	0	0	0	0	0

Bit Name	Description	C Symbol
RTIF	Real-time clock interrupt flag. This bit indicates when a counting overflow (0x00 -> RTCMOD -> 0x00) occurred. It can be cleared by writing "1". Writing "0" produces no effect. 0 – no RTC overflow. 1 – RTC counting overflow occurred.	bRTC_RTIF
RTCLKS	Real-time clock source select bits. These bits select the clock source for the RTC module. Changing the clock source automatically clears the RTC count (RTCCNT).	
	00 – RTC clock source is the 1 kHz low power oscillator (LPO).	RTCLKS_LPO
	01 – RTC clock source is the external clock source (ERCLK).	RTCLKS_EXT
	1x – RTC clock source is the internal clock source (IRCLK).	RTCLKS_INT
RTIE	Real-time clock interrupt enable bit. This bit controls whether the RTIF generates an interrupt to the CPU: 0 – RTC interrupt disabled. 1 – RTC interrupt enabled.	bRTC_RTIE
RTCPS	Real-time clock prescaler selection bits. These four bits select the prescaler dividing factor (the selected clock source signal is divided by this factor before being applied to the RTC counter). Changing the prescaler value automatically clears the RTC count (RTCCNT). Detailed information and C symbols are shown in tables 9.5 and 9.6.	See below

RTCPS	Divide Factor	C Symbol		RTCPS	Divide Factor	C Symbol
0000	RTC disabled	RTC_OFF		1000	1	RTC_PRE1
0001	8	RTC_PRE8		1001	2	RTC_PRE2
0010	32	RTC_PRE32		1010	4	RTC_PRE4
0011	64	RTC_PRE64		1011	10	RTC_PRE10
0100	128	RTC_PRE128		1100	16	RTC_PRE16
0101	256	RTC_PRE256		1101	100	RTC_PRE100
0110	512	RTC_PRE512		1110	500	RTC_PRE500
0111	1024	RTC_PRE1024		1111	1000	RTC_PRE1000

Table 9.5 – RTC prescaler settings when RTCLKS = x0 (LPO or IRCLK clock sources)

RTCPS	Divide Factor	C Symbol		RTCPS	Divide Factor	C Symbol
0000	RTC disabled	RTC_OFF		1000	1000	RTC_PRE_1k
0001	1024	RTC_PRE_1024		1001	2000	RTC_PRE_2k
0010	2048	RTC_PRE_2048		1010	5000	RTC_PRE_5k
0011	4096	RTC_PRE_4096		1011	10000	RTC_PRE_10k
0100	8192	RTC_PRE_8192		1100	20000	RTC_PRE_20k
0101	16384	RTC_PRE_16384		1101	50000	RTC_PRE_50k
0110	32768	RTC_PRE_32768		1110	100000	RTC_PRE_100k
0111	65536	RTC_PRE_65536		1111	200000	RTC_PRE_200k

Table 9.6 – RTC prescaler settings when RTCLKS = x1 (ERCLK or IRCLK clock sources)

9.2.1.2. RTCCNT Register

Name	Model		BIT 7	BIT 6	BIT 5	BIT 4	BIT 3	BIT 2	BIT 1	BIT 0
RTCCNT	Dx, Ex, QB, QE, Sx	Read				RTCCNT				
		Write				-				
		Reset	0	0	0	0	0	0	0	0

The RTCCNT register stores the current counting value of the RTC counter. This register can only be read. Writes have no effect. Any writes to RTCMOD or changes in RTCLKS or RTCPS automatically clear the RTCCNT value.

9.2.1.3. RTCMOD Register

Name	Model		BIT 7	BIT 6	BIT 5	BIT 4	BIT 3	BIT 2	BIT 1	BIT 0
RTCMOD	Dx, Ex, QB, QE, Sx	Read				RTCMOD				
		Write								
		Reset	0	0	0	0	0	0	0	0

This is the modulo counting for the RTC counter. The counter counts from zero up to the value stored into RTCMOD, returning to zero on the next rising edge.

9.2.2. Clock Example with the RTC

The next example shows how to implement a simple real-time clock using the RTC module on QE devices.

This example uses a 32,768 Hz quartz crystal connected on Y1 (DEMO9S08QE128 board; jumpers J17 must be placed and shorted). Do not forget to mount the associated components: Rf, Rs, C8 and C9.

The 1 Hz time base is obtained from the external reference (ERCLK) of the ICS module (the 32768 Hz signal), divided by 1024 (RTC prescaler) and then by 32 (RTC modulo counting value). To achieve a dividing factor of 32, we programmed RTCMOD with 31.

The remaining code is very similar to the one shown for the RTI module.

```
// DEMO9S08QE128 Real-time clock example with RTC

#include <hidef.h>          /* for EnableInterrupts macro */
#include "derivative.h"     /* include peripheral declarations */
#include "hcs08.h"          // This is our definition file!

unsigned char seconds, minutes, hours;

// RTC ISR (implements a real-time clock)
void interrupt VectorNumber_Vrtc rtc_isr(void)
{
  RTCSC_RTIF = 1;            // clear RTIF
  seconds++;                 // increment seconds
  if (seconds>59)           // do we have more than 59 seconds ?
  {
    seconds = 0;             // yes, so we force seconds to zero ...
    minutes++;               // and increment minutes
    if (minutes>59)          // do we have more than 59 minutes ?
    {
      minutes = 0;           // yes, so we force minutes to zero ...
      hours++;               // and increment hours
      if (hours>23) hours = 0;
    }
  }
}
```

```
void main(void)
{
  // configure SOPT1 register, enable STOP instruction and BKGD pin
  SOPT1 = bSTOPE | bBKGDPE;
  // Configure ICS: external crystal, ICSERCLK enabled in stop mode
  // ICSOUT = 32768 * 512 = 16.7 MHz, BUSCLK = 8.4 MHz
  // IRCLK = disabled, ERCLK = 32768 Hz
  ICSC2 = bEREFS | bERCLKEN | bEREFSTEN;
  // Configure ICS: FEE mode, external reference, RDIV = divide by 1
  ICSC1 = ICS_FLL;        // ICS in FEE mode (FLL engaged external because IREFS = 0)
  while (ICSSC_CLKST != ICSC1_CLKS); // wait for the clock change to take effect
  SPMSC1 = 0;             // disable LVD
  SPMSC2 = 0;             // stop3 mode selected
  PTBDD = 0xFF;           // PTB pins as outputs
  PTBD = 0xFF;            // Set all PTB outputs
  // Configure RTC: ERCLK clock source, prescaler dividing by 1024
  RTCMOD = 31;            // counting module = 31+1 = 32
  // RTIF interrupt rate = 32768 / 1024 / 32 = 1 int/sec
  RTCSC = bRTC_RTIE | RTCLKS_EXT | RTC_PRE_1024;
  EnableInterrupts;
  while (1)
  {                    ┌─────────────────────────────────────────────────────────────┐
    //STOP; ◄──────────│ Uncomment this line to enable stop3 mode. The RTI module continues operating in │
  }                    │ stop3 mode provided the external reference clock (ICSERCLK) is enabled │
}                      │ (ERCLKEN = 1) and enabled to operate in stop mode (EREFSTEN = 1). │
                       └─────────────────────────────────────────────────────────────┘
```

Example 9.5 – Real-time clock using the RTC

9.3. MTIM

The modulo timer (MTIM) is a simple 8-bit timer very similar to the RTC module (and to the TIM module of the HC908 devices). The MTIM is found only on QA, QB, QG, SG and SH devices and its block diagram can be seen on figure 9.3.

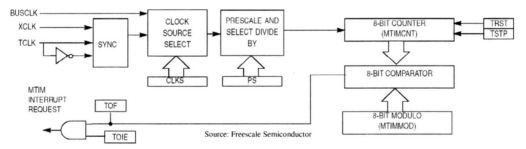

Figure 9.3 – MTIM block diagram

The heart of the module is an 8-bit counter (MTIMCNT). This counter can operate with one of the three possible clock sources: the bus clock signal (BUSCLK), a fixed frequency clock (XCLK) or an external clock (TCLK). The external clock source edge can also be selected: rising or falling edge.

The XCLK input is driven by the fixed frequency output of the ICS module and is equal to the reference clock divided by the RDIV factor.

The TCLK input is multiplexed with the PTA5 pin and can be used to clock the MTIM with an external frequency (not higher than one-fourth of the BUSCLK signal).

The actual clock source is selected by the MTIMCLK:CLKS bits and is then divided by a prescale value selected by the MTIMCLK:PS bits. There are nine possible dividing factors: 1, 2, 4, 8, 16, 32, 64, 128 and 256.

HCS08 Unleashed

The resulting clock signal then feeds the 8-bit binary counter (MTIMCNT) which counts up until it reaches the modulo value programmed into the MTIMMOD register. When a match occurs, the MTIMCNT counter is reset on the next rising edge of its clock signal and the MTIMSC:TOF flag is set. It also generates an interrupt if MTIMSC:TOIE is set.

The frequency of the TOF interrupt can be calculated as follows:

$$F_{TOF} = \frac{F_{Source}}{Prescaler * (MTIMMOD + 1)}$$

In which: F_{TOF} is the frequency of the RTC interrupt; F_{Source} is the frequency of the selected clock source (according to the MTIMCLK:CLKS bits); prescaler is the dividing factor programmed into MTIMCLK:PS bits and MTIMMOD is the value programmed into the MTIMMOD register.

Note that when MTIMMOD = 0, the timer operates in free-running mode (it counts from 0 to 255 and then restarts from 0, a 256 modulo counting). This is equivalent of making MTIMMOD = 255.

The counting of MTIMCNT can be reset by asserting the write-only MTIMSC:TRST bit. Once it is set, MTIMCNT and the internal prescaler counting are reset. TRST is automatically reset by hardware after the operation.

Additionally, the MTIM counting can be stopped by asserting the MTIMSC:TSTP bit. When TSTP = 1, MTIMCNT does not increment, when TSTP = 0, MTIMCNT increments on each rising edge of its input clock signal.

9.3.1. MTIM External Connections

The MTIM has only one external connection: the TCLK input, which is multiplexed with the PTA5 pin.

To configure this pin as TCLK input, all you have to do is select the external clock source on the MTIMCLK:CLKS bits.

9.3.2. MTIM Registers

There are four registers for controlling the MTIM operation:

♦   Modulo TIMer Status and Control (MTIMSC) register.

♦   Modulo TIMer CLocK (MTIMCLK) configuration register.

♦   Modulo TIMer CouNTing (MTIMCNT) register.

♦   Modulo TIMer MODulo (MTIMMOD) register.

9.3.2.1. MTIMCNT Register

Name	Model		BIT 7	BIT 6	BIT 5	BIT 4	BIT 3	BIT 2	BIT 1	BIT 0
MTIMCNT	QA, QB, QG, SG, SH	Read				Count				
		Write								
		Reset	0	0	0	0	0	0	0	0

The MTIMCNT register stores the current counting of the MTIM counter. This register is read-only and writes have no effect. Clearing MTIMCNT is done by a reset, setting the MTIMSC:TRST bit or when the counting reaches the modulo value (MTIMMOD).

9.3.2.2. MTIMSC Register

Name	Model		BIT 7	BIT 6	BIT 5	BIT 4	BIT 3	BIT 2	BIT 1	BIT 0
MTIMSC	QA, QB, QG, SG, SH	Read	TOF	TOIE	0	TSTP	0	0	0	0
		Write	-		TRST		-	-	-	-
		Reset	0	0	0	1	0	0	0	0

Bit Name	Description	C Symbol
TOF	MTIM overflow flag. This bit indicates the overflow of the MTIM counting. It is set on the next clock edge after MTIMCNT = MTIMMOD. Clearing TOF is done by writing "0" into it.	**bTOF**
TOIE	MTIM overflow interrupt enable bit. This bit controls the interrupt request when TOF is set: 0 – no interrupt is requested when TOF = 1. 1 – an interrupt is requested when TOF = 1.	**bTOIE**
TRST	MTIM counter reset bit. Setting TRST clears the MTIMCNT register (and the current internal prescaler counting). Once set, TRST is automatically cleared by hardware.	**bTRST**
TSTP	MTIM counter stop bit. This bit controls whether the MTIM counting is enabled or disabled. Note that enabling MTIM counting does not interfere with the previous content of MTIMCNT. 0 – MTIM counting stopped. 1 – MTIM counting enabled.	**bTSTP**

9.3.2.3. MTIMCLK Register

Name	Model		BIT 7	BIT 6	BIT 5	BIT 4	BIT 3	BIT 2	BIT 1	BIT 0
MTIMCLK	QA, QB, QG, SG, SH	Read	0	0	CLKS		PS			
		Write								
		Reset	0	0	0	0	0	0	0	0

Bit Name	Description	C Symbol
CLKS	Clock source selection bits. These bits select the clock source for the MTIM:	
	00 – BUSCLK.	**MTIM_BUSCLK**
	01 – Fixed-frequency clock (XCLK).	**MTIM_XCLK**
	10 – External source (TCLK pin), falling edge sensitivity.	**MTIM_EXT_FALLING_EDGE**
	11 – External source (TCLK pin), rising edge sensitivity.	**MTIM_EXT_RISING_EDGE**
PS	Prescaler selection bits:	
	0000 – Clock source divided by 1.	**MTIM_DIV1**
	0001 – Clock source divided by 2.	**MTIM_DIV2**
	0010 – Clock source divided by 4.	**MTIM_DIV4**
	0011 – Clock source divided by 8.	**MTIM_DIV8**
	0100 – Clock source divided by 16.	**MTIM_DIV16**
	0101 – Clock source divided by 32.	**MTIM_DIV32**
	0110 – Clock source divided by 64.	**MTIM_DIV64**
	0111 – Clock source divided by 128.	**MTIM_DIV128**
	1000 ... 1111 – Clock source divided by 256.	**MTIM_DIV256**

9.3.2.4. MTIMMOD Register

Name	Model		BIT 7	BIT 6	BIT 5	BIT 4	BIT 3	BIT 2	BIT 1	BIT 0
MTIMMOD	QA, QB, QG, SG, SH	Read				Modulo				
		Write								
		Reset	0	0	0	0	0	0	0	0

This register stores the modulo value for the MTIM counter. The counter counts up to the value stored into this register and then rolls back to zero. MTIM modulo is equal to MTIMMOD + 1.

9.3.3. Periodic Interrupts with MTIM

The following example shows how to implement a timer-based software delay. The variable "timer1" is decremented by one at a one millisecond rate. The main loop stores a value into this variable and then waits for it to go to zero. Storing 50 into timer1 and waiting for it to go down to zero produces a 50ms delay.

```
// DEMO9S08QG8 MTIM example
// LED1 cathode is connected to PTB6

#include <hidef.h>         /* for EnableInterrupts macro */
#include "derivative.h"    /* include peripheral declarations */
#include "hcs08.h"         // This is our definition file!

#define LED1  PTBD_PTBD6

unsigned int timer1;

// MTIM ISR (one interrupt at every millisecond)
void interrupt VectorNumber_Vmtim mtim_isr(void)
{
  MTIMSC_TOF = 0;          // clear TOF
  if (timer1) timer1--;    // if timer1 is greater than zero, decrement it
  // insert other software timers here...
}

void main(void)
{
  // configure SOPT1 register, enable the BKGD pin
  SOPT1 = bBKGDPE;
  // ICS module is operating at 8MHz after reset (BUSCLK = 4MHz)
  PTBDD = 0xFF;            // PTB pins as outputs
  PTBD = 0xFF;             // Set all PTB outputs
  // Configure MTIM: interrupt enabled, MTIM clock = BUSCLK/32, modulo = 125
  // Ttof = 4MHz/(32*125) = 1000
  MTIMSC = bTOIE;
  MTIMCLK = MTIM_DIV32;
  MTIMMOD = 124;           // MTIM modulo = 124+1 = 125
  EnableInterrupts;        // Enable interrupts
  while (1)
  {
    LED1 = 0;              // turn LED1 on
    timer1 = 100;          // initialize timer1 (100ms)
    while (timer1);        // wait for timer1 timeout
    LED1 = 1;              // turn LED1 off
    timer1 = 200;          // initialize timer1 (200ms)
    while (timer1);        // wait for timer1 timeout
  }
}
```

Example 9.6 – MTIM example

This approach can be used with any timer, all you have to do is insert as many if (timerx) timerx--; lines as the number of timers needed (timerx is the desired timer name). Note that each timer spends a small amount of time to be processed within the ISR. The actual type of the timer variable plays a major rule on the total time spent to process each line. Listings 9.4A and 9.4B show the assembly code generated for an 8-bit timer (**char**) and for a 16-bit timer (**int**).

```
LDA    timerx    ; 3 or 4 cycles        LDHX   timerx    ; 4 or 5 cycles
BEQ    *+6       ; 3 cycles             BEQ    *+12      ; 3 cycles
LDHX   #timerx   ; 3 cycles             LDHX   #timerx   ; 3 cycles
DEC    ,X        ; 4 cycles             TST    1,X       ; 4 cycles
                                        BNE    *+3       ; 3 cycles
                                        DEC    ,X        ; 4 cycles
                                        DEC    1,X       ; 5 cycles
```

Listing 9.4A – 8-bit timer	Listing 9.4B – 16-bit timer

The 8-bit timer executes at 13 (DIR addressing mode) or 14 cycles (EXT addressing mode) when timerx is different from zero and 6 (DIR) or 7 cycles (EXT) when it is zero.

The 16-bit timer executes at 26 (DIR) or 27 cycles (EXT) when timerx is different from zero and 7 (DIR) or 8 cycles (EXT) when it is zero.

Note that placing the timer variable into the direct page area causes a reduction of the execution time in one cycle (due to the use of the direct addressing mode on the first instruction shown in listings 9.4A and 9.4B).

 *Using **long** variables for the timer is possible but not recommended, as it generates a larger and slower code!*

The next example shows how to use finite state machines (FSMs) to control two leds blinking at slightly different frequencies.

Using FSMs is a nice way to control the state of leds, buzzers and displays, especially when multiple states are possible (on, off, blinking fast, blinking slowly, and so on). By using a state dedicated to delay control, we also ensure that these delays do not affect the performance of the application (as no code section loops waiting for a delay to finish).

```
// DEMO9S08QG8 MTIM example (finite state machine)
// LED1 cathode is connected to PTB6
// LED2 cathode is connected to PTB7

#include <hidef.h>          /* for EnableInterrupts macro */
#include "derivative.h"     /* include peripheral declarations */
#include "hcs08.h"          // This is our definition file!

#define LED1       PTBD_PTBD6
#define LED2       PTBD_PTBD7
#define LED1_TON   200       // LED1 on time = 200ms
#define LED1_TOFF  220       // LED1 off time = 220ms
#define LED2_TON   220       // LED2 on time = 220ms
#define LED2_TOFF  220       // LED2 off time = 220ms

unsigned int timer1, timer2;

enum eled_fsm
{
    LED_ON, LED_OFF, LED_DELAY
};

// MTIM ISR (one interrupt at every millisecond)
void interrupt VectorNumber_Vmtim mtim_isr(void)
{
    MTIMSC_TOF = 0;        // clear TOF
    if (timer1) timer1--;  // if timer1 is greater than zero, decrement it
    if (timer2) timer2--;  // if timer2 is greater than zero, decrement it
}
```

The three possible states for the led are:
LED_ON – the led is turned on
LED_OFF – the led is turned off
LED_DELAY – the led is waiting for the delay to finish

```c
void led1_state_machine(void)
{
  static enum eled_fsm led_state, next_state;
  switch (led_state)
  {
    case LED_ON:
      LED1 = 0;                 // turn LED1 on
      timer1 = LED1_TON;        // set timer1 (on delay)
      led_state = LED_DELAY;    // change led state to delay
      next_state = LED_OFF;     // next state is "off"
      break;
    case LED_OFF:
      LED1 = 1;                 // turn LED1 off
      timer1 = LED1_TOFF;       // set timer1 (off delay)
      led_state = LED_DELAY;    // change led state to delay
      next_state = LED_ON;      // next state is "on"
      break;
    case LED_DELAY:             // delay
      // when timer1 is zero, change led state
      if (!timer1) led_state = next_state;
      break;
  }
}

void led2_state_machine(void)
{
  static enum eled_fsm led_state, next_state;
  switch (led_state)
  {
    case LED_ON:
      LED2 = 0;                 // turn LED2 on
      timer2 = LED2_TON;        // set timer2 (on delay)
      led_state = LED_DELAY;    // change led state to delay
      next_state = LED_OFF;     // next state is "off"
      break;
    case LED_OFF:
      LED2 = 1;                 // turn LED2 off
      timer2 = LED2_TOFF;       // set timer2 (off delay)
      led_state = LED_DELAY;    // change led state to delay
      next_state = LED_ON;      // next state is "on"
      break;
    case LED_DELAY:             // delay
      // when timer2 is zero, change led state
      if (!timer2) led_state = next_state;
      break;
  }
}

void mcu_init(void)
{
  // configure SOPT1 register, enable the BKGD pin
  SOPT1 = bBKGDPE;
  PTBDD = 0xFF;               // PTB pins as outputs
  PTBD = 0xFF;               // Set all PTB outputs
  // Configure MTIM: interrupt enabled, MTIM clock = BUSCLK/32, modulo = 125
  // Ttof = 4MHz/(32*125) = 1000
  MTIMSC = bTOIE;
  MTIMCLK = MTIM_DIV32;
  MTIMMOD = 124;             // MTIM modulo = 124+1 = 125
  EnableInterrupts;          // Enable interrupts
}

void main(void)
{
  mcu_init();
  while (1)
  {
    led1_state_machine();
    led2_state_machine();
  }
}
```

Example 9.7 – MTIM example (finite state machine)

9.4. TPM

The timer and PWM module (TPM) is a general purpose timer capable of generating periodic interrupts, delays, measuring external signals period, generating pulses and also generating pulse width modulation (PWM) signals.

Each TPM unit comprises a 16-bit timer (a 16-bit counter and a 16-bit comparator) and up to eight channels, each one capable of performing capture, compare or PWM signal generation.

9.4.1. Timer Structure

The main timer structure comprises a 16-bit up/down counter (TPMCNT), a 16-bit modulo register (TPMMOD) and a digital comparator.

By default (TPMSC:CPWMS = 0), the counter counts up, until it reaches the modulo value. When TPMCNT = TPMMOD, TPMCNT is reset on the next clock and TPMSC:TOF is set. It also generates an interrupt request if TMSC:TOIE = 1. Clearing TOF is done by writing "0" into it.

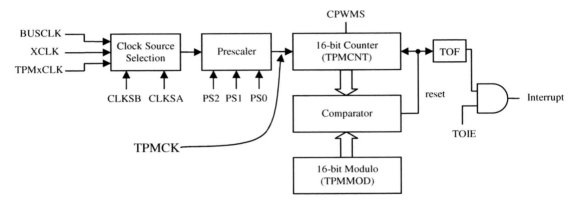

Figure 9.4 – Simplified TPM timer block diagram

The counter can be clocked by one of the three available sources: the BUSCLK signal, the XCLK (external ICS or ICG clock) or the TPMCLK (external TPM clock). The clock source is selected by CLKSB and CLKSA bits (located in the TPMSC register), table 9.7 shows the available options.

CLKSB	CLKSA	Clock Source
0	0	No clock (TPM is stopped)
0	1	TPM clock source is the BUSCLK
1	0	TPM clock source is the XCLK
1	1	TPM clock source is the external TPMCLK pin

Table 9.7

 The external clock input pin option is not present on GB, GT, RC, RD, RE and RG devices!

When an external clock source (XCLK or TPMCLK) is selected, the clock signal passes through a synchronizer circuitry and the maximum clock signal frequency is then limited to one-fourth of the BUSCLK frequency.

The selected clock signal also passes through a prescaler, where it is divided by a selectable factor of 1, 2, 4, 8, 16, 32, 64 or 128. The actual dividing factor is selected by the PS2, PS1 and PS0 bits (all located into TPMSC register). We refer to the clock signal that drives the TPM counter as TPMCK.

The frequency of the TPMCK clock (F_{TPMCK}) is calculated using the following formula:

$$F_{TPMCK} = \frac{F_{Source}}{Prescaler}$$

In which: F_{TPMCK} is the clock signal in the TPMCNT.

F_{Source} is the clock signal from the selected source.

Prescaler is the dividing factor programmed on the TPMSC:PS bits.

We can easily calculate the TOF interrupt rate using the following formula:

$$F_{TOF} = \frac{F_{Source}}{Prescaler * (TPMMOD + 1)}$$

In which: F_{TOF} is the frequency of TPM overflow (in Hz).

F_{Source} is the clock signal from the selected source.

Prescaler is the dividing factor programmed on the TPMSC:PS bits.

TPMMOD is the modulo value programmed in the TPMMOD register.

The current TPMCNT value can be read through TPMCNTL (lower byte) and TPMCNTH (higher byte) registers. Writing in any of these registers clears TPMCNT. To avoid data corruption, there is a "coherency mechanism" that will be discussed later in this chapter.

```
char lower_TPMCNT, higher_TPMCNT;
...
lower_TPMCNT = TPMCNTL;
higher_TPMCNT = TPMCNTH;
```

Thanks to the special placement of the TPMCNTL and TPMCNTH registers, it is possible to read/write into them using 16-bit data movement instructions (such as LDHX and STHX):

```
unsigned int my_TPMCNT;
...
my_TPMCNT = TPMCNT;
```

More efficient because it uses LDHX and STHX instructions!

The modulo value (TPMMOD) can be modified by writing into the TPMMODL (lower byte) and TPMMODH (higher byte) registers. It is also possible to read/write into the TPMMOD register (a 16-bit operation).

```
TPMMODH = 0x12;
TPMMODL = 0x34;
```

or:
```
TPMMOD = 0x1234;
```

More efficient!

When TPMSC:CPWMS = 1, the TPM counter operates as an up/down counter, starting from zero, up to the modulo value and then reverting count direction down to zero, when the process restarts.

This mode (up/down) is especially useful for center-aligned PWM generation, as we will see later.

Table 9.8 shows the available TPM units and capture/compare/PWM channels on most HCS08 devices.

Device	TPM Id	Channels		Device	TPM Id	Channels
AW	TPM1	6		LL	TPM1	2
AW	TPM2	2		LL	TPM2	2
AC	TPM1	4		QA	TPM	1
AC	TPM2	2		QB	TPM	1
AC	TPM3	2		QD	TPM1	2
DN, DV, DZ	TPM1	6		QD	TPM2	1
DN, DV, DZ	TPM2	2		QE4, QE8	TPM1	3
EL	TPM1	4		QE4, QE8	TPM2	3
EL	TPM2	2		QE16/32/64/96/128	TPM1	3
EN	TPM1	4		QE16/32/64/96/128	TPM2	3
GB(A)	TPM1	3		QE16/32/64/96/128	TPM3	6
GT(A)	TPM2	5		QG	TPM	2
JM	TPM1	6		RC, RD, RE, RG	TPM1	2
JM	TPM2	2		SG, SH, SL	TPM1	2
LC	TPM1	2		SG, SH, SL	TPM2	2
LC	TPM2	2				

Table 9.8 – TPM channels availability

9.4.1.1. External Clock Input

When TPMSC:CLKSB and TPMSC:CLKSA are both set, the TPM counter is driven by a clock signal input in the TPMCLK pin (the pin is automatically configured as an input, regardless of the state of the respective PTxDD register). Table 9.9 shows the TPM input pins for some of the HCS08 devices.

Device	TPM Clock Input	Multiplexed Pin		Device	TPM Clock Input	Multiplexed Pin
AW	TPM1CLK	PTD6		LC	TPM1CLK	PTC7
AW	TPM2CLK	PTD4		LC	TPM2CLK	PTC7
AC	TPM1CLK	IRQ/TPMCLK		LL	TCLK	PTC7
AC	TPM2CLK	IRQ/TPMCLK		QA	TCLK	PTA5
AC	TPM3CLK	IRQ/TPMCLK		QD	TPM1CLK	PTA2
DN, DV, DZ	TPM1CLK	PTF2		QD	TPM2CLK	PTA3
DN, DV, DZ	TPM2CLK	PTF3		QE4, QE8	TPM1CLK	PTA5
EL	TPM1CLK	PTA0		QE4, QE8	TPM2CLK	
EL	TPM2CLK	PTA0		QE16/32/64 QE96/128	TPM1CLK	PTA5
EN	TPM1CLK	PTF2		QE16/32/64 QE96/128	TPM2CLK	PTE0
				QE16/32/64 QE96/128	TPM3CLK	PTE7
GB(A), GT(A)	TPM1CLK	-		QG	TPMCLK	PTA5
GB(A), GT(A)	TPM2CLK	-		RC,RD, RE,RG	TPM1CLK	-
JM	TPM1CLK	IRQ/TPMCLK		SG, SH, SL	TPM1CLK	PTA5
JM	TPM2CLK	IRQ/TPMCLK		SG, SH, SL	TPM2CLK	PTA5

Table 9.9 – TPM external clock input pin

9.4.2. Capture/Compare/PWM Channels

Each TPM module can have up to eight independent channels for capture, compare or PWM generation.

Each channel comprises a 16-bit register (TPMCxV) that stores the current counting of TPMCNT when triggered by an external event (in capture mode) or generates a compare event when the value of the register is equal to the current TPMCNT (in compare and PWM modes).

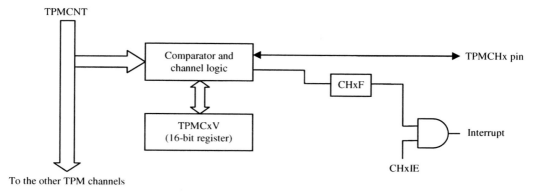

Figure 9.5 – Simplified TPM channel block diagram

The actual channel operating mode selection is controlled by the TPMSC:CPWMS bit and four bits of the TPMCxSC register (MSxB, MSxA, ELSxB and ELSxA), according to table 9.10.

CPWMS	MSxB	MSxA	ELSxB	ELSxA	Channel Mode
x	x	x	0	0	Channel disabled, associated pin (TPMCHx pin) in I/O mode
0	0	0	0	1	Capture mode (capture on rising edge of the TPMCHx pin)
0	0	0	1	0	Capture mode (capture on falling edge of the TPMCHx pin)
0	0	0	1	1	Capture mode (capture on both TPMCHx pin edges)
0	0	1	0	0	Compare mode (interrupt mode, only sets CHxF)
0	0	1	0	1	Compare mode (associated TPMCHx pin is toggled, CHxF is set)
0	0	1	1	0	Compare mode (associated TPMCHx pin is cleared, CHxF is set)
0	0	1	1	1	Compare mode (associated TPMCHx pin is set, CHxF is set)
0	1	x	1	0	Edge-aligned PWM mode (active high PWM)
0	1	x	x	1	Edge-aligned PWM mode (active low PWM)
1	1	x	1	0	Center-aligned PWM mode (active high PWM)
1	1	x	x	1	Center-aligned PWM mode (active low PWM)

Table 9.10 – TPM channel operating modes

9.4.2.1. Capture Mode

The capture mode was designed to allow precise measurements of the period of external signals. The signal is applied into the channel input pin (TPMCHx input) and triggers a capture event on a rising edge, falling edge or on both edges of the input signal (according to the channel configuration).

On each capture event, the current TPMCNT value is copied to the channel value register (TPMCxV) and the channel interrupt flag (TPMCxSC:CHxF) is set. This flag can generate an interrupt request if the channel interrupt is enabled (TPMCxSC:CHxIE = 1).

 The associated input (TPMCHx input) is automatically configured as an input (regardless of the respective PTxDD configuration) when the channel is operating in capture mode.

The difference between two consecutive captures is the period of the external signal (except when both edges are captured).

Once the TPM clock rate is known, the following formula can be used to calculate the period or the frequency of the input signal:

$$T_{CAPTURE} = \frac{CAPTURED\_VALUE}{F_{TPMCK}}$$

In which: $T_{CAPTURE}$ is the period of the input signal.

CAPTURED_VALUE is the value captured into the TPMCxV register.

F_{TPMCK} is the clock signal input into the TPM counter.

Figure 9.6 shows an example of the operation principles of the capture mode (channel set to capture every rising edge of the TPMCHx input).

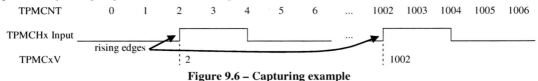

Figure 9.6 – Capturing example

Supposing a BUSCLK = 8MHz and TPMCK = 1MHz (prescaler dividing by 8), we know that each TPMCNT increment equals to 1µs. Calculating the signal period for figure 9.6 is very straightforward: all we need to do is subtract the last capture (1002 on the figure) from the previous capture (2 on the figure). The result (1000) is the period of the input signal. 1000 * 1µs = 1000µs or 1ms!

The next example shows how to configure a TPM channel to capture the total time a key is pressed. This example was designed for the DEMO9S08LC60 board, as it is the only board with a key connected to a TPM input. By adding an external key and making some minor changes in the software, it is possible to run this example on other boards.

The operation principle is very simple: channel 1 of TPM1 is connected to the PTC3 pin that is connected to the SW302 key. By configuring channel 1 to capture falling edges, we can effectively capture the TPM1 counting when the key is pressed. The ISR also changes the capture edge so the next capture is triggered on the rising edge of PTC3 (the exact instant SW302 is released).

Calculating the total time SW302 is pressed is very simple: just subtract one captured value from the first one and we have the total TPM1 counts corresponding to the key press time. This is stored in the "key_time" variable.

Knowing that TPM1 is running from the BUSCLK (which is equal to 5MHz) divided by 128, we get a TPMCLK of 39,062.5 Hz (each TPM1 count is equal to 25.6µs).

```
// DEMO9S08LC60 - TPM capture example
// SW302 is connected to PTC3

#include <hidef.h>          /* for EnableInterrupts macro */
#include "derivative.h"     /* include peripheral declarations */
#include "hcs08.h"          // This is our definition file!

unsigned int key_time;

void interrupt VectorNumber_Vtpm1ch1 tpm1ch1_isr(void)
{
  static unsigned int last_capture;
  static char mode;
  TPM1C1SC_CH1F = 0;        // clear interrupt flag

  if (!mode)
  {
    // if mode=0, we have a falling edge capture (the key was just pressed)
    last_capture = TPM1C1V; // store the current capture
    // change channel mode to capture the key release (rising edge)
    TPM1C1SC = bCHIE | TPM_CAPTURE_RISING_EDGE;
    mode++;                 // mode = 1;
  }
```

```
        else
        {
            // if mode<>0, we have a rising edge capture (the key was just released)
            key_time = TPM1C1V - last_capture; // calculate key press time
            // change channel mode to capture the next key press (falling edge)
            TPM1C1SC = bCHIE | TPM_CAPTURE_FALLING_EDGE;
            mode = 0;                          // mode = 0
        }
}

void init(void)
{
    SOPT1 = bBKGDPE;    // configure SOPT1 register, enable pin BKGD for BDM
    // ICG configuration (ICGOUT = 10MHz, BUSCLK = 5MHz):
    ICGTRM = NVICGTRM;                 // trim IRG oscillator
    ICGC2 = MFDx18 | RFD_DIV4;         // MFD = 111b (N = 18), RFD = 010b (R = 4)
    ICGC1 = ICG_FEI;                   // ICG in FEI mode (CLKS = 01b)
    PTCDD_PTCDD3 = 0;                  // PTC3 as an input
    TPM1SC = TPM_BUSCLK | TPM_DIV128;  // TPM clock = BUSCLK / 128 = 39,062.5Hz
    // Configure channel 1 to capture the next key press (falling edge)
    TPM1C1SC = bCHIE | TPM_CAPTURE_FALLING_EDGE;
    EnableInterrupts;
}

void main(void)
{
    init();
    while (1);
}
```

Example 9.8 – Capture example

On devices with an internal analog comparator, it is possible to use it as the source for capture on channel 0 or channel 1. This option is controlled by the bits ACIC (on devices with only one comparator) or ACIC1 and ACIC2 (on devices with two comparators). These bits are located on the SOPT2 register.

Once the ACIC bit is set and the channel is configured for capture mode, each programmed edge of the comparator triggers a capture event. This is a welcome feature for timing measurements on analog signals.

Table 9.11 shows the devices and TPM channels available for this mode.

Comparator	Device						
	EL, SL	**JM**	**LC**	**LL**	**QE**	**QA, QB, QG**	**SG, SH**
ACMP	TPM1CH0	TPM1CH0	TPM1CH0	TPM2CH0	-	TPMCH0	TPM1CH0
ACMP1	-	-	-		TPM1CH0	-	-
ACMP2	-	-	-		TPM2CH0	-	-

Table 9.11 – ACMP-TPM connections

Refer to topic 5.5.2 for further information on the SOPT2 register and ACIC configuration.

9.4.2.2. Compare Mode

The compare mode can be thought as the opposite of the capture mode. When a TPM channel is operating in this mode, a compare event is generated by the channel hardware every time the TPMCNT value is equal to the channel register value (TPMCxV). The compare event sets the channel interrupt flag (TPMCxSC:CHxF) and optionally changes the state of the output pin associated with the channel. This way it is possible to set, reset or toggle the channel output on each compare event (enabling signal generation under software control).

A typical use of the compare mode is on the generation of waveforms and/or periodical interrupt generation. The compare interval can be easily calculated:

$$T_{COMPARE} = \frac{COMPARE\_VALUE}{F_{TPMCK}}$$

In which: $T_{COMPARE}$ is the period of a half cycle of the output signal.

COMPARE_VALUE is the value stored into the TPMCxV register.

F_{TPMCK} is the clock of the TPM.

Note that for the formula above to be valid, it is necessary that the desired interval value be added to TPMCxV on each compare event. Failing do to so results in a compare event at each 65536 TPMCNT counts!

 The associated channel pin (TPMCHx) is automatically configured as an output (regardless of the respective PTxDD configuration) when the channel is operating in compare toggle, compare set or compare clear mode.

Figure 9.9 shows a waveform generated using the algorithm stated above. The following example demonstrates it:

```
// DEMO9S08AW60 - TPM compare example (toggle output)
// LED connected to PTF0

#include <hidef.h>          /* for EnableInterrupts macro */
#include "derivative.h"     /* include peripheral declarations */
#include "hcs08.h"          // This is our definition file!

#define NVICGTRM (*(const char * __far)0x0000FFBE)

void interrupt VectorNumber_Vtpm1ch2 tpm1ch2_isr(void)
{
  TPM1C2SC_CH2F = 0;        // clear interrupt flag
  TPM1C2V += 1000;          // adds 1000 to TPM1C2V (sets next compare)
}

void main(void)
{
  SOPT = bBKGDPE;           // configure SOPT1 register, enable pin BKGD for BDM
  // ICG configuration (ICGOUT = 40MHz, BUSCLK = 20MHz):
  ICGTRM = NVICGTRM;        // trim IRG oscillator
  ICGC2 = MFDx18;           // MFD = 111b (N = 18), RFD = 000b (R = 1)
  ICGC1 = ICG_FEI;          // ICG in FEI mode (CLKS = 01b)
  TPM1SC = TPM_BUSCLK | TPM_DIV128;  // TPM clock = BUSCLK / 128 = 156250Hz
  // Configure channel 2 to compare mode (toggle output)
  TPM1C2SC = bCHIE | TPM_COMPARE_TOGGLE;
  EnableInterrupts;         // enable interrupts
  while (1);
}
```

This line may be necessary if your version of Codewarrior does not include the NVICGTRM definition for the AW devices. Remove it if you have any compiling error.

Example 9.9 – TPM compare mode example

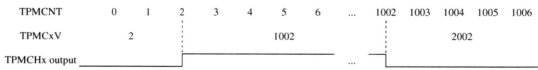

TPMCNT	0	1	2	3	4	5	6	...	1002	1003	1004	1005	1006
TPMCxV		2				1002					2002		

TPMCHx output

Figure 9.7 – Compare mode example

Running the example on the actual hardware (DEMO9S08AW60 board) results in the LED's being steadily lit. The output frequency generated by the TPM can be calculated with the following formula:

$$F_{COMPARE} = \frac{F_{TPMCK}}{2*1000} = 78.125\text{Hz}$$

That is too high for our eyes to notice it is blinking (we can only notice frequencies below ~48Hz). To "help" our eyes, we can use a 10000 counts interval, which produces an output frequency of 7.8Hz!

The next example shows how to use the capture mode to measure the time a key is pressed. The application also uses another channel operating in compare mode to blink a led (the duty cycle is controlled by the captured value). The more time the key is held pressed, the lower the blink rate.

```c
// DEMO9S08LC60 - TPM capture/compare example
// SW302 is connected to PTC3

#include <hidef.h>          /* for EnableInterrupts macro */
#include "derivative.h"     /* include peripheral declarations */
#include "hcs08.h"          // This is our definition file!

#define LED PTBD_PTBD7
unsigned int key_time;

// TPM1 channel 1 isr (process the channel captures)
void interrupt VectorNumber_Vtpm1ch1 tpm1ch1_isr(void)
{
  static unsigned int last_capture;
  static char mode;
  TPM1C1SC_CH1F = 0;          // clear interrupt flag
  if (!mode)
  { // if mode=0 then we have a falling edge capture (the key was just pressed)
    last_capture = TPM1C1V; // store the current capture
    // change channel mode to capture the key release (rising edge)
    TPM1C1SC = bCHIE | TPM_CAPTURE_RISING_EDGE;
    mode++;                   // mode = 1;
  } else
  { // if mode<>0 then we have a rising edge capture (the key was just released)
    key_time = TPM1C1V - last_capture;  // calculate key press time
    // change channel mode to capture the next key press (falling edge)
    TPM1C1SC = bCHIE | TPM_CAPTURE_FALLING_EDGE;
    mode = 0;
  }
}

void interrupt VectorNumber_Vtpm1ch0 tpm1ch0_isr(void)
{
  TPM1C0SC_CH0F = 0;        // clear interrupt flag
  TPM1C0V += key_time;      // update the new cycle
  LED = !LED;              // toggle led state
}

void init(void)
{
  SOPT1 = bBKGDPE;            // configure SOPT1 register, enable pin BKGD for BDM
  // ICG configuration (ICGOUT = 10MHz, BUSCLK = 5MHz):
  ICGTRM = NVICGTRM;         // trim IRG oscillator
  ICGC2 = MFDx18 | RFD_DIV4; // MFD = 111b (N = 18), RFD = 010b (R = 4)
  ICGC1 = ICG_FEI;           // ICG in FEI mode (CLKS = 01b)
  PTCDD_PTCDD3 = 0;          // PTC3 as an input
  PTBDD_PTBDD7 = 1;          // PTB7 as an output
  TPM1SC = TPM_BUSCLK | TPM_DIV128;  // TPM clock = BUSCLK / 128 = 39,062.5Hz
  // Configure channel 0 to compare mode (interrupt only)
  TPM1C0SC = bCHIE | TPM_COMPARE_INT;
  // Configure channel 1 to capture the next key press (falling edge)
  TPM1C1SC = bCHIE | TPM_CAPTURE_FALLING_EDGE;
  EnableInterrupts;          // enable interrupts
}

void main(void)
{
  init();
  while (1);
}
```

Example 9.10 – Capture/compare example

Notice that we are using only the compare interrupt of channel 0. That is because the led is not connected to the TPM1CH0 pin (PTC2) on the DEMO9S08LC60 board. In case a led is externally connected to PTC2, the example can be altered to toggle the pin by hardware. Change the line:

```
// Configure channel 0 to compare mode (interrupt only)
TPM1C0SC = bCHIE | TPM_COMPARE_INT;
```

To:

```
// Configure channel 0 to toggle TPM1CH0 pin on compare (interrupt also enabled)
TPM1C0SC = bCHIE | TPM_COMPARE_TOGGLE;
```

And suppress the line:

```
LED = !LED;            // toggle led state
```

9.4.2.3. PWM Mode

PWM (Pulse Width Modulation) is a special signal largely used in power control. It typically comprises a digital waveform with a fixed period and a variable duty cycle (variable frequency PWM is also commonly used but it is not covered in this book).

The basic idea behind PWM is that maximum voltage is applied in the load, but this only happens for a fraction of time (proportional to the duty cycle). The average voltage on the load over the time is directly proportional to the duty cycle of the waveform.

$$V_{OUT} = D*V_{MAX}$$

In which: V_{OUT} it is the average output voltage.

D it is the duty cycle percentage (0 to 100%).

V_{MAX} it is the maximum voltage applied to the load.

On the HCS08 devices, the PWM signal is generated by using the TPM channels (one for each PWM signal).

Figure 9.8 shows the simplified block diagram for a TPM channel operating in PWM mode.

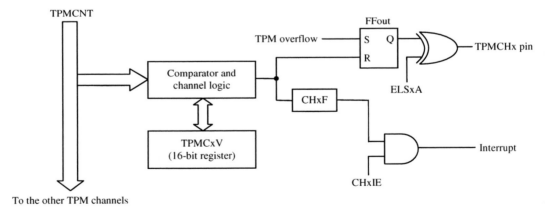

Figure 9.8 – PWM channel simplified block diagram

Edge-aligned PWM

When generating an edge-aligned PWM (TPMSC:CPWMS = 0), the TPM counter starts counting up until it reaches the TPMCxV value, when the flip-flop (FFout) is set (Q=1). This also sets the output pin (TPMCHx) if ELSxA = 0, or clears the output pin if ELSxA = 1.

The TPM counter continues counting until it overflows, when its count is cleared and the flip-flop (FFout) is reset (Q=0). This also clears the output pin (TPMCHx) if ELSxA = 0, or sets the output pin if ELSxA = 1.

The frequency of the PWM signal can be calculated by using the following formula:

$$F_{PWM} = \frac{F_{TPM}}{TPMMOD + 1}$$

In which: F_{PWM} is the frequency of the PWM signal (in Hz).

F_{TPMCK} is the frequency of the TPM increments (TPMCK) in Hertz.

TPMMOD is the modulo value stored into the TPMMOD register.

The pulse width (in seconds) can be calculated using the following formula:

$$T_{PULSE\_WIDTH} = \frac{F_{TPMCK}}{TPMCxV}$$

In which: T_{PULSE_WIDTH} is the pulse width in seconds.

F_{TPMCK} is the frequency of the TPM increments (TPMCK) in Hertz.

TPMCxV is the active cycle value stored into the TPMCxV register.

Figure 9.9 shows a PWM signal with a 50% duty cycle (TPMCxV = TPMMOD/2).

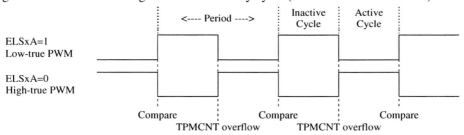

Figure 9.9 – Edge-aligned PWM

To generate the high-true PWM signal shown in figure 9.9, the TPM registers can be configured as follows:

```
TPMSC = TPM_BUSCLK;        // TPM clock = BUSCLK
TPMC0V = 32768;            // Channel value = halt the TPM modulo (65536 by default)
TPMC0SC = TPM_PWM_HIGH;    // channel 0 operating in PWM mode (active high PWM)
```

This generates a 50% duty cycle PWM signal. Considering a BUSCLK frequency of 8MHz, the PWM frequency is equal to:

$$F_{PWM} = \frac{8MHz}{65536} = 122.07Hz$$

The duty cycle is set by the value written into the TPMCxV register: a value of zero sets the duty cycle to 0% and a value equal to or greater than the module (TPMMOD) sets the duty cycle to 100%. The duty cycle is only updated when the TPM count goes from the modulo value (or 0xFFFF) to

0x0000. On TPMv3, this happens when the TPM counter goes from modulo-1 to modulo (or 0xFFFE to 0xFFFF when operating in free-running mode).

By changing the modulo value of the TPM, it is possible to change the frequency of the PWM signal (this is also possible by changing the prescaler of the TPM). Notice, however, that a change into the modulo value also implies in a change in the PWM resolution: when using the PWM on free-running mode (modulo=65536), the cycle value can be anywhere between 0 and 65535 (a 16-bit nominal resolution).

Decreasing the modulo value increases the PWM frequency and reduces the maximum cycle value. For a TPMMOD = 1023 (TPM modulo equal to 1024), the cycle value can be anywhere between 0 and 1023 (a nominal 10-bit resolution).

The highest PWM frequency for an MCU running at BUSCLK = 20 MHz is equal to 305,18Hz (for a 16-bit PWM resolution) or 78.125 kHz (for an 8-bit PWM resolution).

The following example demonstrates the basic operation of the PWM mode on the MC9S08AW60 microcontroller. This first example sets three channels of TPM1 to PWM mode (high-true PWM). The PWM frequency is set through the modulo register to approximately 9.77 kHz (20 MHz / 2048) and the duty cycles are set to 1%, 25% and 75%. The example uses channels 2, 3 and 4 due to their connection to external leds (real-life applications can use any available channel).

```
// DEMO9S08AW60 - TPM PWM example 1
// LEDs connected to PTF0, PTF1 and PTF2

#include <hidef.h>            /* for EnableInterrupts macro */
#include "derivative.h"       /* include peripheral declarations */
#include "hcs08.h"            // This is our definition file!

#define NVICGTRM (*(const char * __far)0x0000FFBE)

void main(void)
{
  SOPT = 0;                   // configure SOPT1 register, disable COP
  // ICG configuration (ICGOUT = 40MHz, BUSCLK = 20MHz):
  ICGTRM = NVICGTRM;          // trim IRG oscillator
  ICGC2 = MFDx18;             // MFD = 111b (N = 18), RFD = 000b (R = 1)
  ICGC1 = ICG_FEI;            // ICG in FEI mode (CLKS = 01b)
  TPM1SC = TPM_BUSCLK;        // TPM clock = BUSCLK = 20MHz
  TPM1MOD = 2047;             // PWM frequency is equal to 9.77 kHz
  // Configure channels 2,3 and 4 to PWM mode (high-true pulses)
  TPM1C2SC = TPM_PWM_HIGH;
  TPM1C3SC = TPM_PWM_HIGH;
  TPM1C4SC = TPM_PWM_HIGH;
  TPM1C2V = 19;               // channel 2 set to 1%
  TPM1C3V = 511;              // channel 3 set to 25%
  TPM1C4V = 1535;             // channel 4 set to 75%
  while (1);
}
```

TPM1 modulo is set to 2048 (2047+1), PWM resolution is 11 bits!

Example 9.11 – PWM example 1

The next example shows a more complex implementation of a dual led flasher. We use the PWM channels to control the brightness of the leds and a compare channel to control the speed of the increment/decrement of the brightness of each led.

```
// DEMO9S08AW60 - TPM PWM example 2
// LEDs connected to PTF0 and PTF1

#include <hidef.h>            /* for EnableInterrupts macro */
#include "derivative.h"       /* include peripheral declarations */
#include "hcs08.h"            // This is our definition file!

#define NVICGTRM (*(const char * __far)0x0000FFBE)

unsigned int new_cycle[2], timer[2];
unsigned char direction[2];
```

```c
void interrupt VectorNumber_Vtpm1ch0 tpm1_channel1_isr(void)
{
  TPM1C0SC_CH0F = 0;              // clear interrupt flag;
  TPM1C0V += 2000;               // next compare in 100us
  TPM1C0V &= 2047;               // adjust TPM1C0V to fit the modulo
  if (timer[0]) timer[0]--; else
  {                              // timer0 timeout
    timer[0] = 15;              // restart timer0
    if (direction[0])          // if direction is 1 (counting down)
    {
      if (new_cycle[0]>12) new_cycle[0] -= 12;  // decrement new_cycle toward 0
        else direction[0] = 0;          // if new_cycle[0] is zero, change direction
    } else     // if direction is 0 (counting up) increment new_cycle toward 2048
    {
      if (new_cycle[0]<2048) new_cycle[0] += 12;
        else direction[0] = 1;         // if new_cycle[0]>=2048, change direction
    }
  }
  if (timer[1]) timer[1]--; else
  {
    // timer1 timeout
    timer[1] = 37;             // restart timer1
    if (direction[1])          // if direction is 1 (counting down)
    {
      if (new_cycle[1]>12) new_cycle[1] -= 12;  // decrement new_cycle toward 0
        else direction[1] = 0;          // if new_cycle[1] is zero, change direction
    } else     // if direction is 0 (counting up) increment new_cycle toward 2048
    {
      if (new_cycle[1]<2048) new_cycle[1] += 12;
        else direction[1] = 1;         // if new_cycle[1]>=2048, change direction
    }
  }
}


void interrupt VectorNumber_Vtpm1ovf tpm1_overflow_isr(void)
{
  TPM1SC_TOF = 0;                // clear interrupt flag
  // update PWM cycles
  TPM1C2V = new_cycle[0];
  TPM1C3V = new_cycle[1];
}
```

> The best moment to update the current duty cycle is on the overflow of the TPM counter!

```c
void MCU_init(void)
{
  SOPT = 0;                      // configure SOPT1 register, disable COP
  // ICG configuration (ICGOUT = 40MHz, BUSCLK = 20MHz):
  ICGTRM = NVICGTRM;             // trim IRG oscillator
  ICGC2 = MFDx18;                // MFD = 111b (N = 18), RFD = 000b (R = 1)
  ICGC1 = ICG_FEI;               // ICG in FEI mode (CLKS = 01b)
  TPM1SC = bTOIE | TPM_BUSCLK;   // TPM clock = BUSCLK = 20MHz
  TPM1MOD = 2047;                // PWM frequency is equal to 9.77kHz
  // Configure channel 0 to compare mode (interrupt only)
  TPM1C0SC = bCHIE | TPM_COMPARE_INT;
  // Configure channels 2 and 3 to PWM mode (high-true pulses)
  TPM1C2SC = TPM_PWM_HIGH;
  TPM1C3SC = TPM_PWM_HIGH;
  TPM1C0V = 1999;                // next compare in 100us ((1999+1)*(1/20MHz))
  TPM1C2V = 0;                   // channel 2 set to 0%
  TPM1C3V = 0;                   // channel 3 set to 0%
  new_cycle[0] = 0;
  new_cycle[1] = 0;
  timer[0] = 15;                 // timer0 timeout is set to 15
  timer[1] = 37;                 // timer1 timeout is set to 37
  direction[0] = 0;
  direction[1] = 0;
  EnableInterrupts;              // enable interrupts
}


void main(void)
{
  MCU_init();
  while (1);
}
```

Example 9.12 – PWM example 2

Center-aligned PWM

Center-aligned PWM is another possible operating mode for the TPM. This mode is characterized by having no aligned transitions when multiple signals (multiple channels) are generated by the TPM. This mode is selected when the CPWMS is set in the TPMSC register (when this bit is set, it is not recommended to operate the channels in compare or capture modes).

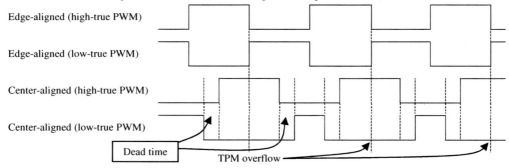

Figure 9.10 – Edge vs. Center-aligned PWM

The center-aligned PWM allows the insertion of dead times between signals, which is a welcome feature when generating complementary PWM signals (usually used when driving half-bridges or full-bridges).

While in this mode, the TPM counter operates as an up/down counter: it counts from 0 up to the modulo value (TPMMOD) and then reverts the count direction, counting towards 0. The channel output is set (or reset) when TPMCNT = TPMCxV and the counter is counting up and reset (or set) when TPMCNT = TPMCxV and the counter is counting down (figure 9.11 shows the resulting waveform).

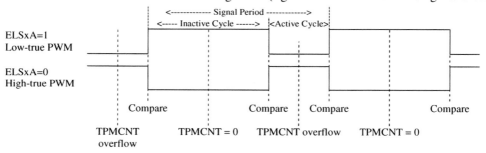

Figure 9.11 – Center-aligned PWM

Using this mode imposes some limits to know:

1. The modulo value must be set between 0x0001 and 0x7FFF (a modulo greater than 0x7FFF does not allow 100% duty cycles).

2. If a 100% duty cycle is necessary, the modulo must not be greater than 0x7FFE.

A 0% duty cycle is generated when the channel value is programmed to 0 (TPMCxV=0) or to a value greater than 0x7FFF (TPMCxV>0x7FFF).

A 100% duty cycle is generated when the channel value is programmed to a value greater than the modulo and lower than 0x8000 (TPMMOD<TPMCxV<0x8000).

9.4.3. Coherency Mechanism

Reading or writing into a 16-bit register cannot be done in a single pass when using an 8-bit CPU such as the HCS08 CPU. This is usually done by two 8-bit load-and-store operations:

```
LDA    #high_byte
STA    TPMMODH
LDA    #low_byte
STA    TPMMODL
```

The same operation can also be performed using LDHX and STHX instructions:

```
LDHX   #16_bit_value
STHX   TPMMOD
```

Even in the above operation, there are four bus transfers: H is loaded with the higher byte, X is loaded with the lower byte, H is stored in the lower address, X is stored in the higher address.

In most cases the extra cycles needed to complete 16-bit transfers are not important and can be disregarded, but when the 16-bit transfer is related to a 16-bit peripheral register, then we must know that the hardware can change the register while the transfer is being performed. Consider the following example:

Now, considering that:

1. TPMCNT is running at a BUSCLK frequency of 8MHz (TPMCNT is incremented at every 125ns).

2. The LDHX TPMCNT instruction runs in four BUSCLK cycles.

3. The instruction performs the two read operations in the two first BUSCLK cycles.

4. TPMCNT = 0x00FF before the instruction is executed.

The first read operation copies the higher byte (0x00) to the H register. This operation takes one BUSCLK cycle (125ns). After that, TPMCNT increments and is now 0x0100. The second read operation takes place: the lower byte of TPMCNT (now = 0x00) is read and stored into the X register. The final result is: H:X = 0x0000 and TPMCNT = 0x0100!

That situation can actually happen on any microcontroller (every time the peripheral register width is larger than the CPU data bus).

Fortunately, Freescale provided the HCS08 devices with a clever logic circuitry to prevent such problems. The so called "coherency mechanism" consists of two 8-bit intermediary latches: when a read operation is requested, the 16-bit value is transferred to the latches in a single operation and can then be read without concerns. The same concept is also valid for writing operations: each 8-bit operation writes in an intermediary 8-bit latch and their content is transferred to the 16-bit target register only when both latches are written.

The following TPM registers implement some kind of coherency mechanism:

TPMCNT – coherency mechanism for reading operations: any read of TPMCNTL or TPMCNTH latches both registers until they are read (TPMCNT continues to be updated). Any writes to TPMCNTH or TPMCNTL clear the TPMCNT counting and resets the coherency mechanism. The mechanism is also reset by writing into the TPMSC register.

TPMMOD – coherency mechanism for writing operations: any writes into TPMMODL or TPMMODH inhibit TOF setting until both registers are written.

TPMCxV – coherency mechanism for reading or writing operations (depending on the channel operating mode): when operating in capture mode, any reads of TPMCxVL or TPMCxVH latch both registers until they are read. When operating in compare or PWM mode, writes into TPMCxVL or

TPMCxVH are latched in a temporary register until both registers are written. Any writes into TPMCxSC reset the mechanism.

9.4.4. TPM External Connections

The TPM module has several external connections for clock input, input capture, output compare and output PWM.

TPM Signal	Device													
	AW	AC	Dx	EL,SL	EN	GB/GT	JM	LC,LL	QA,QB	QD*	QE*	QG	Rx	SG,SH
TPMCLK	-	-	-	-	-	-	-	-	PTA5	-	-	PTA5	-	-
TPM1CLK	PTD6	IRQ	PTF2	PTA0	PTF2	PTD0	IRQ	PTC7	-	PTA2	PTA5	-	PTD6	PTA0
TPM2CLK	PTD4	IRQ	PTF3	PTA0	-	PTD3	IRQ	PTC7	-	PTA3	PTE0	-	-	PTA0
TPM3CLK	-	IRQ	-	-	-	-	-	-	-	-	PTE7	-	-	-
TPMCH0	-	-	-	-	-	-	-	-	PTA0	-	-	PTA0	-	-
TPMCH1	-	-	-	-	-	-	-	-	-	-	-	PTB5	-	-
TPM1CH0	PTE2	PTE2	PTD2	PTA0	PTD2	PTD0	PTE2	PTC2	-	PTA0	PTA0	-	PTD6	PTA0
TPM1CH1	PTE3	PTE3	PTD3	PTB5	PTD3	PTD1	PTE3	PTC3	-	PTA1	PTB5	-	PTB7	PTB5
TPM1CH2	PTF0	PTF0	PTD4	-	PTD4	PTD2	PTF0	-	-	-	PTA6	-	-	-
TPM1CH3	PTF1	PTF1	PTD5	-	PTD5	-	PTF1	-	-	-	-	-	-	-
TPM1CH4	PTF2	-	PTD6	-	-	-	PTF2	-	-	-	-	-	-	-
TPM1CH5	PTF3	-	PTD7	-	-	-	PTF3	-	-	-	-	-	-	-
TPM2CH0	PTF4	PTF4	PTD0	PTA1	-	PTD3	PTF4	PTC4	-	PTA4 PTA5	PTA1	-	-	PTA1
TPM2CH1	PTF5	PTF5	PTD1	PTB4	-	PTD4	PTF5	PTC5	-	-	PTB4	-	-	PTB4
TPM2CH2	-	-	-	-	-	PTD5	-	-	-	-	PTA7	-	-	-
TPM2CH3	-	-	-	-	-	PTD6	-	-	-	-	-	-	-	-
TPM2CH4	-	-	-	-	-	PTD7	-	-	-	-	-	-	-	-
TPM3CH0	-	PTB0	-	-	-	-	-	-	-	-	PTC0	-	-	-
TPM3CH1	-	PTB1	-	-	-	-	-	-	-	-	PTC1	-	-	-
TPM3CH2	-	-	-	-	-	-	-	-	-	-	PTC2	-	-	-
TPM3CH3	-	-	-	-	-	-	-	-	-	-	PTC3	-	-	-
TPM3CH4	-	-	-	-	-	-	-	-	-	-	PTC4	-	-	-
TPM3CH5	-	-	-	-	-	-	-	-	-	-	PTC5	-	-	-

* On QD devices, the TPM2CH0 input is connected to PTA5 and TPM2CH0 output is connected to PTA4
On QE4 and QE8 devices, there is a single input for TPM1 and TPM2, through pin PTA5, TPM3 is not available on these devices

Table 9.12

Some devices allow selection of different pins for some TPM channel input/outputs. This feature is enabled by some bits in the SOPT2 register. Table 9.13 shows the default and alternative connections for some devices.

		Device		
		EL/SL	QB	SG/SH
TPMCH0	default	-	PTA0	-
	alternate	-	PTB5	
TPM1CH0	default	PTA0	-	PTC0
	alternate	PTC0	-	PTA0
TPM1CH1	default	PTB5	-	PTC1
	alternate	PTC1	-	PTB5
TPM2CH0	default	PTA1	-	-
	alternate	PTA6	-	-
TPM2CH1	default	PTB4	-	-
	alternate	PTA7	-	-

Table 9.13

For more information on the pin reallocation, refer to topic 5.5.2 (SOPT2 register).

9.4.5. TPM Module Versions

The following table shows the evolution of TPM modules since version 1:

TPM Version	Devices	Features added/modified
1	GB, GT, Rx	First release.
2	AW, Dx, EL, EN, QA, QD, QG, LC, SG	External clock input option.

3	AC, JM, LL, QE, SH	• Writing to TPMCNTH or TPMCNTL clears both TPMCNT and TPM prescaler.
		• PWM and output compare modes only update cycle/compare value when the TPM count goes from TPMMOD-1 to TPMMOD (or 0xFFFE to 0xFFFF when operating in free-running mode).
		• Improved center-aligned PWM operation:
		  • When TPMCxV=TPMMOD, duty cycle=100%.
		  • When TPMCxV=TPMMOD-1, duty cycle≈100%.
		  • When TPMCxV changes from 0x0000 to a non-zero value, the duty cycle is only updated on the next PWM cycle.
		  • When TPMCxV changes from a non-zero value to 0x0000, the current PWM cycle is finished before the new duty cycle is used.
		• Some improvements for BDM access while in background debug mode.

Table 9.14 – TPM module history

9.4.6. TPM Registers

The TPM module uses five registers for the core timer and three additional registers for each channel (up to eight channels per module):

- TPM status and control (TPMSC) register.
- TPM counter high-byte (TPMCNTH) register.
- TPM counter low-byte (TPMCNTL) register.
- TPM modulo high-byte (TPMMODH) register.
- TPM modulo low-byte (TPMMODL) register.

Each channel comprises three registers:

- TPM channel "x" status and control (TPMCSC) register.
- TPM channel "x" value high-byte (TPMCxVH) register.
- TPM channel "x" value low-byte (TPMCxVL) register.

 TPMCNT, TPMMOD and TPMCxV are 16-bit registers split, each one, into two 8-bit registers. This is necessary as the HCS08 uses an 8-bit CPU!

9.4.6.1. TPMSC Register

Name	Model		BIT 7	BIT 6	BIT 5	BIT 4	BIT 3	BIT 2	BIT 1	BIT 0
TPMSC TPMxSC*	QA, QG all	Read	TOF	TOIE	CPWMS	CLKSB	CLKSA	PS2	PS1	PS0
		Write	-							
		Reset	0	0	0	0	0	0	0	0

* This register is named TPMSC on QA and QG devices, all other devices use TPM1SC, TPM2SC and so on.

Bit Name	Description	C Symbol
TOF	TPM overflow flag. This bit indicates the overflow of the TPM counting. It is set on the next clock edge after TPMCNT = TPMMOD (or when TPMCNT rolls from 0xFFFF to 0x0000 in free-running mode). Clearing TOF is done by writing "0" into it.	bTOF
TOIE	TPM overflow interrupt enable bit. This bit controls the interrupt request when TOF is set: 0 – no interrupt is requested when TOF = 1. 1 – an interrupt is requested when TOF = 1.	bTOIE
CPWMS	Center-aligned PWM mode selection bit. This bit controls the operation of the TPM counter and it is used for center-aligned PWM generation: 0 – edge-aligned PWM mode. TPM operates in up-counting mode only. 1 – center-aligned PWM mode. TPM operates in up/down-counting mode. This mode cannot be used on capture nor compare modes.	bCPWMS

CLKSB CLKSA	TPM clock source selection bits:	
	00 – no clock (TPM disabled).	**TPM_OFF**
	01 – TPM is clocked by BUSCLK.	**TPM_BUSCLK**
	10 – TPM is clocked by fixed system clock (XCLK).	**TPM_XCLK**
	11 – TPM is clocked by the external clock source (TPMxCLK pin)	**TPM_EXT**

PS2 PS1 PS0	TPM prescaler dividing factor selection bits:	
	000 – clock source divided by 1.	**TPM_DIV1**
	001 – clock source divided by 2.	**TPM_DIV2**
	010 – clock source divided by 4.	**TPM_DIV4**
	011 – clock source divided by 8.	**TPM_DIV8**
	100 – clock source divided by 16.	**TPM_DIV16**
	101 – clock source divided by 32.	**TPM_DIV32**
	110 – clock source divided by 64.	**TPM_DIV64**
	111 – clock source divided by 128.	**TPM_DIV128**

9.4.6.2. TPMCNTH and TPMCNTL Registers

Name	Model		BIT 7	BIT 6	BIT 5	BIT 4	BIT 3	BIT 2	BIT 1	BIT 0
TPMCNTH TPMxCNTH	QA, QG all	Read	TPM counting (high byte)							
		Write	Clear TPMCNT							
		Reset	0	0	0	0	0	0	0	0
TPMCNTL TPMxCNTL	QA, QG all	Read	TPM counting (low byte)							
		Write	Clear TPMCNT							
		Reset	0	0	0	0	0	0	0	0

* These registers are named TPMCNTH and TPMCNTL on QA and QG devices, all other devices use TPM1CNTH, TPM1CNTL, and so on.

TPMCNTH and TPMCNTL can be used to read the current TPM count. This register has a coherency mechanism that latches the content of TPMCNT into two 8-bit temporary latches. The mechanism remains locked until both registers are read or a write operation occurs on TPMCNTH, TPMCNTL or TPMSC. Writing on TPMCNTH or TPMCNTL also clears the current count of TPMCNT (on TPMv3 it also clears the prescaler internal counting).

It is also possible to read/write on both registers by using 16-bit transfers (LDHX/STHX instructions or by direct access to TPMCNT in C). Example:

```
var16 = TPMCNT;    // read current TPM count
```

HCS08 Unleashed

9.4.6.3. TPMMODH and TPMMODL Registers

Name	Model		BIT 7	BIT 6	BIT 5	BIT 4	BIT 3	BIT 2	BIT 1	BIT 0
TPMMODH TPMxMODH	QA, QG all	Read	TPM modulo (high byte)							
		Write	TPM modulo (high byte)							
		Reset	0	0	0	0	0	0	0	0
TPMMODL TPMxMODL	QA, QG all	Read	TPM modulo (low byte)							
		Write	TPM modulo (low byte)							
		Reset	0	0	0	0	0	0	0	0

* These registers are named TPMMODH and TPMMODL on QA, QG devices, all other devices use TPM1MODH, TPM1MODL, and so on.

TPMMODH and TPMMODL store the modulo value for the TPM counter. The counter counts up to the modulo value stored on TPMMOD and then rolls to 0x0000 (when CPWMS = 0) or starts counting down (when CPWMS = 1) on the next clock pulse.

This register also implements a coherency mechanism that prevents TOF from being set until both registers are written. This mechanism is reset by writing into both registers (in any order) or by writing into TPMSC.

It is also possible to read/write on both registers by using 16-bit transfers (LDHX/STHX instructions or by direct access to TPMMOD in C). Example:

```
TPMMOD = 1000;    // set modulo to 1000
```

9.4.6.4. TPMCxSC

Name	Model		BIT 7	BIT 6	BIT 5	BIT 4	BIT 3	BIT 2	BIT 1	BIT 0
TPMCxSC TPMxCxSC	QA, QG all	Read	CHxF	CHxIE	MSxB	MSxA	ELSxB	ELSxA	0	0
		Write	CHxF	CHxIE	MSxB	MSxA	ELSxB	ELSxA	-	-
		Reset	0	0	0	0	0	0	0	0

* This register is named TPMCxSC on QA and QG devices, all other devices use TPM1CxSC, TPM2CxSC and so on.

Bit Name	Description	C Symbol
CHxF	Channel x flag. This bit indicates a capture/compare event on channel x: 0 – no capture/compare event. 1 – capture/compare event.	bCHF
CHxIE	Channel x interrupt enable bit: 0 – no interrupt is requested when CHxF is set. 1 – an interrupt is requested when CHxF is set.	bCHIE
MSxB MSxA ELSxB ELSxA	These bits select the current operating mode for channel x:	
	xx00 – channel disabled. The corresponding TPMxCHx pin can be used in I/O mode.	-
	0001 – capture mode (rising edges on TPMxCHx pin).	TPM_CAPTURE_RISING_EDGE
	0010 – capture mode (falling edges on TPMxCHx pin).	TPM_CAPTURE_FALLING_EDGE
	0011 – capture mode (both edges on TPMxCHx pin).	TPM_CAPTURE_BOTH_EDGES
	0100 – compare mode (only set CHxF, the corresponding TPMxCHx pin can be used in I/O mode).	TPM_COMPARE_INT
	0101 – compare mode (toggle TPMxCHx pin state on each compare).	TPM_COMPARE_TOGGLE
	0110 – compare mode (clear pin TPMxCHx pin state on each compare).	TPM_COMPARE_CLEAR
	0111 – compare mode (set pin TPMxCHx pin state on each compare).	TPM_COMPARE_SET
	1x10 – PWM mode (high-true pulses).	TPM_PWM_HIGH
	1xx1 – PWM mode (low-true pulses).	TPM_PWM_LOW

9.4.6.5. TPMCxVH and TPMCxVL Registers

Name	Model		BIT 7	BIT 6	BIT 5	BIT 4	BIT 3	BIT 2	BIT 1	BIT 0
TPMCxVH TPMxCxVH	QA, QG all	Read	TPM channel value (high byte)							
		Write								
		Reset	0	0	0	0	0	0	0	0
TPMCxVL TPMxCxVL	QA, QG all	Read	TPM channel value (low byte)							
		Write								
		Reset	0	0	0	0	0	0	0	0

* These registers are named TPMCxVH and TPMCxVL on QA and QG devices, all other devices use TPM1CxVH, TPM1CxVL, and so on.

TPMCxVH and TPMCxVL allow access to the TPM channel value.

When the channel is set to capture mode, these registers store the last value captured by the channel. While in this mode, writing onto TPMCxVH and TPMCxVL is not recommended.

When the channel is set to compare mode, these registers store the current compare value (to be compared with the TPM counting).

When the channel is set to PWM mode, these registers store the current duty cycle.

These registers include a coherency mechanism to prevent data corruption: when the channel is operating in capture mode, the coherency mechanism latches the 16-bit content of TPMCxV into two temporary 8-bit registers. This allows the application to read both registers (TPMCxVH and TPMCxVL) without worrying about their being updated in the middle of the operation. When the channel is operating in compare or PWM mode, the same mechanism prevents the actual update of TPMCxV content until both registers (TPMCxVH and TPMCxVL) are written. This coherency mechanism is reset on any write into TPMCxSC.

It is also possible to read/write on both registers by using 16-bit transfers (LDHX/STHX instructions or by direct access to TPMCxV in C). Example:

```
TPMC1V = 1000;     // set channel 1 value to 1000
```

9.4.7. PWM Led Dimmer

The next example shows the implementation of a led dimmer. It uses four keys to adjust the brightness of a led by controlling the PWM duty cycle applied on it.

Two keys are used to increment/decrement the brightness in small steps. A third key turns the led off and a fourth one turns the led on fully (100% duty cycle).

The example also includes an interrupt-driven function to periodically read the keys (controlled by the RTI module, with one sample at every 128 ms) and a simple mechanism to implement an auto-increment function (the more time the key is held pressed, the faster the PWM duty cycle is incremented/decremented).

```
// DEMO9S08AW60 - TPM PWM example (led dimmer)
// SW1 connected to PTC2
// SW2 connected to PTC6
// SW3 connected to PTD3
// SW4 connected to PTD2
// LED is connected to PTF0 (TPM1CH2 pin)

#include <hidef.h>              /* for EnableInterrupts macro */
#include "derivative.h"         /* include peripheral declarations */
#include "hcs08.h"              // This is our definition file!

#define NVICGTRM (*(const char * __far)0x0000FFBE)
#define SW1 PTCD_PTCD2
#define SW2 PTCD_PTCD6
#define SW3 PTDD_PTDD3
#define SW4 PTDD_PTDD2

unsigned int duty_cycle;
```

```
// Key sampler function (called automatically by hardware (RTI))
void interrupt VectorNumber_Vrti key_sampler(void)
{
  static char auto_inc=1, key_timer;
  char flag=0;
  SRTISC_RTIACK = 1;               // clear RTIF
  if (!SW1)                        // if SW1 is pressed
  {
    // decrement the duty cycle by the "auto_inc" amount
    if (duty_cycle>auto_inc) duty_cycle -= auto_inc; else duty_cycle = 0;
    key_timer++;
    flag = 1;                      // set the keypressed flag
  }
  if (!SW2)                        // if SW2 is pressed
  {
    // increment the duty cycle by the "auto_inc" amount
    if (duty_cycle<=TPM1MOD)
    {
      duty_cycle += auto_inc;
      if (duty_cycle>TPM1MOD) duty_cycle = TPM1MOD+1;
    }
    key_timer++;
    flag = 1;                      // set the keypressed flag
  }
  if (key_timer>4)                 // if the key is pressed for more than 4 samples
  {
    auto_inc++;                    // increment the auto_inc value
    key_timer=5;                   // limit the key_timer variable to 5
  }
  if (!flag)                       // if SW1 nor SW2 are pressed
  {
    auto_inc = 1;                  // set auto_inc to 1
    key_timer = 0;                 // clear key_timer
  }
  // if SW3 is pressed, set duty cycle to 0% (led = off)
  if (!SW3) duty_cycle = 0;
  // if SW4 is pressed, set duty cycle to 100% (led = full bright)
  if (!SW4) duty_cycle = TPM1MOD+1;
}

void interrupt VectorNumber_Vtpm1ovf tpm1_overflow_isr(void)
{
  TPM1SC_TOF = 0;                  // clear interrupt flag
  TPM1C2V = duty_cycle;            // update PWM cycle
}

void main(void)
{
  SOPT = 0;                        // configure SOPT1 register, disable COP
  // ICG configuration (ICGOUT = 40MHz, BUSCLK = 20MHz):
  ICGTRM = NVICGTRM;               // trim IRG oscillator
  ICGC2 = MFDx18;                  // MFD = 111b (N = 18), RFD = 000b (R = 1)
  ICGC1 = ICG_FEI;                 // ICG in FEI mode (CLKS = 01b)
  TPM1SC = bTOIE | TPM_BUSCLK;     // TPM clock = BUSCLK = 20MHz
  TPM1MOD = 255;                   // PWM frequency is equal to 78.125 kHz
  // Configure channel 2 of TPM1 to PWM mode (active high)
  TPM1C2SC = TPM_PWM_HIGH;         // channel 2 in high-true PWM mode
  TPM1C2V = 0;                     // channel 2 set to 0%
  // Configure the RTI for periodic sampling the keys
  SRTISC = bRTIE | RTI_128ms;
  duty_cycle = 0;
  PTCPE = BIT_2 | BIT_6;
  PTDPE = BIT_2 | BIT_3;
  EnableInterrupts;
  while (1);
}
```

Example 9.13 – LED dimmer

9.4.8. Sound Generation

The next example shows how to use a TPM channel for sound generation. The channel is configured to operate in compare mode, toggling the channel pin on each compare. The channel pin (PTB5) is connected to a piezoelectric buzzer. The example was tested on a DEMO9S08QE128 board.

Figure 9.12 shows the generation of a 1000Hz frequency and some of the related interrupt events.

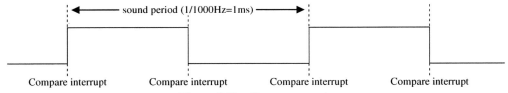

Figure 9.12 – Compare events

The `sound()` function sets the TPM 1 channel 1 to operate in compare mode, toggling the channel pin on each compare event. All you have to do is reload the channel register with a value corresponding to half the sound period.

Note that the higher the sound frequency, the higher the error due to the interrupt latency. This happens because the interrupt latency is fixed (with a small variation due to the different instruction timings) and relatively small when compared to the generated frequencies (at 16MHz BUSCLK speed, the maximum interrupt latency is equal to 1.375μs).

The ISR itself takes the following BUSCLK cycles to run:

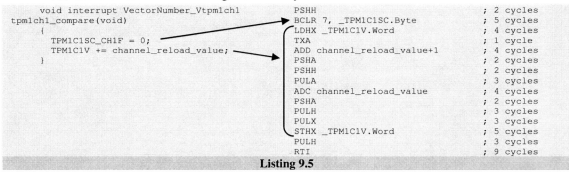

Listing 9.5

It takes 40 BUSCLK cycles to execute the full reload of TPM1C1V (2.5μs). The total latency for the TPM1C1V reload is equal to 22 (maximum ISR entrance latency) + 40 = 62 BUSCLK cycles (or 3.875μs). For a 1 kHz signal generation, the register reloading time adds 0.3875% to the signal period. For a 10 kHz signal, this value increases to 3.875%! It is easy to note that the higher the generated frequency, the higher the error due to the interrupt latency.

There are two ways to minimize the timing impact:

1. Decrease the ISR processing time (not always easy). In the present example, we could store the "channel_reload_value" variable into the direct memory. This would make ADD and ADC instructions a little bit faster (3 BUSCLK cycles instead of 4) as they would be using DIR addressing mode (instead of EXT). The overall gain would be equal to 2 BUSCLK cycles.

2. Increase the CPU clock (not always possible). On QE devices, the maximum BUSCLK frequency is 25MHz. In the present example we are operating at 16MHz. Changing the clock frequency would represent an overall gain of 56%.

```
// DEMO9S08QE128 - TPM example (sound generation)
// TPM1 channel 1 pin (PTB5) connected to a piezoelectric buzzer (mounted in the
// demonstration board)

#include <hidef.h>          /* for EnableInterrupts macro */
#include "derivative.h"     /* include peripheral declarations */
#include "hcs08.h"          // This is our header file!

unsigned int channel_reload_value, duration_timer;
char sound_playing;

// TPM1 overflow interrupt isr: control the duration of the sound
void interrupt VectorNumber_Vtpm1ch0 tpm1ch0_isr(void)
{
  TPM1C0SC_CH0F = 0;                    // clear interrupt flag
  TPM1C0V += 4000;                      // next compare in 1 ms
  if (duration_timer) duration_timer--; // decrement the timer if > 0
  else                                  // if the timer is = 0
  {
    TPM1C1SC = 0;                       // disable TPM1CH1 channel (stops sound)
    sound_playing = 0;                  // clear the flag (no sound is playing)
  }
}

// TPM1 channel1 compare interrupt isr: control the frequency generation
// the channel pin is automatically toggled on each compare
void interrupt VectorNumber_Vtpm1ch1 tpm1ch1_compare(void)
{
  TPM1C1SC_CH1F = 0;                    // clear interrupt flag
  TPM1C1V += channel_reload_value;      // next compare
}

// Outputs a sound through the TPM1CH1 pin (PTB5): freq in Hz, dur in ms
void sound(unsigned int freq, unsigned int dur)
{
  while (sound_playing);               // if a sound is playing, wait until if finishes
  // the channel reload value is equal to half the period of the signal
  channel_reload_value = (4000000/freq)/2;
  duration_timer = dur;                // set the total duration time
  // set channel 1 to compare mode (toggling the channel pin)
  TPM1C1V = channel_reload_value;      // set the first compare value
  TPM1C1SC = bCHIE | TPM_COMPARE_TOGGLE;
  sound_playing = 1;                   // sound is playing (until the flag is cleared)
}

void main(void)
{
  unsigned char song_sel;
  SOPT1 = bBKGDPE;                     // enable the debug pin
  ICSSC = DCO_MID | NVFTRIM;           // configure FTRIM value, select DCO high range
  ICSTRM = NVICSTRM;                   // configure TRIM value
  ICSC1 = ICS_FLL | bIREFS;            // select FEI mode (ICSOUT = DCOOUT = 1024 * IRCLK)
  ICSC2 = BDIV_1;                      // ICSOUT = DCOOUT / 1
  // BUSCLK = 16MHz
  TPM1SC = TPM_BUSCLK | TPM_DIV4;      // TPMCK = 4MHz
  // set channel 0 to compare mode (interrupt only)
  TPM1C0V = 3999;
  TPM1C0SC = bCHIE | TPM_COMPARE_INT;
  sound_playing = 0;
  EnableInterrupts;                    // enable interrupts (CCR:I=0)
  while(1)
  {
    sound(1000,300);                   // output a 1kHz sound for 300ms
    sound(2000,500);                   // output a 2kHz sound for 500ms
  }
}
```

Example 9.14 – Sound generation

 To modify this example to output the signal through any general purpose I/O pin, all you have to do is configure the TPM channel to interrupt-only compare mode TPM1C1SC = bCHIE | TPM_COMPARE_INT; *and insert a command to toggle the desired I/O pin by software (such as:* PTBD_PTBD0 = !PTBD_PTBD0; *to toggle the PTB0 pin)!*

9.4.9. Bit-banged UART Implementation

An interesting use of the TPM compare mode is on the implementation of bit-banged UARTs. This is useful for devices without a hardware UART (such as the QD and Rx devices) or when an additional low-speed UART is needed.

This software UART uses two channels of the TPM: one for transmitting and another for receiving.

In the current implementation (using the MC9S08QG8 on the DEMO9S08QG8 board), the transmit pin (PTB1) is set/reset on each compare event. The transmission timing is controlled by channel 0 of the TPM (configured to compare interrupt-only mode). The channel generates interrupts at a rate equal to the bit time length (104μs for 9600bps).

On each compare event, the tx_state variable is incremented. When tx_state=0, the start bit is generated (TX_PIN is cleared) and the transmission buffer is loaded with the data to be transmitted.

When tx_state is higher than 0 and less than 9, a data bit is shifted through the TX_PIN (starting by the lowest significant bit).

When tx_state=9, a stop bit is generated and when tx_state=10, the transmission is finished (tx_state is set to zero and the tx_enable flag is cleared). The tx_enable flag is used to control whether the transmitter is transmitting (tx_enable=1) or not (tx_enable=0). This flag is set by the transmitting function and automatically reset at the end of each character.

Figure 9.13 shows the transmission of 'A' (0x41 or 65 decimal).

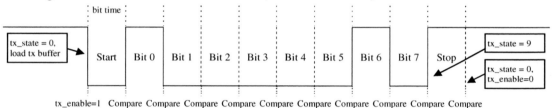

Figure 9.13 – Transmit compare events

The receiver section is controlled by TPM channel 1. When the receiver is idle, the KBI interrupt for the RX pin (PTB0) is enabled. Once a falling edge is detected on the RX pin, the KBI ISR disables the KBI interrupt pin (through the KBIPE register) and enables the TPM channel 1 compare interrupt mode. The first comparing value is set to 0.5 bit times from the current TPMCNT. This makes the first compare happen exactly or nearly in the middle of the start bit.

The RX pin is sampled on the first compare event and if it is found cleared, a valid start bit is detected and the receive state variable (rx_state) is incremented on each compare event. Note that after the first compare event, the channel compare register is reloaded with a full bit time length (104μs for 9600bps).

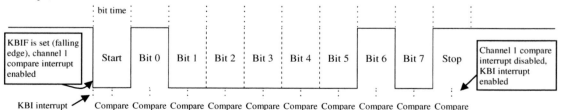

Figure 9.14 – Receive compare events

This operation continues until rx_state=9. In this case, a valid stop bit (high level) is found on the RX pin. This makes the channel 1 register be reset and re-enables the KBI interrupt (waiting for another start bit on the RX pin).

Figure 9.14 depicts the operation of the receiver section and related interrupt events (a 'A' character is being received).

```
// DEMO9S08QG8 - Bit-banged UART
// TxD pin is PTB1
// RxD pin is PTB0
// BUSCLK = 4MHz
// Baud-rate = 9600 bps
// By Fábio Pereira - fabio@sctec.com.br
// www.sctec.com.br

#include <hidef.h>              /* for EnableInterrupts macro */
#include "derivative.h"         /* include peripheral declarations */
#include "hcs08.h"              // This is our definition file!

#define TX_PIN     PTBD_PTBD1   // PTB1 as the TX pin
#define RX_PIN     PTBD_PTBD0   // PTB0 as the RX pin
#define RX_KBI     BIT_4        // PTB0(RX) is the 4th KBI input

#define BAUDRATE   9600         // this is the desired baud-rate
#define TPMCK      1000000      // the TPMCK clock
#define BITTIME    TPMCK/BAUDRATE  // bit time calculation

char tx_data, rx_data;

#pragma DATA_SEG __DIRECT_SEG MY_ZEROPAGE
struct
{
  char tx_enable    : 1;        // transmitter is enabled
  char rx_flag      : 1;        // character received
} __near flags;
#pragma DATA_SEG DEFAULT

// This is the timer isr that generates the transmittion timing
void interrupt VectorNumber_Vtpmch0 tpmch0_isr(void)
{
  static char tx_state;
  #pragma DATA_SEG __DIRECT_SEG MY_ZEROPAGE
  static __near char buffer;
  #pragma DATA_SEG DEFAULT
  TPMC0SC_CH0F = 0;             // clear interrupt flag
  TPMC0V += BITTIME;            // next compare in 104us
  if (flags.tx_enable)
  {
    if (!tx_state)             // its the start bit
    {
      TX_PIN = 0;
      buffer = tx_data;
      tx_state++;
    }
    else
    if (tx_state<9)            // its a data bit
    {
      TX_PIN = buffer & 1;     // output the LSB
      buffer >>= 1;            // shift buffer one bit to the left
      tx_state++;
    }
    else
    if (tx_state==9)           // its the stop bit
    {
      TX_PIN = 1;
      tx_state++;
    }
    else                       // transmission completed
    {
      tx_state = 0;            // restart tx_state
      flags.tx_enable = 0;     // disable the transmitter
    }
  }
}
```

These flags are stored into the direct page for faster access times (by using the DIR addressing mode)! Note that without the __near qualifier the compiler would use EXT addressing mode instead of DIR mode!

This places the next variable into the direct page for faster access times! Note the __near qualifier!

```c
// This is the timer isr that generates the receiver timing
void interrupt VectorNumber_Vtpmch1 tpmch1_isr(void)
{
  static char rx_state;
  #pragma DATA_SEG __DIRECT_SEG MY_ZEROPAGE
  static __near char buffer;
  #pragma DATA_SEG DEFAULT
  TPMC1SC_CH1F = 0;                 // clear interrupt flag
  TPMC1V += BITTIME;                // next compare in 104us
  if (!rx_state)                    // it is the start bit ?
  {
    if (!RX_PIN)
    {
      buffer = 0;                   // it is a start bit
      rx_state++;                   // advance to the next state (bit 0)
    }
    else
    {
      KBIPE = RX_KBI;               // re-enable the RX pin interrupt
      KBISC_KBACK = 1;              // clear interrupt flag
      TPMC1SC = 0;
    }
  }
  else if (rx_state<9)              // its a data bit
  {
    buffer >>= 1;                   // shift the buffer one bit to the left
    if (RX_PIN) buffer |= 128;
    rx_state++;
  }
  else
  {
    if (RX_PIN)                     // it is the stop bit ?
    {
      rx_data = buffer;
      flags.rx_flag = 1;            // set the receiver flag
    }
    KBIPE = RX_KBI;                 // re-enable the RX pin interrupt
    KBISC_KBACK = 1;               // clear interrupt flag
    rx_state = 0;
    TPMC1SC = 0;
  }
}


// This is the timer isr that generates the receiver timing
void interrupt VectorNumber_Vkeyboard kbi_isr(void)
{
  KBISC_KBACK = 1;                  // clear interrupt flag
  TPMC1SC = bCHIE | TPM_COMPARE_INT;
  TPMC1V = TPMCNT + BITTIME/2-1;
  KBIPE_KBIPE4 = 0;                 // disable the RX pin interrupt
}

// This is a simple string to serial function
void send_serial_string(char *string)
{
  while (*string)                   // while the current char of the string is not null
  {
    while (flags.tx_enable);        // wait for the transmitter to be idle
    tx_data = *string;              // write the current char into the transmit buffer
    flags.tx_enable = 1;
    string++;            // increment the current char position within the string
  }
}


void mcu_init(void)
{
  SOPT1 = bBKGDPE;                  // Enable debug pin
  ICSSC = NV_FTRIM;                 // configure FTRIM value
  ICSTRM = NV_ICSTRM;               // configure TRIM value
  ICSC2 = 0;
  // BUSCLK is now 8 MHz
  // Configure the I/O pins
  PTBDD = BIT_1;                    // PTB1 as an output
  // Configure the TPM for the bit time generation:
  TPMSC = TPM_BUSCLK | TPM_DIV8;    // TPMCK = 1MHz
  TPMC0SC = bCHIE | TPM_COMPARE_INT;
```

This places the next variable into the direct page for faster access times!

```
      // Configure the RX pin interrupt (KBI module)
      KBISC = bKBIE;
      KBIPE = RX_KBI;                         // bit 4 = PTB0
      EnableInterrupts;
      flags.tx_enable = 0;
      flags.rx_flag = 0;
      TX_PIN = 1;                             // TX pin is idle
   }


   void main(void)
   {
      mcu_init();
      send_serial_string("This is a test!\r\n");
      while (1)
      {
         if (flags.rx_flag)
         {
            tx_data = rx_data;
            flags.rx_flag = 0;
            flags.tx_enable = 1;
         }
      }
   }
```

Example 9.15 – Bit-banged TX/RX

 To change RX and TX pins it is necessary to modify the #define directives at the beginning of the program. Additionally, the #define RX_KBI must be changed to reflect the KBI input corresponding to the selected RX pin.

Note that the interrupt latency problems described in the previous example are even worse when implementing a bit-banged UART (especially for the receiver section). For a 9600bps receiver, the RX pin must be sampled at a precise rate of 104.17µs.

Using a TPMCK of 1MHz, the TPM counter is incremented at a 1µs rate and thus, the compare value must be equal to 104 (the fractional part of 104.17 is actually lost).

Considering the BUSCLK frequency is equal to 8MHz, the maximum interrupt latency (for interrupt entrance only) is equal to 22 BUSCLK or 2.75µs. Clearing the channel interrupt flag and reloading TPMC0V takes 16 BUSCLK cycles and the initial stacking of H, 2 more BUSCLK cycles (H stacking is automatically handled by the compiler as H:X is used within the ISR). The total latency is equal to 40 BUSCLK cycles (5µs at 8MHz). Refer to listing 9.6 for further details.

```
void interrupt VectorNumber_Vtpmch0
tpmch0_isr(void)
{
   static char tx_state;                          PSHH                    ; 2 cycles
   #pragma DATA_SEG __DIRECT_SEG MY_ZEROPAGE       BCLR 7, _TPM1C0SC.Byte  ; 5 cycles
   static __near char buffer;                      LDHX _TPMC1V.Word       ; 4 cycles
   TPMC0SC_CH0F = 0;                               AIX  #104               ; 2 cycles
   TPMC0V += BITTIME;                              STHX _TPMC1V.Word       ; 5 cycles
   ...                                             ...
```

Listing 9.6

The 5µs latency is very important as it is equivalent to 4.8% of a full bit time length. In serial receiving, this is even worse because the error is accumulated through the received bits. If the accumulated error rate is too high, the last bits (bit 6, bit 7 or the stop bit) can be sampled incorrectly and the received character is compromised. In this case, a fast ISR and a high BUSCLK are very important to achieve higher baud-rates.

 Note that the reloading of TPMC0V takes several cycles less than in the previous example. This is because the reloaded value is a constant value, that enables the compiler to generate a more optimized code!

10

Analog Peripherals

We live in an analog world, but the microcontrollers "live" in a digital world. In many applications, the microcontroller needs to interface somehow to our analog world.

For this kind of interfacing, there are the analog peripherals: they can do this link between a digital microcontroller and an analog quantity.

The HCS08 devices include two analog peripherals: analog comparators and analog to digital converters (ADCs).

The analog comparators can be used, as an example, for measuring analog signal timings (with help of a timer such as the TPM). In this case, the objective is not measuring the amplitude or the level of the signal, but its period or frequency.

For measuring the amplitude of an analog level, the ADC is indicated. It can convert the analog level into a digital value proportional to the input voltage level.

10.1. Analog Comparator (ACMP)

The analog comparator (ACMP) is a basic analog input module designed for simple interface between analog signals and the digital microcontroller.

Each ACMP module comprises a two-input analog comparator, a fixed internal voltage reference and a logic circuitry for interrupt generation and module control.

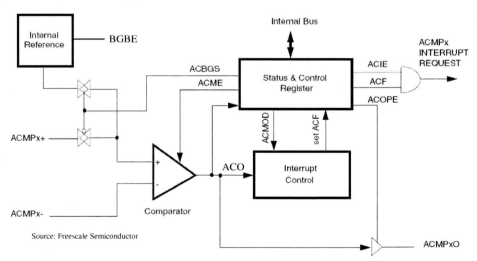

Figure 10.1 – Analog comparator simplified block diagram

As can be seen on figure 10.1, one input of the analog comparator (the inverting one) is always tied to the ACMPx- input and the other comparator input (the non-inverting one) is connected through an analog multiplexer to the ACMPx+ or to the internal bandgap reference (this is selected by the ACMPxSC:ACBGS bit).

If ACBGS = 0, the analog comparator output (ACO) is high if $V_{ACMPx+} > V_{ACMPx-}$. ACO is low if $V_{ACMPx+} < V_{ACMPx-}$.

If ACBGS = 1, ACO is high if the bandgap voltage is higher than the ACMPx- voltage and low if the bandgap voltage is lower than ACMPx-.

 To use the internal bandgap voltage reference, it is necessary to set the BGBE bit in the SPMSC1 register!

On most devices (except LC, RC, RD, RE and RG ones), the analog comparator output (ACO) can be output externally on the ACMPO pin if the ACOPE bit is set.

 When ACOPE is set, the ACMPO pin is configured as an output regardless of the state of the respective PTxDD register!

ACO can also generate an interrupt request if ACIE is set. The ACMOD bits select the event that sets the interrupt flag (ACF) and generate an interrupt request:

ACIE	ACMOD	Interrupt Event
0	xx	No interrupt event (ACF is set, but no interrupt is requested).
1	00	An interrupt is requested (ACF=1) on a falling edge of ACO.
1	01	An interrupt is requested (ACF=1) on a rising edge of ACO.
1	10	An interrupt is requested (ACF=1) on a falling edge of ACO.
1	11	An interrupt is requested (ACF=1) on both edges of ACO.

Table 10.1

Once set, ACF can be cleared by writing "1" on it.

10.1.1. ACMP External Connections and Availability

The ACMP module is available on several HCS08 devices. Table 10.2 presents the devices with built-in ACMP modules as well as the external pins of these modules.

	Dx	EL/SL	EN	JM	LC	LL	QE	QA, QG	Rx	SG	SH
ACMP1+	PTA1	PTA0	PTA1	PTD0	PTA2	PTA6	PTA0	PTA0	PTD5	PTA0	PTA0
ACMP1-	PTA2	PTA1	PTA2	PTD1	PTA3	PTA7	PTA1	PTA1	PTD4	PTA1	PTA1
ACMP1O	PTA3	PTA2	PTA3	PTD2	-	PTC6	PTA4	PTA4	-	PTA2	PTA4
ACMP2+	PTF4	PTC6	-	-	-		PTC6	-	-	-	-
ACMP2-	PTF5	PTC7	-	-	-		PTC7	-	-	-	-
ACMP2O	PTF6	PTC5	-	-	-		PTC5	-	-	-	-

Table 10.2 – ACMP external pin connections

 On devices with only one ACMP module, the module is named ACMP (and not ACMP1). On these devices, the inverting input is named ACMP-, the non-inverting input is named ACMP+ and the comparator output pin is named ACMPO.

10.1.2. ACMP Module Versions

The following table shows the evolution of the ACMP module since version 1:

ACMP Version	Devices	Features added/modified
1	Rx	First release.
2	EL, JM, LL, QA, QG, QE, SG, SH and SL	External comparator output pin.
3	Dx, EN, LC	No noticeable differences since version 2.

Table 10.3 – ACMP module history

 Despite using an ACMPv3, the LC devices do not implement an external output pin for the comparator module!

10.1.3. ACMP Registers

The ACMP module uses one register for controlling the operation and configuration of each analog comparator:

* Analog Comparator Status and Control (ACMPSC) register.

10.1.3.1. ACMPSC Register

Name	Model		BIT 7	BIT 6	BIT 5	BIT 4	BIT 3	BIT 2	BIT 1	BIT 0
ACMPxSC	EL, EN, Dx, JM, LC, LL, QA, QE, QG, Rx, SG, SL, SH	Read	ACME	ACBGS	ACF	ACIE	ACO	ACOPE	ACMOD	
		Write					-			
		Reset	0	0	0	0	0	0	0	0

Bit Name	Description	C Symbol
ACME	Analog comparator enable bit. This bit controls whether the ACMP module is enabled or disabled: 0 – ACMP disabled. 1 – ACMP enabled.	bACME
ACBGS	Reference selection for the non-inverting comparator input: 0 – non-inverting input connected to ACMP+ pin. 1 – non-inverting input connected to the internal bandgap reference.	bACBGS
ACF	Comparison event flag. This bit is set according to the operating mode set by ACMOD bits and it is cleared by writing "1" on it. 0 – no compare event. 1 – a compare event occurred.	bACF
ACIE	ACMP module interrupt enable bit. This bit controls whether an interrupt is requested when ACF is set: 0 – interrupt disabled (no interrupt is requested when ACF is set). 1 – interrupt enabled (an interrupt is requested when ACF is set).	bACIE
ACO	Analog comparator output bit. This bit reflects the current state of the output of the ACMP module: $0 - V_{ACMP-} > V_{ACMP+}$ $1 - V_{ACMP-} < V_{ACMP+}$	bACO
ACOPE	Analog comparator output pin enable bit. This bit controls whether the output of the analog comparator is connected to the ACMPO pin: 0 – external ACMP output disabled (ACMPO pin in I/O mode). 1 – external ACMP output enabled (ACMPO pin is an output and its state is equal to ACO).	bACOPE
ACMOD	Analog comparator operating mode selection:	
	00 – ACF is set on the falling edge of ACO.	ACMOD_0
	01 – ACF is set on the rising edge of ACO.	ACMOD_1
	10 – ACF is set on the falling edge of ACO.	ACMOD_2
	11 – ACF is set on both edges of ACO.	ACMOD_3

10.1.4. ACMP Examples

The next example shows how to use the analog comparator for comparing two analog voltages: one from a light sensor (a phototransistor) and the other from a trimpot. This example uses the DEMO9S08QG8 demonstration board. Figure 10.2 illustrates the schematic of the demonstration board.

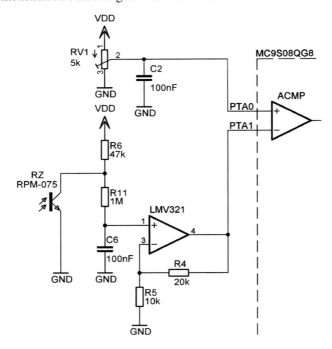

Figure 10.2 – Analog circuitry on DEMO9S08QG8 board

```
// DEMO9S08QG8 - ACMP example (no interrupt)
// LED1 connected to PTB6, LED2 connected to PTB7
// RV1 connected to PTA0 (ACMP+), RZ connected to PTA1 (ACMP-)

#include <hidef.h>           /* for EnableInterrupts macro */
#include "derivative.h"      /* include peripheral declarations */
#include "hcs08.h"           // This is our definition file!

#define LED1_ONPTBD = BIT_6
#define LED2_ONPTBD = BIT_7

void main(void)
{
  SOPT1 = bBKGDPE;           // Enable debug pin
  PTBDD = BIT_7 | BIT_6;     // PTB6 and PTB7 as outputs
  ACMPSC = bACME;            // Enable ACMP
  while(1)
  {
    if (ACMPSC_ACO)
    {                        // if ACO=1 (ACMP+>ACMP-)
      LED1_ON;
    } else
    {                        // if ACO=0 (ACMP+<ACMP-)
      LED2_ON;
    }
  }
}
```

Example 10.1 – ACMP example

Testing the example is very simple: expose RZ1 to a light source and then turn RV1 until LED1 turns on (and LED2 turns off). Now a minimal reduction of light over the light sensor toggles the leds. This could be a good start for an object counter using a light barrier.

The next example shows how to use the interrupts to create a 16-bit counter.

```
// DEMO9S08QG8 - ACMP example (using interrupt)
// LED1 connected to PTB6
// LED2 connected to PTB7
// RV1 connected to PTA0 (ACMP+)
// RZ connected to PTA1 (ACMP-)

#include <hidef.h>          /* for EnableInterrupts macro */
#include "derivative.h"     /* include peripheral declarations */
#include "hcs08.h"          // This is our definition file!

#define LED1  PTBD_PTBD6
#define LED2  PTBD_PTBD7

unsigned int counter;

void interrupt VectorNumber_Vacmp acmp_isr(void)
{
  ACMPSC_ACF = 1;           // clear interrupt flag
  counter++;
}

void main(void)
{
  SOPT1 = bBKGDPE;          // Enable debug pin
  PTBD = BIT_7 | BIT_6;     // turn leds off
  PTBDD = BIT_7 | BIT_6;    // PTB6 and PTB7 as outputs
  // Enable ACMP, ACMP interrupts, ACF is set on falling edges of ACO
  ACMPSC = bACME | bACIE | ACMOD_0;
  counter = 0;
  EnableInterrupts;
  while(1)
  {
    if (counter>10) LED1 = 0; // turn LED1 on when counter>10
    if (counter>20) LED2 = 0; // turn LED2 on when counter>20
  }
}
```

Example 10.2 – ACMP example (with interrupts)

 You can watch the counter incrementing by using the debug application and the periodical visualization! Refer to topic 3.6.1 for further information.

The next example is a more robust version of the example 10.2, despite doing the same work. This new version includes a digital filter to reject short pulses, thus, reducing counting error.

The example uses TPM channel 0 operating in compare mode to generate a time base of 1ms. This time base increments a 16-bit counter ("ms_timer") when its flag is set.

The analog comparator ISR is called by the ACMP hardware when a rising or a falling edge occurs on the comparator output (ACO). When a falling edge is detected (ACO = 0), the ms_timer counter is enabled (flag = 1) and cleared (ms_timer = 0) and starts counting up. When a rising edge is detected (ACO = 1), the current counting of ms_timer is stored into the "capture" variable and checked against the threshold value (TTIME). If "capture" is higher than "TTIME", the "counter" variable is incremented by one.

This simple mechanism rejects any pulse shorter than TTIME milliseconds and it is a nice way to implement a good object counter or a passageway counter.

```
// DEMO9S08QG8 - ACMP example 3 (object counter)
// LED1 connected to PTB6, LED2 connected to PTB7
// RV1 connected to PTA0 (ACMP+), RZ connected to PTA1 (ACMP-)

#include <hidef.h>              /* for EnableInterrupts macro */
#include "derivative.h"         /* include peripheral declarations */
#include "hcs08.h"              // This is our definition file!

#define LED1    PTBD_PTBD6
#define LED2    PTBD_PTBD7
#define CTIME 300

unsigned int ms_timer, counter, capture;
char flag;

void interrupt VectorNumber_Vtpmch0 timer_isr(void)
{
  TPMC0SC_CH0F = 0;        // clear interrupt flag
  TPMC0V += 1000;          // set the next compare to 1ms ahead
  if (flag) ms_timer++;    // if flag is set, increment ms_timer
}

void interrupt VectorNumber_Vacmp comparator_isr()
{
  ACMPSC_ACF = 1;          // clear interrupt flag
  if (!ACMPSC_ACO)
    {                      // if ACO = 0 (ACMP+ < ACMP-)
    flag = 1;              // set flag (start the ms_timer counter)
    ms_timer = 0;          // clear the ms_timer counter
    }
  else
    {                      // if ACO = 1 (ACMP+ > ACMP-)
    capture = ms_timer;    // store the current ms_timer count
    flag = 0;              // stop the ms_timer counter
    // if the captured value is greater than CTIME, increment counter
    if (capture>CTIME) counter++;
    }
}

void main(void)
{
  SOPT1 = bBKGDPE;         // Enable debug pin
  ICSSC = NV_FTRIM;        // configure FTRIM value
  ICSTRM = NV_ICSTRM;      // configure TRIM value
  ICSC2 = 0;               // ICSOUT = DCOOUT / 1
  // BUSCLK is now 8 MHz
  PTBD = BIT_7 | BIT_6;    // turn leds off
  PTBDD = BIT_7 | BIT_6;   // PTB6 and PTB7 as outputs
  ACMPSC = bACME | bACIE | ACMOD_3; // Enable ACMP, ACMOD = 3 (both edges)
  TPMSC = TPM_BUSCLK | TPM_DIV8;
  TPMC0SC = bCHIE | TPM_COMPARE_INT;
  TPMC0V = 999;
  counter = 0;
  flag = 0;
  EnableInterrupts;
  while(1)
    {
    if (counter>10) LED1 = 0; // turn LED1 on when counter>10
    if (counter>20) LED2 = 0; // turn LED2 on when counter>20
    }
}
```

Example 10.3 – Object counter

 For other interesting uses of the analog comparator module, take a look at the application note AN3552 available at Freescale's website. This application note shows how to use the ACMP module as a standalone operational amplifier for multiple purposes.

10.2. ADC

The ADC module is more sophisticated than the analog comparator. It can convert an analog voltage into its digital representation, allowing precise measurements of analog quantities. Of course these analog quantities must be an analog voltage (as the ADC can only convert voltages to digital values).

The converter is built around a Successive Approximation Register (SAR), allowing fast (up to 470,000 conversions per second) and precise conversion (up to 12 bits on some devices). The following devices include an ADC module: Ax, Dx, Ex, JM, Lx, Qx and Sx.

 GB/GT devices do not implement an ADC converter as the one presented in this topic. Instead, they implement another analog-to-digital converter called ATD, which is not studied in this book.

The high speed (high sampling-rate, technically speaking) allows conversion of high frequency signals (up to 235 kHz, according to the Nyquist-Shannon theorem[41]), whereas the 12-bit resolution allows higher precision on measurements.

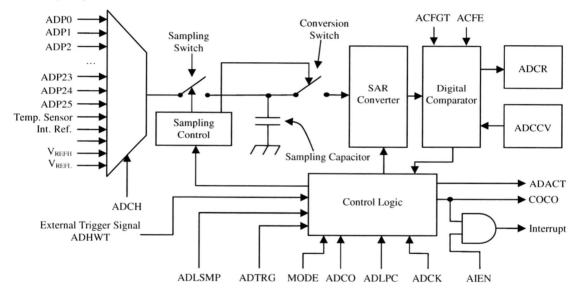

Figure 10.3 – ADC simplified block diagram

The ADC operating principle is very simple: the desired channel is selected through ADCSC1:ADCH bits. Once the channel is selected, it is sampled by the sample-and-hold circuit. This charges the sampling capacitor with the same voltage level of the input voltage. Note that the typical value for the internal sampling capacitor is 4.5pF and the ADC input impedance is 10kΩ (maximum).

The total sampling time is selected by the ADCCFG:ADLSMP bit. When ADLSMP = 0, the sampling capacitor is connected to the input channel for 3.5 ADC clock cycles (this is the standard sampling time). When ADLSMP = 1, the sampling time is extended to 23.5 ADC clock cycles.

Once sampling is complete, the sampling switch opens and the conversion switch closes, connecting the sampling capacitor to the SAR converter.

The SAR converter then starts converting the value. It takes 16.5 ADCK cycles to complete a conversion. During the conversion, the activity indicator bit (ADCSC2:ADACT) remains set. It is automatically cleared by the hardware when the conversion ends.

There are three possible operating modes for the SAR converter: 8-bit, 10-bit or 12-bit. The mode selection is made through the ADCCFG:MODE bits. Notice that the 12-bit mode is only available on JM, LC, LL and QE devices.

Once the conversion is complete, the resulting value can be directly stored into the ADCR registers, or pass through the digital comparator circuitry. The ADCSC2:ACFE bit controls whether the comparator is enabled or disabled.

When ACFE = 0, the digital comparator is disabled and the conversion result is directly stored into the ADCRL (lower byte of the result) and ADCRH (higher byte of the result). The ADCSC1:COCO bit is set (indicating the conversion is complete) and can generate an interrupt request if ADCSC1:AIEN is set.

Notice that the conversion range depends on the voltage applied to the V_{REFH} and V_{REFL} converter inputs.

The V_{REFH} voltage sets the upper limit for the input voltage that generates a full-scale result (0x3FF for a 10-bit converter). The V_{REFH} voltage must be higher or equal to 1.8 Volts (for 1.8 to 3.3V devices) or 2.7 Volts (for 2.7 to 5.5V devices). V_{REFH} cannot be higher than the V_{DDAD} voltage.

The V_{REFL} voltage sets the lower limit for the input voltage that generates a result equal to zero. The V_{REFL} voltage must be always equal to the V_{SSAD} voltage.

When V_{REFH} = V_{DDAD} and V_{REFL} = V_{SSAD}, the digital value (ADCR) for an analog input value (V_{IN}) can be calculated as follows (the resolution can be 8, 10 or 12 bits):

$$ADCR = \frac{V_{IN} * 2^{Resolution}}{V_{DDAD}}$$

 For improved precision, an external precision reference source can be connected to the V_{REFH} input!

When ACFE = 1, the digital comparator is enabled and the conversion result is compared to the ADCCV value (stored into the ADCCVH and ADCCHL registers). The digital comparator can operate in two modes (controlled by the ADCSC2:ACFGT bit):

"Less than" mode – when ACFGT = 0, the comparator compares the result with the ADCCV value and sets the conversion complete flag (COCO) only if the result is less than the ADCCV value.

"Greater than or equal" mode – when ACFGT = 1, the comparator compares the result with the ADCCV value and sets the COCO flag if the result is greater than or equal to the ADCCV value.

In both cases, the value actually stored into ADCR is:

$$ADCR = result - ADCCV$$

 When ACFE=1, the ADCRL and ADCRH registers are updated only when the compare function results true!

The digital comparator can be used to reduce CPU load when monitoring an analog signal. An interrupt is generated only when the analog voltage crosses a specific threshold value. This enables the

CPU to be used for other tasks or even to enter stop mode, waiting for a wake up event (such as the crossing of the voltage threshold).

 The COCO flag is automatically cleared when the ADCSC1 register is written or when the ADCRL register is read.

After a conversion is completed, the converter does not initiate a new conversion. A new conversion is started by a write into the ADCSC1 register (unless the ADCH bits are set to all ones), or by a new trigger signal on the ADHWT input when the trigger selection bit (ADCSC2:ADTRG) is set. When the converter is operating in continuous mode (ADCSC1:ADCO = 1), a new conversion is started right after the previous one is completed.

Note that the ADC includes a blocking mechanism to avoid data corruption when reading the ADCRH and ADCRL registers: reading ADCRH automatically latches ADCRL and blocks new conversions until ADCRL is read.

10.2.1. ADC Clocking Circuitry

The ADC can operate with one of four possible sources: the BUSCLK frequency, the BUSCLK/2 frequency, the alternate clock (which varies from one chip to another) or the ADC internal asynchronous oscillator (ADACK). The clock source is selected by the ADCCFG:ADICLK bits, according to the following table:

ADICLK	ADC Clock Source
00	BUSCLK
01	BUSCLK/2
10	Alternate clock (ALTCLK)
11	Internal oscillator (ADACK)

Table 10.4 – ADC clock source selection

The selected clock source can also be divided by a selectable factor of 1, 2, 4 or 8. This is selected by the ADCCFG:ADIV bits.

The resulting clock signal (ADCK) feeds the ADC control logic and the SAR converter. The manufacturer states that the maximum ADCK frequency is limited to 8MHz and the minimum ADCK frequency is 400 kHz.

The converter also implements a low power mode which allows reduced power consumption but limits the maximum ADCK clock frequency to 4MHz. This low power mode can be enabled by setting the ADCCFG:ADLPC bit.

The internal asynchronous oscillator (ADACK) typically operates at 3.3MHz (when ADLPC=0) or 2MHz (when ADLPC=1) and is a good option when it is necessary to let the ADC operating while the CPU is in a stop mode.

The alternate clock (ALTCLK) is connected to the external reference clock signal. On devices with the ICS clock module, the ALTCLK input is connected to the ICSERCLK signal, on devices with the ICG clock module, the ALTCLK input is connected to the ICGERCLK signal and on devices with the MCG clock module, the ALTCLK input is connected to the MCGERCLK signal.

 On QG devices, the alternate clock source (ALTCLK) signal is not available and must not be used!

Note that the total conversion time depends on the ADC clock source and ADC operating mode. Table 10.5 shows the conversion time for all ADC modes.

Conversion Type	ADLSMP	Maximum Conversion Time
Single/first continuous 8-bit*	0	20 ADCK + 5 BUSCLK
Single/first continuous 10-bit/12-bit*	0	23 ADCK + 5 BUSCLK
Single/first continuous 8-bit*	1	40 ADCK + 5 BUSCLK
Single/first continuous 10-bit/12-bit*	1	43 ADCK + 5 BUSCLK
Subsequent continuous 8-bit ($f_{BUS} \geq f_{ADCK}$)	0	17 ADCK
Subsequent continuous 10-bit/12-bit ($f_{BUS} \geq f_{ADCK}$)	0	20 ADCK
Subsequent continuous 8-bit ($f_{BUS} \geq (f_{ADCK})/11$)	1	37 ADCK
Subsequent continuous 10-bit/12-bit ($f_{BUS} \geq (f_{ADCK})/11$)	1	40 ADCK

* Note that when the ADC clock source is the internal ADC oscillator (ADICLK=11b), there is an additional delay of 5µs in the first conversion!

Table 10.5 – ADC conversion time

10.2.2. Initiating/Aborting Conversions

A new conversion is started by one of the following events:

1. A valid channel selection in the ADCSC1 register.

2. An external trigger (rising edge of the ADHWT input) assertion (when ADCSC2:ADTRG=1).

3. After a previous conversion is completed and the ADC is operating in continuous mode (ADCSC1:ADCO = 1).

The external trigger input (ADHWT) is usually connected to the RTI or RTC output, but some devices (EL, DN, DV and DZ) allow using the IRQ pin as an alternate trigger source. On these devices, the actual trigger source is selected by the SOPT2:ADHTS bit (refer to the topic 5.5.2 for further information about the SOPT2 register). The external trigger can be used to start a conversion when the chip is in a low power mode (provided that the ADC clock source is still running). The ADC conversion complete interrupt then wakes up the device once the conversion is completed.

When a conversion is in progress, writing into the ADCSC1, ADCSC2, ADCCFG, ADCCVH or ADCCVL registers aborts the current conversion (the content of ADCRL and ADCRH is no longer valid).

10.2.3. Input Channel Selection

Writing a value different from 11111b into the ADCH bits (register ADCSC1) starts a new conversion on the selected channel (writing 11111b into ADCH disables the ADC module).

The ADC module includes some special registers to disable the digital circuitry associated with the pin, thus, reducing the leakage currents which can influence the ADC measurements. Each bit of the APCTL registers control whether the pin is operating as an analog input or in digital mode.

From the thirty-one possible channels, four are connected to specific internal sources: channel 26 is connected to the internal temperature sensor, channel 27 is connected to the internal bandgap reference, channel 29 is connected to the V_{REFH} voltage and channel 30 is connected to the V_{REHL} voltage.

Internal Temperature Sensor

When the channel 26 is selected as the source for the ADC converter, the internal temperature sensor output is converted.

The internal temperature sensor outputs a DC voltage proportional to the chip internal temperature (typically 701.2mV at 25°C). **The manufacturer advises using long sampling mode and a maximum of 1 MHz ADCK.**

The current temperature can be calculated by using the following formula:

$$Temp(^oC) = 25 - \frac{V_{sensor} - V_{25degrees}}{m}$$

In which: V_{sensor} is the current voltage output of the temperature sensor.

$V_{25degrees}$ is the output voltage when the sensor is at 25°C (typically 701.2mV).

m is the temperature slope (m=1.646mV/°C for temperatures among -40 and +25°C, m=1.769mV/°C for temperatures among +25 and +85°C).

Note that the uncalibrated accuracy of the temperature sensor is approximately ±12°C. By calibrating at three points (-40, 25 and 85°C) it is possible to improve the accuracy to ±2.5°C. The temperature slopes will need to be recalculated accordingly.

The example 13.2 presents an application that measures ambient temperature using the ADC internal temperature sensor.

Internal Bandgap Reference

When the channel 27 is selected, the output of the internal bandgap voltage reference is converted. Typically, the bandgap reference outputs a voltage equal to 1.20V.

The BGBE bit (register SPMSC1) must be set in order to read the bandgap output voltage!

An interesting use of the bandgap reference is in measuring the V_{REFH} voltage supplied to the ADC converter. On most devices, this voltage is directly derived from the supply voltage (V_{DD}).

The procedure for calculating the V_{REFH} value is very simple. The following formula can be used to calculate V_{REFH} (or V_{DD}) from the bandgap conversion result:

$$V_{REFH} = \frac{1.2 * 2^{ADC\_res}}{ADCR_{BANDGAP}}$$

In which: V_{REFH} is the positive reference voltage

ADC_res is the ADC resolution in bits (8, 10 or 12).

$ADCR_{BANDGAP}$ is the conversion result when the selected channel is the bandgap.

Example 13.2 presents an application that measures ambient temperature using the ADC internal temperature sensor. The bandgap voltage is used to calibrate the correct value of V_{DD}.

10.2.4. ADC External Connections

The ADC module is connected to several input pins, most of them multiplexed with general I/O functions. Table 10.6 shows the ADC external connections for some HCS08 devices.

ADC signal	AC	AW	Dx	EL/SL	EN	JM	LC	LL	QA, QD	QE	QG	SG/SH
ADP0	PTB0	PTB0	PTA0	PTA0	PTA0	PTB0	PTA0	PTA0	PTA0	PTA0	PTA0	PTA0
ADP1	PTB1	PTB1	PTA1	PTA1	PTA1	PTB1	PTA1	PTA1	PTA1	PTA1	PTA1	PTA1
ADP2	PTB2	PTB2	PTA2	PTA2	PTA2	PTB2	PTA2	PTA2	PTA2	PTA2	PTA2	PTA2
ADP3	PTB3	PTB3	PTA3	PTA3	PTA3	PTB3	PTA3	PTA3	PTA3	PTA3	PTA3	PTA3
ADP4	V_{REFL}	PTB4	PTA4	PTB0	PTA4	PTB4	PTA4	PTA4	V_{SS}	PTB0	PTB0	PTB0
ADP5	V_{REFL}	PTB5	PTA5	PTB1	PTA5	PTB5	PTA5	PTA5	V_{SS}	PTB1	PTB1	PTB1
ADP6	V_{REFL}	PTB6	PTA6	PTB2	PTA6	PTB6	PTA6	PTA6	V_{SS}	PTB2	PTB2	PTB2
ADP7	V_{REFL}	PTB7	PTA7	PTB3	PTA7	PTB7	PTA7	PTA7	V_{SS}	PTB3	PTB3	PTB3

Table 10.6 – ADC external connections (continued)

ADC signal	AC	AW	Dx	EL/SL	EN	JM	LC	LL	QA, QD	QE	QG	SG/SH
ADP8	PTD0	PTD0	PTB0	PTC0	PTB0	PTD0	V_{REFL}	-	V_{SS}	PTA6	V_{SS}	PTC0
ADP9	PTD1	PTD1	PTB1	PTC1	PTB1	PTD1	V_{REFL}	-	V_{SS}	PTA7	V_{SS}	PTC1
ADP10	PTD2	PTD2	PTB2	PTC2	PTB2	PTD2	V_{REFL}	-	V_{SS}	PTF0	V_{SS}	PTC2
ADP11	PTD3	PTD3	PTB3	PTC3	PTB3	PTD3	V_{REFL}	-	V_{SS}	PTF1	V_{SS}	PTC3
ADP12	V_{REFL}	PTD4	PTB4	PTC4	-	PTD4	V_{REFL}	-	V_{SS}	PTF2	V_{SS}	V_{SS}
ADP13	V_{REFL}	PTD5	PTB5	PTC5	-	V_{REFL}	V_{REFL}	-	V_{SS}	PTF3	V_{SS}	V_{SS}
ADP14	V_{REFL}	PTD6	PTB6	PTC6	-	V_{REFL}	V_{REFL}	-	V_{SS}	PTF4	V_{SS}	V_{SS}
ADP15	V_{REFL}	PTD7	PTB7	PTC7	-	V_{REFL}	V_{REFL}	-	V_{SS}	PTF5	V_{SS}	V_{SS}
ADP16	V_{REFL}	V_{REFL}	PTC0	V_{REFL}	-	V_{REFL}	V_{REFL}	-	V_{SS}	PTF6	V_{SS}	V_{SS}
ADP17	V_{REFL}	V_{REFL}	PTC1	V_{REFL}	-	V_{REFL}	V_{REFL}	-	V_{SS}	PTF7	V_{SS}	V_{SS}
ADP18	V_{REFL}	V_{REFL}	PTC2	V_{REFL}	-	V_{REFL}	V_{REFL}	-	V_{SS}	PTG2	V_{SS}	V_{SS}
ADP19	V_{REFL}	V_{REFL}	PTC3	V_{REFL}	-	V_{REFL}	V_{SS}	-	V_{SS}	PTG3	V_{SS}	V_{SS}
ADP20	V_{REFL}	V_{REFL}	PTC4	V_{REFL}	-	V_{REFL}	V_{LCD}	V_{LCD}	V_{SS}	PTG4	V_{SS}	V_{SS}
ADP21	V_{REFL}	V_{REFL}	PTC5	V_{REFL}	-	-	V_{LL1}	V_{LL1}	V_{SS}	PTG5	V_{SS}	-
ADP22	-	-	PTC6	V_{REFL}	-	-	V_{DDASW}	V_{REFL}	-	PTG6	-	-
ADP23	-	-	PTC7	V_{REFL}	-	-	V_{DDA}	-	-	PTG7	-	-
ADP24	-	-	-	-	-	-	V_{DDSW}	-	-	-	-	-
ADP25	-	-	-	-	-	-	V_{DD}	-	-	-	-	-
ADP29(V_{REFH})	V_{REFH}	V_{REFH}	V_{REFH}	V_{REFH}	V_{REFH}	V_{REFH}	V_{REFH}	V_{REFH}	V_{DD}	V_{DD}	V_{DD}	V_{DD}
ADP30(V_{REFL})	V_{REFL}	V_{REFL}	V_{REFL}	V_{REFL}	V_{REFL}	V_{REFL}	V_{REFL}	V_{REFL}	V_{SS}	V_{SS}	V_{SS}	V_{SS}

Table 10.6 – ADC external connections

10.2.5. ADC Registers

The ADC module includes up to ten registers for controlling its operations:

- ADC Status and Control 1 (ADCSC1) register.
- ADC Status and Control 2 (ADCSC2) register.
- ADC Configuration (ADCCFG) register.
- ADC Result (ADCR) register:
  - ADC Result High byte (ADCRH).
  - ADC Result Low byte (ADCRL).
- ADC Compare Value (ADCCV) register:
  - ADC Compare Value High byte (ADCCVH).
  - ADC Compare Value Low byte (ADCCVL).
- ADC Pin Control 1, 2 and 3 (APCTL1, APCTL2 and APCTL3) registers.

10.2.5.1. ADCSC1 Register

Name	Model		BIT 7	BIT 6	BIT 5	BIT 4	BIT 3	BIT 2	BIT 1	BIT 0
ADCSC1	all (except GB, GT, and Rx)	Read	COCO	AIEN	ADCO	ADCH				
		Write	-	AIEN	ADCO	ADCH				
		Reset	0	0	0	1	1	1	1	1

Bit Name	Description	C Symbol
COCO	ADC conversion complete. This read-only bit can be cleared by writing in the ADCSC1 register or by reading ADCR. 0 – conversion not completed. 1 – conversion completed.	bCOCO
AIEN	ADC interrupt enable bit. This bit controls whether an interrupt is requested when COCO is set. 0 – no interrupt is requested when COCO is set. 1 – an interrupt is requested when COCO is set.	bAIEN

ADCO	Continuous conversion mode: 0 – single conversion mode. The converter does only one conversion per trigger signal. 1 – continuous conversion mode. The converter automatically starts a new conversion after the previous one is completed.	bADCO
ADCH	ADC input channel selection bits. These bits control which channel is selected for the next analog-to-digital conversion.	
	00000 – channel 0 (ADP0).	ADC_CH0
	00001 – channel 1 (ADP1).	ADC_CH1
	…	…
	11001 – channel 25 (ADP25).	ADC_CH25
	11010 – channel 26: internal temperature sensor.	ADCH_TEMP_SENSOR
	11011 – channel 27: internal bandgap reference.	ADCH_BANDGAP
	11100 – channel 28: reserved.	-
	11101 – channel 29: V_{REFH}.	ADCH_REFH
	11110 – channel 30: V_{REFL}.	ADCH_REFL
	11111 – ADC disabled.	ADC_OFF

10.2.5.2. ADCSC2 Register

Name	Model		BIT 7	BIT 6	BIT 5	BIT 4	BIT 3	BIT 2	BIT 1	BIT 0
ADCSC2	all (except GB, GT, and Rx)	Read	ADACT	ADTRG	ACFE	ACFGT	0	0	0	0
		Write	-				-	-		
		Reset	0	0	0	0	0	0	0	0

Bit Name	Description	C Symbol
ADACT	Analog to digital conversion in progress indicator bit (this is a read-only bit): 0 – no conversion in progress. 1 – a conversion is in progress.	bADACT
ADTRG	ADC conversion trigger source selection: 0 – a new conversion is started by writing into ADCSC1. 1 – a new conversion is started by the ADHWT input.	bADTRG
ACFE	ADC digital comparator enable bit: 0 – digital comparator disabled. 1 – digital comparator enabled.	bACFE
ACFGT	ADC digital comparator operation mode (valid only when ACFE=1): 0 – COCO is set when the result is less than the comparison value in ADCCV. 1 – COCO is set when the result is equal to or greater than the comparison value in ADCCV.	bACFGT

10.2.5.3. ADCCFG Register

Name	Model		BIT 7	BIT 6	BIT 5	BIT 4	BIT 3	BIT 2	BIT 1	BIT 0
ADCCFG	all (except GB, GT, and Rx)	Read	ADLPC	ADIV		ADLSMP	MODE		ADICLK	
		Write								
		Reset	0	0	0	0	0	0	0	0

Bit Name	Description	C Symbol
ADLPC	Low power ADC operating mode selection: 0 – full power mode (high speed). 1 – low power mode (low speed).	bADLPC
ADIV	ADC clock divider:	
	00 – clock divided by 1.	ADIV_1
	01 – clock divided by 2.	ADIV_2
	10 – clock divided by 4.	ADIV_4
	11 – clock divided by 8.	ADIV_8

ADLSMP	Sampling time selection bit: 0 – standard sampling time (3.5 ADC clocks). 1 – long sampling time (23.5 ADC clocks).	bADLSMP
MODE	ADC operating mode:	
	00 – 8-bit mode (results stored in ADCRL).	ADC_8BITS
	01 – 12-bit mode.	ADC_12BITS
	10 – 10-bit mode.	ADC_10BITS
	11 – reserved.	-
ADICLK	ADC clock source selection bits:	
	00 – ADC clocked by BUSCLK.	ADC_BUSCLK
	01 – ADC clocked by BUSCLK/2.	ADC_BUSCLK_DIV2
	10 – ADC clocked by the alternate clock (ALTCLK)	ADC_ALTCLK
	11 – ADC clocked by the internal ADC oscillator (ADACK).	ADC_INTCLK

10.2.5.4. ADCR Registers (ADCRH and ADCRL)

Name	Model		BIT 7	BIT 6	BIT 5	BIT 4	BIT 3	BIT 2	BIT 1	BIT 0
ADCRH	all (except GB, GT, and Rx)	Read	0	0	0	0	ADR11	ADR10	ADR9	ADR8
		Write	-	-	-	-	-	-	-	-
		Reset	0	0	0	0	0	0	0	0
ADCRL	all (except GB, GT, and Rx)	Read	ADR7	ADR6	ADR5	ADR4	ADR3	ADR2	ADR1	ADR0
		Write	-	-	-	-	-	-	-	-
		Reset	0	0	0	0	0	0	0	0

When the converter is operating in 10 or 12-bit mode, the most significant bits of the result are stored into ADCRH and the least significant byte is stored into ADCRL. In this case, ADCRH must be read first and then ADCRL. Reading ADCRH latches ADCRL until it is read (this prevents the result changing while the read operating is in progress). When the converter is operating in 8-bit mode, the result is stored into ADCRL.

In case the ADC digital comparator is enabled, the result stored into ADCR is equal to:

$$ADCR = \text{Conversion result} - ADCCV$$

10.2.5.5. ADCCV Registers (ADCCVH and ADCCVL)

Name	Model		BIT 7	BIT 6	BIT 5	BIT 4	BIT 3	BIT 2	BIT 1	BIT 0
ADCCVH	all (except GB, GT, and Rx)	Read	0	0	0	0	ADCV11	ADCV10	ADCV9	ADCV8
		Write	-	-	-	-	-	-	-	-
		Reset	0	0	0	0	0	0	0	0
ADCCVL	all (except GB, GT, and Rx)	Read	ADCV7	ADCV6	ADCV5	ADCV4	ADCV3	ADCV2	ADCV1	ADCV0
		Write	-	-	-	-	-	-	-	-
		Reset	0	0	0	0	0	0	0	0

The ADCCV register stores the comparing value used when the ADC digital comparator is enabled.

10.2.5.6. APCTL1, APCTL2 and APCTL3 Registers

Name	Model		BIT 7	BIT 6	BIT 5	BIT 4	BIT 3	BIT 2	BIT 1	BIT 0
APCTL1	all (except GB, GT, and Rx)	Read	ADPC7	ADPC6	ADPC5	ADPC4	ADPC3	ADPC2	ADPC1	ADPC0
		Write								
		Reset	0	0	0	0	0	0	0	0
APCTL2	all (except GB, GT, and Rx)	Read	ADPC15	ADPC14	ADPC13	ADPC12	ADPC11	ADPC10	ADPC9	ADPC8
		Write								
		Reset	0	0	0	0	0	0	0	0
APCTL3	all (except GB, GT, and Rx)	Read	ADPC23	ADPC22	ADPC21	ADPC20	ADPC19	ADPC18	ADPC17	ADPC16
		Write								
		Reset	0	0	0	0	0	0	0	0

Bit Name	Description	C Symbol
ADPCx	Each ADPCx bit controls whether the respective ADC input pin operates in analog or digital mode. 0 – digital mode (the digital input buffer is not disabled and can interfere in the result of the conversion. 1 – analog mode (the digital input buffer is disabled and cannot interfere in the result of the conversion.	bADPC0 bADPC1 ... bADPC22 bADPC23

10.2.6. ADC Examples

The first ADC example simply shows how to set up the ADC module to perform continuous conversions. It monitors the trimpot RV1 on the PTA0 pin. This example was tested on a DEMO9S08QG8 board.

LED1 lights when the voltage on PTA0 pin is higher than:

$$V_{IN} = \frac{ADCR * V_{DD}}{1023} = \frac{512 * 3.3}{1023} \cong 1.65V$$

LED2 lights when the voltage on PTA0 pin is higher than:

$$V_{IN} = \frac{ADCR * V_{DD}}{1023} = \frac{768 * 3.3}{1023} \cong 2.48V$$

```
// DEMO9S08QG8 - ADC example 1
// LED1 connected to PTB6
// LED2 connected to PTB7
// RV1 connected to PTA0
// RZ connected to PTA1

#include <hidef.h>          /* for EnableInterrupts macro */
#include "derivative.h"     /* include peripheral declarations */
#include "hcs08.h"          // This is our definition file!

#define LED1  PTBD_PTBD6
#define LED2  PTBD_PTBD7

unsigned int result;

void main(void)
{
  SOPT1 = bBKGDPE;      // configure SOPT1 register, enable pin BKGD for BDM
  ICSSC = NV_FTRIM;             // configure FTRIM value
  ICSTRM = NV_ICSTRM;          // configure TRIM value
  ICSC2 = 0;                    // ICSOUT = DCOOUT / 1
  ADCCFG = bADLSMP | ADC_10BITS; // enable long sampling, 10-bit mode
  APCTL1 = BIT_0;               // ADP0 in analog mode
```

```
      ADCSC1 = bADCO | ADC_CH0;          // ADC in continuous mode, channel 0
      PTBDD = BIT_6 | BIT_7;             // PTB6 and PTB7 as outputs
      while (1)
      {
        if (ADCSC1_COCO)                 // if a new conversion is complete
        {
          result = ADCR;                 // read the result (clear COCO)
          if (result>512) LED1 = 0; else LED1 = 1;
          if (result>768) LED2 = 0; else LED2 = 1;
        }
      }
}
```

Example 10.4 – ADC example 1

 You can watch the "result" variable in real-time by using the debug application and the periodical visualization! Refer to topic 3.6.1 for further information.

Now, let us calculate the sampling-rate for the example above. Considering we are using the BUSCLK as the clock source for the ADC and BUSCLK = 8MHz, we have an ADCK = 8MHz (ADICLK = 00b). This means that an ADCK cycle takes 125ns.

According to the table 10.5, the first conversion takes 43 ADCK + 5 BUSCLK = 48 BUSCLK = 6µs. The next conversions take 5µs (40 BUSCLK cycles). This leads us to a performance of 200,000 samples (conversions) per second or 200ksps.

Note that the manufacturer states that when long sampling is enabled, f_{BUS} should be preferably eleven times higher than f_{ADCK} (to guarantee the correct sampling time). The previous example can be modified to reduce the ADCK, by using the BUSCLK/2 as ADC clock source and by dividing down this signal by 8 (ADICLK = 11b). To do this, simply modify the ADCCFG configuration to the following one:

```
      ADCCFG = bADLSMP | ADC_10BITS | ADC_BUSCLK_DIV2 | ADIV_8;
```

 The higher the analog source series impedance, the slower the sampling capacitor charges. If the analog source has a high internal output impedance, consider using an analog buffer (with a low output impedance) between the analog source and the ADC input!

The next example shows how to use the ADC interrupts to perform the same operation seen in example 10.4.

```
// DEMO9S08QG8 - ADC example 2 (interrupts)
// LED1 connected to PTB6, LED2 connected to PTB7
// RV1 connected to PTA0, RZ connected to PTA1

#include <hidef.h>           /* for EnableInterrupts macro */
#include "derivative.h"      /* include peripheral declarations */
#include "hcs08.h"           // This is our definition file!

#define LED1  PTBD_PTBD6
#define LED2  PTBD_PTBD7

unsigned int result;

// ADC ISR (store the last conversion result)
void interrupt VectorNumber_Vadc adc_isr(void)
{
  result = ADCR;             // read the result (clear interrupt flag)
}

void main(void)
{
  SOPT1 = bBKGDPE;           // configure SOPT1 register, enable pin BKGD for BDM
  ICSSC = NV_FTRIM;          // configure FTRIM value
```

```
    ICSTRM = NV_ICSTRM;        // configure TRIM value
    ICSC2 = 0;                 // ICSOUT = DCOOUT / 1
    // enable long sampling, 10-bit mode, ADICLK = 11b, ADCK = BUSCLK/2
    ADCCFG = bADLSMP | ADC_10BITS | ADC_BUSCLK_DIV2 | ADIV_8;
    APCTL1 = BIT_0;            // ADP0 in analog mode
    ADCSC1 = bAIEN | bADCO | ADC_CH0;// ADC in continuous mode, channel 0, interrupts enabled
    PTBDD = BIT_6 | BIT_7;     // PTB6 and PTB7 as outputs
    EnableInterrupts;          // Enable interrupts (CCR:I=0)
    while (1)
    {
      if (result>512) LED1 = 0; else LED1 = 1;
      if (result>768) LED2 = 0; else LED2 = 1;
    }
}
```

Example 10.5 – ADC example (with interrupts)

The next example shows how to perform sequential conversions with the ADC module. In this example, channel 0 and channel 1 are converted sequentially. On each conversion, the interrupt servicing function reads the result and change the input channel to the next channel in the sequence.

The example reads the trimpot RV1, connected to the PTA0/ADP0 input and the light sensor RZ1, connected to the PTA1/ADP1 input.

The main loop compares the two results and lights LED1 when the voltage on the trimpot RV1 is higher than 1.65V (ADC result is higher than 512). Otherwise, the led is turned off. LED2 is turned on if the voltage on the light sensor (RZ1) is higher than the voltage set on RV1.

The application was tested on the DEMO9S08QG8 board.

```
// DEMO9S08QG8 - ADC example 3 (sequential readings)
// LED1 connected to PTB6, LED2 connected to PTB7
// RV1 connected to PTA0, RZ connected to PTA1
#include <hidef.h>           /* for EnableInterrupts macro */
#include "derivative.h"      /* include peripheral declarations */
#include "hcs08.h"           // This is our definition file!

#define LED1  PTBD_PTBD6
#define LED2  PTBD_PTBD7

unsigned int trimpot, sensor;

// ADC ISR (store the conversion result for RV1 or RZ)
void interrupt VectorNumber_Vadc adc_isr(void)
{
  if (ADCSC1 & 0x1F)
  {
    sensor = ADCR;             // read the channel 1 result
    ADCSC1 = bAIEN | ADC_CH0; // next channel = 0
  }
  else
  {
    trimpot = ADCR;            // read the channel 0 result
    ADCSC1 = bAIEN | ADC_CH1; // next channel = 1
  }
}

void main(void)
{
  SOPT1 = bBKGDPE;            // configure SOPT1 register, enable pin BKGD for BDM
  ICSSC = NV_FTRIM;          // configure FTRIM value
  ICSTRM = NV_ICSTRM;        // configure TRIM value
  ICSC2 = 0;                 // ICSOUT = DCOOUT / 1
  // enable long sampling, 10-bit mode, ADICLK = 11b, ADCK = BUSCLK/2
  ADCCFG = bADLSMP | ADC_10BITS | ADC_BUSCLK_DIV2 | ADIV_8;
  APCTL1 = BIT_0 | BIT_1;    // ADP0 and ADP1 in analog mode
  // ADC channel 0, interrupts enabled
  ADCSC1 = bAIEN | ADC_CH0;
  PTBDD = BIT_6 | BIT_7;     // PTB6 and PTB7 as outputs
  EnableInterrupts;          // Enable interrupts (CCR:I=0)
```

```
   while (1)
   {
     if (trimpot>512) LED1 = 0; else LED1 = 1;
     if (sensor>trimpot) LED2 = 0; else LED2 = 1;
   }
}
```

Example 10.6 – ADC example (sequential reading)

You can watch the "trimpot" and "sensor" variables in real-time by using the debug application and the periodical visualization! Refer to topic 3.6.1 for further information.

The next example shows how to use the ADC digital comparator function. The example monitors sensor RZ1 and turn LED1 on when the sensor is in the dark.

The example also shows how to use the ADC in stop3 mode with the ADC clock sourced by the internal asynchronous oscillator (ADACK).

```
// DEMO9S08QG8 - ADC example 4 (ADC comparator and stop mode)
// LED1 connected to PTB6, LED2 connected to PTB7
// RV1 connected to PTA0, RZ connected to PTA1

#include <hidef.h>          /* for EnableInterrupts macro */
#include "derivative.h"     /* include peripheral declarations */
#include "hcs08.h"          // This is our definition file!

#define LED1  PTBD_PTBD6

unsigned int sensor;

// ADC ISR
void interrupt VectorNumber_Vadc adc_isr(void)
{
  sensor = ADCR;
  if (ADCSC2_ACFGT)    // if adc comparator mode is higher than
  {
    LED1 = 0;          // turn LED1 on
    ADCSC2_ACFGT = 0;  // set ADC comparator to less than mode
  }
  else                 // if adc comparator mode is less than
  {
    LED1 = 1;          // turn LED1 off
    ADCSC2_ACFGT = 1;  // set ADC comparator to greater than or equal to mode
  }
  ADCSC1_ADCH = 1;     // restart ADC conversion
}

void main(void)
{
  // configure SOPT1 register, enable pin BKGD for BDM and STOP instruction
  SOPT1 = bSTOPE | bBKGDPE;
  ICSSC = NV_FTRIM;          // configure FTRIM value
  ICSTRM = NV_ICSTRM;        // configure TRIM value
  ICSC2 = 0;                 // ICSOUT = DCOOUT / 1
  SPMSC2 = 0;                // stop3 mode selected
  // enable long sampling, 10-bit mode, ADICLK = 11b, ADCK = ADACK
  ADCCFG = bADLSMP | ADC_10BITS | ADC_INTCLK | ADIV_8;
  ADCCV = 900;               // compare value for the ADC
  ADCSC2 = bACFE | bACFGT;   // ADC comparator enabled (greater than or equal to)
  APCTL1 = BIT_1;            // ADP1 in analog mode
  // ADC in continuous mode, channel 1, interrupts enabled
  ADCSC1 = bAIEN | bADCO | ADC_CH1;
  PTBDD = BIT_6;             // PTB6 as output
  LED1 = 1;                  // turn LED1 off
  EnableInterrupts;          // Enable interrupts (CCR:I=0)
  while (1)
  {
    STOP;                    // enter stop3 mode
  }
}
```

After changing the ADC comparator mode, it is necessary to restart the conversion by writing into the ADCSC1 register!

Example 10.7 – ADC example (digital comparator)

The ADC module can also be used in an unusual way: generation of periodical interrupts!

The next example shows how to set up the ADC to accomplish this task. By using an internal channel (such as temperature sensor, bandgap reference, REFH or REFL), no external pin is used!

Note that using a long sampling time (ADLSMP=1) and the ADC clock derived from the BUSCLK divided by sixteen (ADIV=11b and ADICLK=01b), each conversion (except the first one) takes 80µs at an 8MHz BUSCLK. Using a short sampling time (ADLSMP=0), each conversion takes 40µs to complete.

While it is possible to produce even faster interrupt rates (less than 40µs), this may lead us to an incredibly high number of interrupt requests per second and can interfere, and be interfered with by other interrupt processes (in case other peripheral interrupts are enabled).

```c
// DEMO9S08QG8 - ADC example 5 (using the ADC as a simple timer)

#include <hidef.h>          /* for EnableInterrupts macro */
#include "derivative.h"     /* include peripheral declarations */
#include "hcs08.h"          // This is our definition file!

#define LED1  PTBD_PTBD6

unsigned int timeout;

// ADC interrupt
void interrupt VectorNumber_Vadc adc_isr(void)
{
  char temp;
  temp = ADCRL;    // dummy read of ADCR (only the low byte) to clear the interrupt flag
  if (timeout) timeout--;
}

// ADC delay function (each "delay" unit is equal to 80us)
void adc_delay(unsigned int delay)
{
  timeout = delay;
  while (timeout);
}

void main(void)
{
  // configure SOPT1 register, enable pin BKGD
  SOPT1 = bBKGDPE;
  ICSSC = NV_FTRIM;          // configure FTRIM value
  ICSTRM = NV_ICSTRM;        // configure TRIM value
  ICSC2 = 0;                 // ICSOUT = DCOOUT / 1
  // enable long sampling, 10-bit mode, ADICLK = 11b, ADCK = BUSCLK
  // ADC clock = 500 kHz
  ADCCFG = bADLSMP | ADC_10BITS | ADC_BUSCLK_DIV2 | ADIV_8;
  // ADC in continuous mode, channel 30 (VREFL), interrupts enabled
  // one conversion each 40 ADC clocks = 80us
  ADCSC1 = bAIEN | bADCO | ADCH_REFL;
  PTBDD = BIT_6;             // PTB6 as output
  EnableInterrupts;          // Enable interrupts (CCR:I=0)
  while (1)
  {
    LED1 = 0;          // LED1 on
    adc_delay(6250);   // wait for 500ms
    LED1 = 1;          // LED1 off
    adc_delay(6250);   // wait for 500ms
  }
}
```

Example 10.8 – ADC example (timer)

11

Communication Peripherals

In this chapter we take a look at some of the communication peripherals available on HCS08 devices. Three basic interfaces are covered: the Serial Peripheral Interface (SPI), the Inter-Integrated Communication (IIC or I^2C) and the Serial Communication Interface (SCI, also known as asynchronous interface or UART - Universal Asynchronous Receiver-Transmitter).

Some newer devices may include Controller Area Network (CAN) interfaces (DN, DV and DZ families), or Local Interconnect Network (LIN) interfaces (EL devices) or Universal Serial Bus (USB) interfaces (JM family), but these peripherals are not discussed in this book due to the lack of testing hardware at the time of writing.

11.1. SPI

The serial peripheral interface (SPI) protocol was developed by Motorola to enable easy interfacing between microprocessors and support ICs (such as analog to digital converters, memories, etc.).

This is a very simple synchronous, master-slave protocol based on four lines: a clock line, a serial output, a serial input and a selection (enable) line. The master device always initiates the communication (by asserting the enable/selection line). After enabling the slave device, the master starts clocking out the data (through its output line). For each clock edge, one bit is output and one input. After eight clocks the transfer is done: the master sent one byte of data to the slave device and received on byte of data from it.

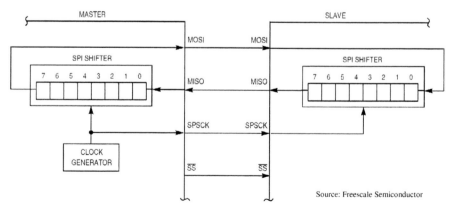

Figure 11.1 – SPI communication block diagram

Thanks to this simple construction, the SPI can even communicate with "dumb" devices such as serial shift registers.

The SPI interface in the HCS08 microcontrollers has a simple and efficient design and is found on most devices (except the QA, QD and 8-pin versions of the QG). There is an 8-bit shift register, a clock generation circuit (used only when the interface operates as a master) and additional circuits to control the interface pins. Each module can be enabled/disabled by using the SPIC1:SPE bit.

 On JM devices, the SPI module can work in 8 or 16-bit modes. The SPI16V1 module of these devices has some differences from the SPI module seen in this topic but when operating in 8-bit mode (default mode) it is very similar to the SPI module of other HCS08 devices.

When the module is operating in master mode (SPIC1:MSTR = 1), the clock output is done through the SPSCK pin, the data input is done through the MISO pin and the data output is done through the MOSI pin. The slave selection pin (SS) is used to enable the slave device (the target of the communication). It is also possible to configure the SS pin to operate as an input (in this case, it is possible to detect another master device trying to control the slave device).

Before a transmission takes place, it is necessary to check if the transmit buffer (SPID register) is empty. This is indicated by the SPIS:SPTEF flag's being set. A write operation into the SPID register starts a transmission (the SPTEF flag is cleared in the same time, indicating the transmission buffer is empty).

Once the transmission is started, the shift register starts shifting the data, starting by the most significant bit (MSB). Each bit is placed on the MOSI output on each edge of the SPSCK signal. At the same time, the MISO input is read and stored into the receive buffer. If the SPIC1:LSBFE bit is set, the transmission starts by the least significant bit (LSB).

After eight clock pulses, the slave device received one byte of data from the master and the master received one byte of data from the slave. The new data just received sets the "reception buffer full" flag (SPIS:SPRF).

Notice that the slave selection (SS) pin must be driven low ("0") during the whole transfer. This control can be done through the master SS pin (when SPIC2:MODFEN=1 and SPIC1:SSOE=0) or through a general-purpose I/O pin. In this case the software must enable (low level) the slave device before starting the transmission and disable it (high level) once the transmission is done.

A typical communication between a master and a slave SPI device comprises several steps:

1. The master device sends one byte of data, typically a command that the slave device must interpret and process.

2. If the command has additional bytes (such as parameters), they are transmitted in sequence.

3. In case the command implies a response from the slave device (such as an ADC conversion result); the master sends a dummy byte and simultaneously receives a byte corresponding to the desired data from the slave. As an example, let us see a 16-bit analog to digital converter: to read a conversion result, the master first sends a dummy byte and stores the byte read from the slave. Then the master sends another dummy byte and receives the second byte of data. These two bytes of data comprise the 16-bit result of the conversion.

It is important to note that there is no way to only receive data in SPI: data reception is always associated with data writing.

The SPI clock (which synchronizes data transmission and reception), is generated from the bus clock (BUSCLK) after being divided by two factors located in the SPIBR register: SPPR, selectable among 1, 2, 3, 4, 5, 6, 7 or 8 and SPR, selectable among 2, 4, 8, 16, 32, 64, 128 or 256. This way, the maximum clock rate is BUSCLK/2 (up to 12.5MHz on 50MHz devices) and the minimum clock rate is BUSCLK/2048.

It is also possible to select the clock polarity and clock phase in which the data shift actually occurs. These selections are performed by the CPOL and CPHA bits (both located in the SPIC1 register). The following table shows the possible combinations.

SPI Mode	CPOL	CPHA	Description
0	0	0	clock active high ("1"), the data is stored in the rising edge of the SPSCK
1	0	1	clock active high ("1"), the data is stored in the falling edge of the SPSCK
2	1	0	clock active low ("0"), the data is stored in the falling edge of the SPSCK
3	1	1	clock active low ("0"), the data is stored in the rising edge of the SPSCK

Table 11.1- SPI modes

Figure 11.2 shows the signal diagram for the transmission of 0x15 using the four SPI formats available (the slave output is not pictured in the figure).

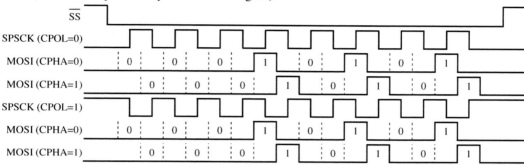

Figure 11.2 – SPI formats

When the SPI interface is operating in slave mode (SPIC1:MSTR = 0), the clock signal is supplied by the master device. While in this mode, the interface can continue operating in wait mode if the SPIC2:SPISWAI bit is set and the interface is enabled and configured before entering wait mode.

Note that while the interface is operating in slave mode, the SPSCK pin operates as a clock input, the MOSI pin operates as a data input, the MISO pin operates as an output and the SS pin operates as a selection input (active in low level).

11.1.1. SPI Transfers

As we said earlier, each SPI transfer comprises a transmission and a reception. The master device clocks the slave device and for each bit sent (through the MOSI output) another one is read (through the MISO input).

The SPID register operates as a read/write double buffer: data written to SPID is transferred to the internal shift register at the beginning of the transfer (this sets the transmitter buffer empty (SPIS:SPTEF) flag). When the transfer finishes, the received data (stored in the SPI internal shift register) is transferred to the SPID register and the SPI read buffer full (SPIS:SPRF) flag is set.

These two flags can be used to control and monitor SPI transfers. SPRF, when set, generates an interrupt request if SPIC1:SPIE = 1. SPRF is cleared by first reading SPIS (when SPRF is set) and then reading the SPID register.

SPTEF, when set, generates an interrupt request if SPIC1:SPTIE = 1. It is cleared by reading SPIS when SPTEF is set and then writing into SPID.

 It is necessary to read SPIS:SPTEF before writing data to SPID. Failing to do so causes the write to be ignored by the SPI interface and no data is transmitted!

11.1.2. Slave Selection Pin

When the interface is operating in slave mode, the SS pin is configured as an input and enables the communication when its logic level is "0".

When the interface is operating in master mode, the behavior of the SS pin is controlled by two bits: SPIC1:SSOE and SPIC2:MODFEN.

If MODFEN = 0, the pin reverts to general purpose I/O mode and is not used by the interface (slave device selection must be done by the application, using I/O pins).

If MODFEN = 1, the pin is controlled by the SPI module but its actual function depends on the SSOE bit:

When SPIC1:SSOE = 1, the pin operates as a single slave selection output (it is cleared during a transfer, otherwise it is set).

When SSOE = 0, the pin operates as a mode fault detection input. This feature allows detection of another master device communicating with the same slave device.

The mode fault pin must stay high for the interface to operate regularly. If SS is found low during a transfer, a mode fault error arises and the SPIS:MODF flag is set. This flag generates an interrupt request if SPIC1:SPIE = 1.

Once MODF is set, the application must abort the transfer an retry it later (aborting a transfer can be done at any time by clearing the SPIC1:SPE bit). MODF is cleared by reading SPIS (when MODF is set) and then writing into SPIC1.

 Note that when using the mode fault detection feature, the slave selection must be done by using general purpose I/O pins (the slave selection signal is no longer automatically generated)!

11.1.3. Single Wire Data Mode

One interesting feature of the SPI interface on the HCS08 devices is the single wire capability. It enables the SPI to operate with only three wires: bidirectional data, clock and slave selection.

The single wire mode is selected by setting the SPIC2:SPC0 bit. While in this mode, the MOSI pin becomes MOMI (master in, master out) and the MISO pin becomes SISO (slave in, slave out).

A master device communicates through its MOMI pin and a slave device communicates through its SISO pin.

The direction of the data pin (input or output) is controlled by the SPIC2:BIDIROE bit. When BIDIROE = 0, the data pin is an input; when BIDIROE = 1, the data pin is an output.

11.1.4. SPI External Connections

The SPI modules are connected to general purpose I/O pins. Tables 11.2 and 11.3 show the connections for some HCS08 devices.

SPI Signal	Ax	Dx, EN	EL, SL	GB, GT	LL*	QE4/8	QG	Rx	SG/SH
MOSI(MOMI)	PTE6	PTE4	PTB3	PTE4	PTA3	PTB3	PTB3	PTC4	PTB3
MISO(SISO)	PTE5	PTE5	PTB4	PTE3	PTA2	PTB4	PTB4	PTC5	PTB4
SPSCK	PTE7	PTE3	PTB2	PTE5	PTA1	PTB2	PTB2	PTC6	PTB2
SS	PTE4	PTE2	PTB5	PTE2	PTA0	PTB5	PTB5	PTC7	PTB5

* When SOPT2:SPIPS=1, MOSI = PTB5, MISO = PTB4, SPSCK = PTB6 and SS = PTB7

Table 11.2

SPI Signal	JM	LC	QE64/96/128
MOSI1(MOMI1)	PTB1	PTB5	PTE1
MISO1(SISO1)	PTB0	PTB4	PTE2
SPSCK1	PTB2	PTB6	PTE0
$\overline{SS1}$	PTB2	PTB7	PTE3
MOSI2(MOMI2)	PTE5	PTC0	PTD1
MISO2(SISO2)	PTE4	PTC1	PTD2
SPSCK2	PTE6	PTC2	PTD0
$\overline{SS2}$	PTE7	PTC3	PTD3

Table 11.3

11.1.5. SPI Registers

Each SPI module includes five registers:

♦ SPI Control 1 (SPIC1) register.

♦ SPI Control2 (SPIC2) register.

♦ SPI Status (SPIS) register.

♦ SPI Data (SPID) register.

♦ SPI Baud Rate (SPIBR) register.

11.1.5.1. SPIC1 Register

Name	Model		BIT 7	BIT 6	BIT 5	BIT 4	BIT 3	BIT 2	BIT 1	BIT 0
SPIC1	All*	Read	SPIE	SPE	SPTIE	MSTR	CPOL	CPHA	SSOE	LSBFE
		Write								
		Reset	0	0	0	0	0	0	0	0

* The SPI is not available on QA and QD devices. On JM, LC and QE64/96/128 devices, there are two SPI interfaces and two SPIC1 registers: SPI1C1 and SPI2C1.

Bit Name	Description	C Symbol
SPIE	SPI interrupt enable bit. This bit enables interrupt request for the SPI receive buffer full (SPRF) and mode fault (MODF) events. 0 – SPRF and MODF interrupts disabled. 1 – SPRF and MODF interrupts enabled.	bSPIE
SPE	SPI system enable bit. This bit controls whether the SPI interface is enabled or disabled. It also initializes internal state machines and clears the data buffers. 0 – SPI interface disabled. 1 – SPI interface enabled.	bSPE
SPTIE	SPI transfer interrupt enable bit. This bit controls whether the transmit buffer empty flag (SPTEF) generates an interrupt request when set. 0 – SPTEF interrupt disabled. 1 – SPTEF interrupt enabled.	bSPTIE
MSTR	Master/slave mode selection bit. This bit controls the operating mode for the SPI interface: 0 – Slave mode. 1 – Master mode.	bMSTR
CPOL	Clock polarity selection bit. This bit controls the polarity of the clock signal: 0 – SPI clock active high (idles in low state). 1 – SPI clock active low (idles in high state).	bCPOL
CPHA	Clock phase selection bit. This bit controls the phase of the clock signal: 0 – First edge of clock occurs at the middle of the first cycle of an 8-bit data transfer. 1 – First edge of clock occurs at the start of the first cycle of an 8-bit data transfer.	bCPHA
SSOE	Slave select output enable bit. This bit controls the operation of the SS pin when MODFEN=1 in master mode (MSTR=1): 0 – SS is an input and detects when another master tries to select the slave device. 1 – SS is an output and is used to select the slave device.	bSSOE
LSBFE	LSB first enable bit. This bit controls the direction of the SPI shift register: 0 – SPI transfers initiate by the most significant bit (MSB). 1 – SPI transfers initiate by the least significant bit (LSB).	bLSBFE

11.1.5.2. SPIC2 Register

Name	Model		BIT 7	BIT 6	BIT 5	BIT 4	BIT 3	BIT 2	BIT 1	BIT 0
SPIC2	All*	Read	0	0	0	MODFEN	BIDIROE	0	SPISWAI	SPC0
		Write	-	-	-			-		
		Reset	0	0	0	0	0	0	0	0

* The SPI is not available on QA and QD devices. On JM, LC and QE64/96/128 devices, there are two SPI interfaces and two SPIC2 registers: SPI1C2 and SPI2C2.

Bit Name	Description	C Symbol
MODFEN	Master mode-fault function enable bit. This bit controls the function of the slave select (SS) pin. 0 – SS pin not controlled by the SPI interface. 1 – SS pin controlled by the SPI interface (according to the SSOE configuration).	bMODFEN
BIDIROE	Bidirectional mode output enable bit. This bit controls the direction of the data pin (MOMI or SOSI) when the SPI interface is configured for single wire mode (SPC0 = 1). 0 – Output driver disabled (data pin as an input). 1 – Output driver enabled (data pin as an output).	bBIDIROE
SPISWAI	SPI stop in wait mode. This bit controls whether the SPI interface continue operating in wait mode. 0 – SPI interface continue operating in wait mode. 1 – SPI interface does not operate in wait mode.	bSPISWAI
SPC0	SPI single wire mode select bit. This bit selects the single wire operation for the SPI interface. 0 – The SPI interface uses separate pins for data input and output. 1 – The SPI interface uses a single pin for data input and output. In slave mode (SPIC1:MSTR=0), the data input/output is the MISO/SISO pin, in master mode (SPIC1:MSTR=1), the data input/output is the MOSI/MOMI pin.	bSPC0

11.1.5.3. SPIS Register

Name	Model		BIT 7	BIT 6	BIT 5	BIT 4	BIT 3	BIT 2	BIT 1	BIT 0
SPIS	All*	Read	SPRF	0	SPTEF	MODF	0	0	0	0
		Write	-	-	-	-	-	-	-	-
		Reset	0	0	0	0	0	0	0	0

* The SPI is not available on QA and QD devices. On JM, LC and QE64/96/128 devices, there are two SPI interfaces and two SPIS registers: SPI1S and SPI2S.

Bit Name	Description	C Symbol
SPRF	SPI read buffer full flag. This bit indicates when a full byte is received by the SPI interface and is available in the buffer (SPID). This bit is cleared by reading SPIS (when SPRF is set) and then reading SPID. 0 – No data available in the SPI receive buffer. 1 – A byte of data is available in the SPI receive buffer (SPID).	bSPRF
SPTEF	SPI transmit buffer empty flag. This bit indicates when the SPI transmit buffer is empty and ready for a new byte of data. It is cleared by reading SPIS (when SPTEF is set) and then writing into SPID. This flag generates an interrupt request when SPIC1:SPTIE=1. 0 – SPI transmit buffer is not empty. 1 – SPI transmit buffer is empty. *** In order to successfully transmit a byte of data, the application needs to read SPIS and then write the data into SPID. Failing to do so discards the data written into SPID and no transmission occurs!**	bSPTEF
MODF	Master mode fault flag. This bit indicates when the SPI interface is operating in master mode and the slave select input (SS) is externally driven low (indicating another master is accessing the same slave device). MODF is cleared by reading SPIS (when MODF is set) and then writing into SPIC1. 0 – No mode fault error. 1 – Mode fault error detected.	bMODF

11.1.5.4. SPID Register

Name	Model		BIT 7	BIT 6	BIT 5	BIT 4	BIT 3	BIT 2	BIT 1	BIT 0
SPID	All*	Read	\multicolumn{8}{c}{Return the last byte received by the SPI interface}							
		Write	\multicolumn{8}{c}{Write new data into the SPI data buffer}							
		Reset	0	0	0	0	0	0	0	0

* The SPI is not available on QA and QD devices. On JM, LC and QE64/96/128 devices, there are two SPI interfaces and two SPID registers: SPI1D and SPI2D.

Writing into SPID causes the SPI to start transmitting (in another transmission was not already in progress). Reading SPID returns the last character received by the SPI.

11.1.5.5. SPIBR Register

Name	Model		BIT 7	BIT 6	BIT 5	BIT 4	BIT 3	BIT 2	BIT 1	BIT 0
SPIBR	All*	Read	0	SPPR2	SPPR1	SPPR0	0	SPR2	SPR1	SPR0
		Write	-				-			
		Reset	0	0	0	0	0	0	0	0

* The SPI is not available on QA and QD devices. On JM, LC and QE64/96/128 devices, there are two SPI interfaces and two SPIBR registers: SPI1BR and SPI2BR.

Bit Name	Description	C Symbol
SPPR2 **SPPR1** **SPPR0**	Prescaler divisor select bits. These bits select the prescaler divisor for the SPI clock generator.	
	000 – Divide by 1.	**SPI_PRE1**
	001 – Divide by 2.	**SPI_PRE2**
	010 – Divide by 3.	**SPI_PRE3**
	011 – Divide by 4.	**SPI_PRE4**
	100 – Divide by 5.	**SPI_PRE5**
	101 – Divide by 6.	**SPI_PRE6**
	110 – Divide by 7.	**SPI_PRE7**
	111 – Divide by 8.	**SPI_PRE8**
SPR2 **SPR1** **SPR0**	SPI baud rate divisor bits.	
	000 – Divide by 2.	**SPI_DIV2**
	001 – Divide by 4.	**SPI_DIV4**
	010 – Divide by 8.	**SPI_DIV8**
	011 – Divide by 16.	**SPI_DIV16**
	100 – Divide by 32.	**SPI_DIV32**
	101 – Divide by 64.	**SPI_DIV64**
	110 – Divide by 128.	**SPI_DIV128**
	111 – Divide by 256.	**SPI_DIV256**

11.1.6. SPI Examples

The first example shows how to interface a serial-in parallel-out CD4094 shift register to the SPI module of an MC9S08QG8 device. This is an easy and inexpensive way to increase the total number of outputs of a microcontroller.

The CD4094 is an 8-bit serial-in parallel-out shift register with a serial data in pin (D), a clock input pin (CLK), an enable pin (STR) and an output enable pin (OE).

The shift register stores serial data when the STR pin is low. The stored data is output when OE=1. The output pins (Q1 through Q8) are in high impedance state when OE=0.

In our example, the OE pin is directly connected to V_{DD} and the state of the output pins reflect the data stored into the shift register.

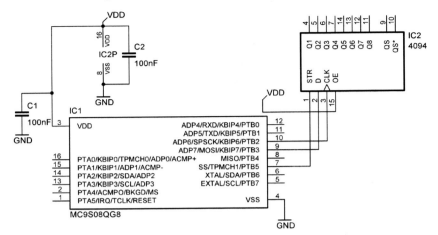

Figure 11.3

```
// DEMO9S08QG8 - SPI example 1 (driving a 4094 shift register)
// STROBE connected to PTB5/SS
// DATA connected to PTB3/MOSI
// CLOCK connected to PTB2/SPSCK

#include <hidef.h>          /* for EnableInterrupts macro */
#include "derivative.h"     /* include peripheral declarations */
#include "hcs08.h"          // This is our definition file!

char counter;

void mcu_init(void)
{
  SOPT1 = bBKGDPE;          // Enable debug pin
  ICSSC = NV_FTRIM;         // configure FTRIM value
  ICSTRM = NV_ICSTRM;       // configure TRIM value
  // BUSCLK is now 8 MHz
  // Now we configure the SPI interface: Master mode, SS pin as slave select output
  // SPI clock = BUSCLK/4 = 1MHz
  SPIC1 = bSPE | bMSTR | bSSOE;
  SPIC2 = bMODFEN;
  SPIBR = SPI_PRE1 | SPI_DIV4;  // SPI clock = BUSCLK/4 = 1MHz
}

void main(void)
{
  unsigned int temp;
  mcu_init();               // initialize the MCU
  counter = 0;
  while(1)
  {
    temp = SPIS;            // read the SPI flags
    SPID = counter++;       // write "counter" through the SPI
    for (temp=20000; temp; temp--); // wait for a while
  }
}
```

We are not testing SPTEF because the big delay added by the **for** loop makes it unnecessary!

Example 11.1 – SPI example 1 (driving one 4094)

The example implements a simple 8-bit counter and outputs through the shift register the current counting.

The SS pin is directly controlled by the SPI module. This works when there is just a single slave device, but when there are multiple slave devices, it is necessary to have multiple SS pins.

HCS08 Unleashed

The next example shows how to use the SPI module to interface with two CD4094 shift registers. The data inputs (D) are both connected to PTB3/MOSI, the clock inputs (CLK) are both connected to PTB2/SPSCK, the strobe inputs (STR) are connected to PTB5/SS (IC2) and to PTB6 (IC3).

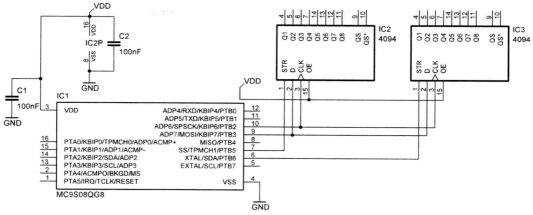

Figure 11.4

 The 4094 shift registers allow cascading multiple devices using the same chip select and SPI connection. The objective of this example is only show how to address two different slave devices with the same SPI port.

The slave selection pins are controlled by software. Each SS pin is activated at the beginning of the transfer and deactivated after the transfer is complete.

The SPI receive buffer full flag (SPIS:SPRF) is used to detect the end of the transfer. This is necessary because the SPI transmit buffer empty flag (SPIS:SPTEF) cannot be used to diagnose the end of the transmission (it is set when the transmit buffer is emptied and that happens immediately after the beginning of the transmission).

```
// DEMO9S08QG8 - SPI example 2 (driving two 4094 shift registers)
// DATA connected to PTB3/MOSI
// CLOCK connected to PTB2/SPSCK
// STROBE (IC2) connected to PTB5/SS
// STROBE (IC3) connected to PTB6

#include <hidef.h>          /* for EnableInterrupts macro */
#include "derivative.h"     /* include peripheral declarations */
#include "hcs08.h"          // This is our definition file!

#define SS1 PTBD_PTBD5
#define SS2 PTBD_PTBD6

char counter;
char spi_busy;

void interrupt VectorNumber_Vspi spi_receive_isr(void)
{
  unsigned char temp;
  temp = SPIS;       // read SPIS
  temp = SPID;       // read received data (dummy read, clear SPRF)
  SS1 = 1;           // disable SS1 (IC2)
  SS2 = 1;           // disable SS2 (IC3)
  spi_busy = 0;      // clear busy flag (SPI is idle)
}

// MCU initialization
void mcu_init(void)
{
  SOPT1 = bBKGDPE;          // Enable debug pin
  ICSSC = NV_FTRIM;         // configure FTRIM value
```

```
        ICSTRM = NV_ICSTRM;        // configure TRIM value
        // BUSCLK is now 4 MHz
        // Now we configure the SPI interface:
        // Master mode, SS pin as slave select output, SPI clock = BUSCLK/4 = 1MHz
        SPIC1 = bSPIE | bSPE | bMSTR;
        SPIBR = SPI_PRE1 | SPI_DIV4;          // SPI clock = BUSCLK/4 = 1MHz
        // Configure Port B pins:
        PTBDD = BIT_5 | BIT_6;                // PTB5 and PTB6 as outputs
        EnableInterrupts;
}

void spi_write(char data, char device)
{
    char temp;
    while (spi_busy);        // verify if the SPI is idle (wait until it is free)
    spi_busy = 1;            // SPI is busy from now on
    temp = SPIS;             // read the SPI flags
    if (device) SS2 = 0; else SS1 = 0;  // enable the desired 4094
    SPID = data;             // transmit data through SPI
}

void main(void)
{
    unsigned int temp;
    mcu_init();              // initialize the MCU
    counter = 0;
    while(1)
    {
      spi_write(counter++,0);          // write to the first 4094 (IC2)
      spi_write(~counter,1);           // write to the second 4094 (IC3)
      for (temp=20000; temp; temp--);  // wait for a while
    }
}
```

Example 11.2 – SPI example 2 (driving two 4094)

The next example shows an implementation of a software (bit-banged) SPI. The whole transfer is controlled by the spi_write() function. Note that the example implements SPI mode 0.

```
// DEMO9S08QG8 - SPI example 3 (bit-banged  SPI)
// DATA connected to PTB3/MOSI
// CLOCK connected to PTB2/SPSCK
// STROBE connected to PTB5/SS

#include <hidef.h>           /* for EnableInterrupts macro */
#include "derivative.h"      /* include peripheral declarations */
#include "hcs08.h"           // This is our definition file!

#define DATA    PTBD_PTBD3
#define CLOCK   PTBD_PTBD2
#define STROBE  PTBD_PTBD5

char counter;

// MCU initialization
void mcu_init(void)
{
  SOPT1 = bBKGDPE;                  // Enable debug pin
  ICSSC = NV_FTRIM;                 // configure FTRIM value
  ICSTRM = NV_ICSTRM;               // configure TRIM value
  // BUSCLK is now 4 MHz
  // Configure Port B pins:
  PTBDD = BIT_2 | BIT_3 | BIT_5;    // PTB2, PTB3 and PTB5 as outputs
  EnableInterrupts;
}

// This is the bit-banged SPI function. It emulates SPI master (output only) in SPI mode 0
void spi_write(char data)
{
  unsigned temp;
  STROBE = 0;                       // enable the 4094
  for (temp=8; temp; temp--)        // repeat eight times
  { // if the msb of data = 1, force the DATA pin to 1, else force the DATA pin to 0
    if (data & BIT_7) DATA = 1; else DATA = 0;
```

```
          CLOCK = 1;                    // set the CLOCK pin
          data = data << 1;             // shift data one bit to the left
          CLOCK = 0;                    // clear the CLOCK pin
      }
    STROBE = 1;                         // disable the 4094
}


void main(void)
{
    unsigned int temp;
    mcu_init();                         // initialize the MCU
    counter = 0;
    while(1)
    {
      spi_write(counter++);             // write to the 4094
      for (temp=10000; temp; temp--);   // wait for a while
    }
}
```

Example 11.3 – Bit-banged SPI

11.2. I²C Interface

The I²C interface was formerly developed by Royal Philips Electronics to aid reducing cost and increasing the modularity of TV sets.

The I²C (Inter-Integrated Communication) is a synchronous master-slave protocol based on a two-wire interface: the SDA (Serial Data) line and the SCL (Serial Clock) line. The protocol also specifies that both lines are open-drain (external pull up devices are needed for proper operation).

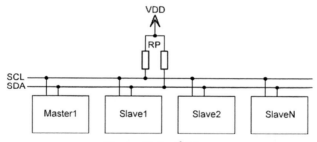

Figure 11.5 – I²C bus

The I²C protocol, as the SPI, is largely used for intra-circuit communication: several A/D converters, D/A converters, serial memories, real-time clocks, displays, etc.

While slower than the SPI, the I²C protocol has the advantages of lower communication lines (only two) and higher number of devices in the same bus (up to 127 devices can share the same I²C bus).

The I²C interface included on the HCS08 devices (except QA, QD and 8-pin QG devices) is capable of operating in master or slave modes.

11.2.1. I²C Basics

The main idea behind the protocol is using as few as possible communication lines and being as simple as possible.

The protocol is master-slave synchronous and it uses a clock line (SCL) to synchronize the slave with the master device. A single bidirectional data line (SDA) is used to interchange data between the master and the slave device.

Four simple rules govern the I²C communication:

1. The data line (SDA) and clock line (SCL) cannot be actively driven high by any I²C device (either master or slave device). I²C devices must use open-drain (or open-collector) drivers. High logic level is obtained by using external pull up devices (usually resistors).

2. The information on SDA is only read on the high phase of the clock line (SCL).

3. Changing the level of SDA is only allowed in the low phase of the clock line (SCL).

4. When the bus is not in use, the SDA and SCL lines must be left in off state (forced by the pull ups to "1").

The protocol also states that the start and the end of the transmission is signaled by violating the third rule. The start of the transmission is signaled by a falling edge on SDA when SCL is high (this is called a start condition). The end of the transmission is signaled by a rising edge on SDA when SCL is high (this is called a stop condition). Note that both the start and the stop conditions are always generated by the master device.

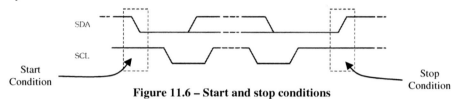

Figure 11.6 – Start and stop conditions

After the start condition, eight bits of data are transmitted, starting from the MSB. Following the 8 bits of data, the receiver generates an acknowledge bit (ACK) by forcing down the SDA line on the ninth clock pulse of the SCL line.

If the receiver does not recognize the received data, it generates a not-acknowledge bit (NACK) by not driving the SDA line low on the ninth clock pulse of the SCL line. In this case, the transmitter aborts the current transmission (generating a stop condition) and retries it later. Figure 11.7 shows a typical I²C frame (the data is 0x26 and an ACK is present).

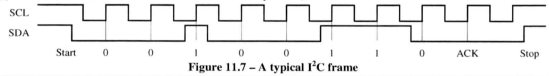

Figure 11.7 – A typical I²C frame

Rising and falling edges are usually not as sharp as shown in figure 11.7 due to the line capacitance and pull up devices. The larger the capacitance and the pull up resistance, the higher are the rise and fall times. High rise and fall times can lead to communication errors.

When designing I²C buses, it is important to consider the number of devices, their input capacitances, the line capacitance and the value of the pull up devices. Some semiconductor manufacturers offer special devices to extend bus capability and provide special active pull up devices.

Note that the maximum allowed bus capacitance is 400pF.

An additional feature of the I²C protocol is the ability of the slave devices to slow down the master device while a transfer is in progress. Once the SCL line is low, the slave device can actively force it down, until it has time to continue with the transfer, releasing the SCL line, allowing the master to continue. This feature is called "clock stretching".

To allow sharing the bus among multiple devices, the I²C protocol states that each device (master or slave) must have its own unique address. An address, in I²C terms, can be seven or ten bits long.

Figure 11.8 shows the structure of a 7-bit addressing frame. There are seven bits for the slave address, one bit for the R/W bit and the ACK/NACK bit. The addressing frame is always generated by the master device and the ACK bit is generated by the slave device whose address matches the address specified in the frame.

	Bit 7	Bit 6	Bit 5	Bit 4	Bit 3	Bit 2	Bit 1	Bit 0	
START	A6	A5	A4	A3	A2	A1	A0	R/W	ACK
			Slave address (7 *bits*)					0-write	0-ACK
								1-read	1-NACK

Figure 11.8 – I²C addressing frame

Typically, a 7-bit address comprises two distinct fields: the first four bits (A6 through A3) specify the class of the slave device and the 3 last bits (A2, A1 and A0) specify one among up to eight devices of that class.

Table 11.4 shows some classes of slave devices and other special I²C addresses.

Address	R/W	Description
0000000	0	General call address
0000000	1	Start byte (helps slow slave devices to identify the beginning of a transfer)
0000001	x	CBUS address
0000010	x	Reserved for different bus formats
0000011	x	Reserved for future purposes
00001xx	x	High-speed master mode (up to 3.4Mbps). This mode is not available on HCS08 devices.
0010xxx	x	Speech synthesizers
0011xxx	x	PCM audio interfaces
0100xxx	x	Audio-tone generators
0111xxx	x	LED/LCD displays
1000xxx	x	Video interfaces
1001xxx	x	A/D and D/A converters
1010xxx	x	Memories
1100xxx	x	RF synthesizers
1101xxx	x	Clock/calendars
11111xx	x	Reserved for future purposes
11110xx	x	10-bit addressing mode

Table 11.4 – Some special reserved I²C addresses

The R/W bit indicates whether the master is going to write data into the slave (R/W=0) or write data from the slave (R/W=1).

The addressing procedure is defined as follows:

1. The master device generates a start condition.

2. The master sends an addressing byte.

3. According to the R/W bit of the addressing byte, the master sends data to the slave (when R/W=0) or reads data from the slave (when R/W=1).

4. Once all data is transferred, the master device generates a stop condition.

 The protocol allows a special condition called "repeated start" in which the master device sends another start condition and another addressing byte without first generating a stop condition. This is useful for changing the transfer direction.

11.2.2. I²C Interface Basics

The I²C module has a very simple operation and is based on five registers. The core operation is controlled by the IICC register.

The interface is enabled by the IICC:IICEN bit. Once IICEN is set, the SDA and SCL functions are controlled by the I^2C module.

The IICC:MST bit controls whether the interface operates in master (MST=1) or slave (MST=0) mode. This bit also controls the start and stop conditions generation. A transition from "0" to "1" of MST (master mode selection) generates a start condition on the bus. A transition from "1" to "0" of MST (slave mode selection) generates a stop condition on the bus.

Another bit, IICC:TX, controls whether the interface is in transmit (TX=1) or receive (TX=0) mode. This selection is made regardless of the interface being in master or slave mode.

The data to be transmitted is written in the IICD register. This register also stores data received by the interface. That means that you cannot read the data you have just written.

There are also three flags for signaling internal I^2C states (all located in the IICS register):

- TCF – when set, indicates a transfer was completed.

- IAAS – when set, indicates the interface was addressed as a slave device.

- ARBL – when set, indicates the interface, operating in master mode, has lost the bus arbitration and switched to slave mode.

When any of these flags is set, the I^2C interrupt flag (IICS:IICIF) is set and generates an interrupt request if IICC:IICIE is set.

Clearing ARBL is done by writing "1" in it, the IAAS flag is cleared by writing into the IICC register and the TCF flag is cleared by reading IICD (when IICC:TX=0) or by writing into IICD (when IICC:TX=1).

11.2.3. Master Mode Operation

To configure the I^2C interface for master operation, some steps must be followed:

1. The communication speed must be properly set. This is done using the IICF register. The clock multiplier factor M is set through the IICF:MULT bits and the divisor factor D is set through the IICF:ICR bits. The communication speed is calculated through the following formula:

$$I^2C_{CLOCK} = \frac{BUSCLK}{M * D}$$

The maximum baud-rate for the HCS08 I^2C interface is 100kbps. Higher speeds are allowed if the bus capacitance is below the 400pF maximum.

2. The I^2C module must be enabled by setting the IICC:IICEN bit.

3. The master mode must be selected by setting the IICC:MST bit. This also generates a start condition on the bus.

4. Transmit mode must be selected. This is done by setting the IICC:TX bit.

5. The addressing byte must be written into the IICD register. The slave device whose own address matches the addressing byte generates an ACK bit. The address byte must follow the standard shown in figure 11.7.

6. After writing to the IICD register, the master must wait for the transfer to be completed. This is signaled by the IICS:TCF being set.

7. Once the transfer is completed, the master must verify the IICS:RXAK flag. This flag reflects the state of the ACK bit. The ACK bit is always generated by the receiver (the slave device in this case). If RXAK = 0, the master received an ACK bit. If RXAK=1, the master

received a NACK, this can indicate there is no slave device with such address present in the bus. After a NACK, the master must send a stop condition and, if it is the case, retry the operation later.

8. After receiving an ACK from the slave, the master can start sending or receiving bytes from the slave, depending on the operation signaled by the R/W bit of the address byte sent on step 5.

9. To transmit a byte, the IICC:TX bit must stay set and the new data must be written into the IICD register. After each byte is transmitted, the slave sends an ACK or NACK. After transmitting all data (or after receiving a NACK), the master must send a stop condition, by clearing IICC:MST.

10. To receive a byte from the slave device, the master must clear the IICC:TX bit and read the IICD register. This is a dummy read operation and it is needed to start I^2C data receiving.

11. After each transfer (sent or receive), the IICS:TCF flag is set. If the I^2C interface is in transmitting mode the TCF flag is cleared by writing into the IICD register (each new transmission clears the TCF flag). If the I^2C interface is in receive mode, the TCF flag is cleared by reading the IICD register.

12. While in receive mode, the master device must generate the acknowledge bit. This is controlled by the IICC:TXAK bit. When TXAK=0, an ACK bit is generated after receiving a byte of data. When TXAK=1, a NACK is generated after receiving a byte of data. Notice that the desired acknowledgement bit is always generated on the next transfer following its change.

13. To finish a data transfer, the master must generate a stop condition. This is done by clearing the IICC:MST bit, which also switches the interface to slave mode.

11.2.4. Slave Mode Operation

The operation of the I^2C interface in slave mode is very similar to the master mode. The steps for slave communication are:

1. The interface slave address must be set in the IICA register.

2. The slave device must wait until it gets addressed by a master device. This event is signaled by the IICS:IAAS flag's being set.

3. The software must read the IICS:SRW bit. This bit reflects the state of the R/W bit of the address byte. When SRW=0, the master writes into the slave device. When SRW=1, the master reads from the slave device.

4. If SRW=1, the application software must set the IICC:TX bit. This configures the interface for transmitting mode (the slave sends data to the master). Once TX is set, the first byte of data must be written into the IICD register.

5. After a transfer is completed, the IICS:TCF flag is set and the application needs to read the IICS:RXAK bit to verify if the master acknowledged the transfer. If RXAK=0, the master acknowledged the transfer and another byte can be written to the IICD register. In case RXAK=1, the master did not acknowledge the transfer. In this case, the application must set the interface to receive mode (IICC:TX=0) and perform a dummy read of the IICD register (to receive data on the next transfer).

6. If SRW=0 after step 3, the application must set the interface to receive mode (IICC:TX=0) and perform a dummy read of the IICD register to start the data reception.

7. After a transfer is completed, the application must read the received data in the IICD register. On each transfer, the slave interface generates an acknowledgement bit depending

on the state of the IICC:TXAK bit. If TXAK=0, an ACK is sent. If TXAK=1, a NACK is sent.

11.2.5. I²C Bus Arbitration Loss

When several master devices share the same I²C bus, it is possible that two or more masters try to use the bus at the same time.

When such condition occurs, a process called "bus arbitration" begins. In this process, each master interface monitors its SDA line. Every master that writes a "1" into the bus and reads a "0" instead, automatically switches to slave receiver mode.

After the arbitration process takes place, only one master controls the bus and all others are operating as slave receivers.

The HCS08 I²C interface signalizes the arbitration loss event by setting the IICS:ARBL flag. Note that the following events can lead to an arbitration loss:

1. The SDA line reads "0" when the master is writing "1" on a master transmission.

2. The SDA line reads "0" when the master is writing "1" on a NACK phase.

3. Attempts to generate a start condition when the bus is busy (IICS:BUSY=1). Note that the bus is considered busy when a start condition is detected and is considered idle after a stop condition is detected.

4. Attempts to generate a repeated start condition when the interface is operating in slave mode.

5. Detection of a stop condition not generated by the master interface.

11.2.6. I²C External Connections

The I²C modules are connected to general purpose I/O pins. Table 11.5 shows the I²C connections for some HCS08 devices.

I²C Signal	AC/AW	Dx	EL/SL	GB/GT	JM	LC/LL	QE*	QG	SG/SH
SDA	PTC1	PTF3	PTA2	PTC2	PTC1	PTB4	PTA2/PTH7	PTA2	PTA2
SCL	PTC0	PTF2	PTA3	PTC3	PTC0	PTB5	PTA3/PTH6	PTA3	PTA3

* Some QE devices include two I²C interfaces (IIC1 and IIC2). SDA1 is connected to PTA2, SDA2 is connected to PTH7, SCL1 is connected to PTA3 and SCL2 is connected to PTH6.

Table 11.5 – I²C pins

Some devices allow alternate pins for the I²C interface. This selection is made through the SOPT1 or SOPT2 register, according to each device. Refer to topic 5.5 for further information on the option registers.

I²C Signal	Dx	EL/SL*	LL	QE	QG	SG/SH
SDA	PTE5	PTB6 PTB2	PTA2	PTB6	PTB6	PTB6
SCL	PTE4	PTB7 PTB3	PTA3	PTB7	PTB7	PTB7

* These devices allow three connection options for the I²C interface.

Table 11.6 – Alternate I²C pins

11.2.7. I²C Module Versions

The following table shows the evolution of I²C modules since version 1:

I²C Version	Devices	Features added/modified
1	AW, GB, GT, LC, QG	First release.
2	AC, Dx, EL, JM, LL, QE, SG, SH, SL	General call address, 10-bit slave address

Table 11.7 – I²C module history

11.2.8. I²C Registers

The I²C interface includes some registers for controlling its operation and status:

- ◆ IIC Control (IICC) register.
- ◆ IIC Control 2 (IICC2) register (only on AC, Dx, EL, JM, QE, SG, SH and SL devices).
- ◆ IIC Status (IICS) register.
- ◆ IIC Frequency Divider (IICF) register.
- ◆ IIC Address (IICA) register.
- ◆ IIC Data (IICD) register.

11.2.8.1. IICC Register

Name	Model		BIT 7	BIT 6	BIT 5	BIT 4	BIT 3	BIT 2	BIT 1	BIT 0
IICC	AC, AW, Dx, EL,	Read	IICEN	IICIE	MST	TX	TXAK	0	0	0
IIC1C*	GB, GT, JM, LC,	Write						RSTA	-	-
IIC2C*	LL, QE, QG, SG, SH, SL	Reset	0	0	0	0	0	0	0	0

* On AC, AW, GB and GT devices, this register is named IIC1C. On QE devices the I²C control registers are named IIC1C and IIC2C.

Bit Name	Description	C Symbol
IICEN	I²C interface enable bit. This bit controls whether the I²C interface is enabled or disabled. 0 – I²C interface disabled. 1 – I²C interface enabled.	bIICEN
IICIE	I²C interrupt enable bit. 0 – I²C interrupt disabled. 1 – I²C interrupt enabled.	bIICIE
MST	Master mode selection bit. This bit selects whether the I²C interface operates in master or slave mode. When MST changes from "0" to "1", a start condition is generated. When MST changes from "1" to "0", a stop condition is generated. 0 – Slave mode. 1 – Master mode.	bMST
TX	Transmit mode selection bit. This bit controls the direction of the SDA pin. 0 – Receive mode. 1 – Transmit mode.	bTX
TXAK	Transmit acknowledge bit. This bit controls the acknowledge generation when the interface is receiving data (either in master or slave mode). 0 – An acknowledge bit is sent upon receiving a byte of data. 1 – A not-acknowledge bit is sent upon receiving a byte of data.	bTXAK
RSTA	Repeated start generation. When set, in master mode, this bit causes the generation of a repeated start condition. This is a write-only bit and is always read as "0".	bRSTA

11.2.8.2. IICS Register

Name	Model		BIT 7	BIT 6	BIT 5	BIT 4	BIT 3	BIT 2	BIT 1	BIT 0
IICS IIC1S* IIC2S*	AC, AW, Dx, EL, GB, GT, JM, LC, LL, QE, QG, SG, SH, SL	Read	TCF	IAAS	BUSY	ARBL	0	SRW	IICIF	RXAK
		Write	-		-		-	-		-
		Reset	0	0	0	0	0	0	0	0

* On AC, AW, GB and GT devices, this register is named IIC1S. On QE devices the I²C control registers are named IIC1S and IIC2S.

Bit Name	Description	C Symbol
TCF	Transfer complete flag. This bit indicates the completion of a byte transfer. The TCF flag is cleared by reading IICD (when in receive mode) or by writing a new byte of data into IICD (when in transmit mode). 0 – Transfer in progress. 1 – Transfer completed.	bTCF
IAAS	Addresses as a slave flag. This bit is set when the I²C interface (operating as a slave device) is successfully addressed (received address match IICA). This flag is cleared by writing into the IICC register.	bIAAS
BUSY	Bus busy flag. This bit indicates whether the I²C bus is busy or not (regardless of the I²C interface mode). The busy flag is set when a start condition is detected and cleared when a stop condition is detected. 0 – I²C bus is free. 1 – I²C bus is busy.	bBUSY
ARBL	Arbitration lost flag. This bit indicates when the interface has lost the arbitration of the bus. ARBL is cleared by writing "1" into it. 0 – Standard bus operation. 1 – Arbitration lost. The device must wait and retry the operation later.	bARBL
SRW	Slave read/write bit. This bit reflects the state of the R/W bit of the address field. 0 – Slave receive. The master is writing into the slave device. 1 – Slave transmit. The master is reading from the slave device.	bSRW
IICIF	I²C interrupt flag. This bit, when set, indicates an I²C interrupt (TCF, IAAS and ARBL) is pending. IICIF is cleared by writing "1" into it. 0 – No interrupt pending. 1 – An I²C interrupt is pending.	bIICIF
RXAK	Receive acknowledge flag. This flag reflects the acknowledge bit received by the transmitter after transferring a byte of data. 0 – Transmitted data acknowledged by the receiver. 1 – Transmitted data not-acknowledged by the receiver.	bRXAK

11.2.8.3. IICA Register

Name	Model		BIT 7	BIT 6	BIT 5	BIT 4	BIT 3	BIT 2	BIT 1	BIT 0
IICA IIC1A* IIC2A*	AC, AW, Dx, EL, GB, GT, JM, LC, LL, QE, QG, SG, SH, SL	Read				ADDR				0
		Write								-
		Reset	0	0	0	0	0	0	0	0

* On AC, AW, GB and GT devices, this register is named IIC1A. On QE devices the I²C control registers are named IIC1A and IIC2A.

The IICA register stores the 7-bit slave address of the I²C interface. This address is automatically compared with the address field received by the I²C interface when addressed as a slave device and it they are equal, the IICS:IAAS bit is set.

11.2.8.4. IICF Register

Name	Model		BIT 7	BIT 6	BIT 5	BIT 4	BIT 3	BIT 2	BIT 1	BIT 0
IICF	AC, AW, Dx, EL,	Read	MULT		ICR					
IIC1F*	GB, GT, JM, LC,	Write								
IIC2F*	LL,QE, QG, SG, SH, SL	Reset	0	0	0	0	0	0	0	0

* On AC, AW, GB and GT devices, this register is named IIC1F. On QE devices the I²C control registers are named IIC1F and IIC2F.

Bit Name	Description	C Symbol

MULT These bits define the multiplication factor (M) which is used to calculate the speed (baud rate) of the I^2C interface:

00 – M = 1	**I2C_M1**
01 – M = 2	**I2C_M2**
10 – M = 4	**I2C_M4**
11 – reserved.	-

ICR I^2C clock rate. These bits select the divisor factor used to calculate the I^2C clock frequency and the SDA holding value, which is the delay between the falling edge of SCL and the change in the SDA line. The table 11.8 shows the possible ICR values for some SCL divisors and SDA hold. —

ICR	SCL Divisor	SDA hold (μs)	ICR	SCL Divisor	SDA hold (μs)
0x00	20	7	0x20	160	17
0x01	22	7	0x21	192	17
0x02	24	8	0x22	224	33
0x03	26	8	0x23	256	33
0x04	28	9	0x24	288	49
0x05	30	9	0x25	320	49
0x06	34	10	0x26	384	65
0x07	40	10	0x27	480	65
0x08	28	7	0x28	320	33
0x09	32	7	0x29	384	33
0x0A	36	9	0x2A	448	65
0x0B	40	9	0x2B	512	65
0x0C	44	11	0x2C	576	97
0x0D	48	11	0x2D	640	97
0x0E	56	13	0x2E	768	129
0x0F	68	13	0x2F	960	129
0x10	48	9	0x30	640	65
0x11	56	9	0x31	768	65
0x12	64	13	0x32	896	129
0x13	72	13	0x33	1024	129
0x14	80	17	0x34	1152	193
0x15	88	17	0x35	1280	193
0x16	104	21	0x36	1536	257
0x17	128	21	0x37	1920	257
0x18	80	9	0x38	1280	129
0x19	96	9	0x39	1536	129
0x1A	112	17	0x3A	1792	257
0x1B	128	17	0x3B	2048	257
0x1C	144	25	0x3C	2304	385
0x1D	160	25	0x3D	2560	385
0x1E	192	33	0x3E	3072	513
0x1F	240	33	0x3F	3840	513

Table 11.8

11.2.8.5. IICD Register

Name	Model		BIT 7	BIT 6	BIT 5	BIT 4	BIT 3	BIT 2	BIT 1	BIT 0
IICD IIC1D* IIC2D*	AC, AW, Dx, EL, GB, GT, JM, LC, LL,QE, QG, SG, SH, SL	Read				I²C Data				
		Write								
		Reset	0	0	0	0	0	0	0	0

* On AC, AW, GB and GT devices, this register is named IIC1D. On QE devices the I²C control registers are named IIC1D and IIC2D.

In master mode, writing to the IICD register causes the data to be transmitted through the I²C bus. Reading IICD initiates the receiving of the next byte of data. Each read of the IICD retrieves the last byte received by the interface.

The first byte written to IICD after a start condition is sent is used as the address field. The application must properly configure this byte to meet the I²C standards.

In slave mode, writing to IICD queues the data to be transmitted the next time the device is read by the master. Reading IICD returns the last byte of data received by the master.

11.2.8.6. IICC2 Register

Name	Model		BIT 7	BIT 6	BIT 5	BIT 4	BIT 3	BIT 2	BIT 1	BIT 0
IICC2 IIC1C2* IIC2C2*	AC, Dx, EL, JM, LL, QE, SG, SH, SL	Read	GCAEN	ADEXT	0	0	0	AD10	AD9	AD8
		Write			-	-	-			
		Reset	0	0	0	0	0	0	0	0

* On AC devices, this register is named IIC1C2. On QE devices the I²C control registers are named IIC1C2 and IIC2C2.

Bit Name	Description	C Symbol
GCAEN	General Call Address Enable bit. This bit allows the interface to respond to a general call address. 0 – The interface does not respond to a general call. 1 – The interface responds to a general call.	bGCAEN
ADEXT	Address Extension enable bit. This bit controls whether the slave address is 7 or 10 bits wide. 0 – Slave address is 7 bits. 1 – Slave address is 10 bits.	bADEXT
AD10 AD9 AD8	Three upper bits of the slave address. The 10-bit slave address is formed by these bits and the seven bits of IICA. These bits are only valid when ADEXT=1.	-

11.2.9. Interfacing to an External I²C EEPROM

The next examples show how to interface a 24LC256 I²C EEPROM memory to the MC9S08QG8 through the I²C module.

Figure 11.8 shows the basic schematic for the test circuit. The EEPROM was connected to a DEMO9S08QG8 board through the expansion connector.

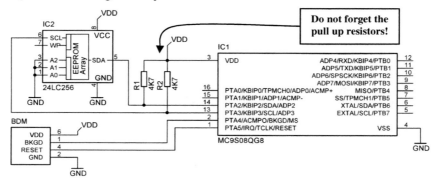

Figure 11.9

Before continuing, it is important to understand some basic principles on the I²C EEPROM operation.

1. These EEPROM devices use the 0xA0 slave address. The last three bits of their address is controlled by the A0, A1 and A2 lines. This allows up to eight devices to share the same bus (this leads to slave addresses ranging from 0xA0 (A0=A1=A2=0) to 0xAE (A0=A1=A2=1)).

2. A write operation is performed by sending the address byte (with R/W=0), which is followed by the EEPROM 2-byte address (a 24LC256 memory has 32,768 bytes or 32,768 addresses).

   The data to be written follows the EEPROM address. Some memories allow several sequential bytes to be written, but in our case, let us consider only a single byte write.

   After the data is sent, a stop condition starts the memory programming process.

3. A read operation is performed much like the write operation, but after the address is sent, the master must send a repeated start condition with R/W=1 (indicating the read operation).

   The EEPROM starts sending the content of the address. On each read operation, the content of the next address is sent. After the desired information is read, the master generates a stop condition and the transfer is finished.

The first example is based on polling the interface flags. This is a simpler approach that allows a better understanding of the interface operation.

```
// DEMO9S08QG8 - I2C example 1 (I2C EEPROM (no interrupts))
// EEPROM SDA pin connected to PTA2/SDA
// EEPROM SCL pin connected to PTA3/SCL

#include <hidef.h>              /* for EnableInterrupts macro */
#include "derivative.h"         /* include peripheral declarations */
#include "hcs08.h"              // This is our definition file!

#define EEPROM_ID   0xA0
#define EEPROM_ADR  0
#define RD          1
#define WR          0

unsigned int eeprom_address;
char eeprom_data, operation;


// MCU initialization
void mcu_init(void)
{
  SOPT1 = bBKGDPE;              // Enable debug pin
  ICSSC = NV_FTRIM;             // configure FTRIM value
  ICSTRM = NV_ICSTRM;           // configure TRIM value
  // BUSCLK is now 4 MHz
  // Now we configure the I2C interface:
  // I2C clock is 4MHz*1/40 = 100kHz, SDA hold = 9/4MHz=2.25us
  IICF = I2C_M1 | 0x0B;
  IICC = bIICEN;                // enable the I2C interface
}


// This is the I2C EEPROM byte writing function
void eeprom_bytewrite(unsigned int address, char data)
{
  char temp;
  do  // first we send the address field
  {
    // enter master transmit mode and send a start
    IICC = bIICEN | bTX | bMST;
    // send the address field (R/W = write)
    IICD = EEPROM_ID | EEPROM_ADR | WR;
    for (temp=40; temp; temp--);          // wait for a while
    while (!IICS_TCF);                     // wait until the transmission ends
```

```
      if (IICS_RXAK)                      // was an ACK received from the memory ?
      {
        IICC = bIICEN;                    // if no ACK was received, send a stop
        for (temp=40; temp; temp--);      // wait for a while
      }
    } while (IICS_RXAK);                   // if no ACK was received, resend the address field
    IICD = address >> 8;                   // send the higher byte of the address
    while (!IICS_TCF);                     // wait until the transmission ends
    IICD = address & 0xFF;                 // send the lower byte of the address
    while (!IICS_TCF);                     // wait until the transmission ends
    IICD = data;                           // send the data to be written
    for (temp=40; temp; temp--);           // wait for a while
    while (!IICS_TCF);                     // wait until the transmission ends
    // send stop, exit master mode (the memory starts programming)
    IICC = bIICEN;
}

// This is the I2C EEPROM byte reading function
char eeprom_byteread(unsigned int address)
{
    char temp, data;
    while (IICS_BUSY);
    IICS_ARBL = 1;                         // clear ARBL flag
    IICC = bIICEN | bTX | bMST;            // enter master transmitter mode
    IICD = EEPROM_ID | EEPROM_ADR | WR;
    for (temp=40; temp; temp--);           // wait for a while
    while (!IICS_TCF);                     // wait until the transmission ends
    IICD = address >> 8;                   // send the higher byte of the address
    while (!IICS_TCF);                     // wait until the transmission ends
    IICD = address & 0xFF;                 // send the lower byte of the address
    while (!IICS_TCF);                     // wait until the transmission ends
    IICC_RSTA = 1;                         // send a repeated start
    IICD = EEPROM_ID | EEPROM_ADR | RD;    // read operation
    for (temp=40; temp; temp--);           // wait for a while
    while (!IICS_TCF);                     // wait until the transmission ends
    IICC_TX = 0;                           // receive mode
    data = IICD;                           // dummy read of IICD (start the reception)
    for (temp=40; temp; temp--);           // wait for a while
    while (!IICS_TCF);                     // wait until the transmission ends
    IICC_TXAK = 1;                         // send NACK in the next read
    data = IICD;                           // read data from EEPROM
    while (!IICS_TCF);                     // wait until the transmission ends
    IICC = bIICEN;                         // send a stop, leave master mode
    for (temp=100; temp; temp--);          // wait for a while
    return(data);
}

void main(void)
{
    mcu_init();                            // initialize the MCU
    eeprom_address = 0;
    eeprom_data = 0;
    operation = 0;
    while(1)
    {
        if (operation==1)                  // write operation
        {
            eeprom_bytewrite(eeprom_address, eeprom_data);
            operation = 0;
        }
        if (operation==2)                  // read operation
        {
            eeprom_data = eeprom_byteread(eeprom_address);
            operation = 0;
        }
    }
}
```

Example 11.4 – Communicating with an I²C EEPROM (without interrupts)

Testing the application is very simple: by using the real-time debug, all you have to do is add three variables ("eeprom_address", "eeprom_data" and "operation") to the data watch window and set the window to update periodically (200ms for example).

By running the application, set the desired address (in the "eeprom_address" variable) and the value to be written (in the "eeprom_data" variable) and write "1" into the "operation" variable. This sends a write command to the EEPROM and performs the desired writing operation.

Reading the EEPROM is also very simple: set the desired address (in the "eeprom_address" variable) and write "2" into the "operation" variable. This sends a read command to the EEPROM and performs the read of the selected address. The content of that address is stored into the "eeprom_data" variable.

The next example shows how to perform the same task using the interrupts available on the I^2C module.

```c
// DEMO9S08QG8 - I2C example 2 (I2C EEPROM interface using interrupts and FSM)
// EEPROM SDA pin connected to PTA2/SDA
// EEPROM SCL pin connected to PTA3/SCL
// By Fábio Pereira - fabio@sctec.com.br (01/10/08)
// Joinville - SC - Brasil

#include <hidef.h>          /* for EnableInterrupts macro */
#include "derivative.h"     /* include peripheral declarations */
#include "hcs08.h"          // This is our definition file!

#define EEPROM_ID    0xA0
#define EEPROM_ADR   0
#define RD           1
#define WR           0

enum ei2c_states
{
  I2C_IDLE,
  I2C_START, I2C_REPEATED_START,
  I2C_WRITE_HIGHER_ADDRESS, I2C_WRITE_LOWER_ADDRESS,
  I2C_WRITE_BYTE,
  I2C_DUMMY_READ, I2C_READ_BYTE,
  I2C_STOP,
  I2C_EEPROM_WR, I2C_EEPROM_RD
};

unsigned int eeprom_address;
char eeprom_data, operation;
char i2c_buffer[4], command_ready;

enum ei2c_states i2c_fsm(char new_state);

// MCU initialization
void mcu_init(void)
{
  SOPT1 = bBKGDPE;         // Enable debug pin
  ICSSC = NV_FTRIM;        // configure FTRIM value
  ICSTRM = NV_ICSTRM;      // configure TRIM value
  // BUSCLK is now 4 MHz
  // Now we configure the I2C interface:
  // I2C clock is 4MHz*1/40 = 100kHz, SDA hold = 9/4MHz=1.125us
  IICF = I2C_M1 | 0x0B;
  IICC = bIICEN | bIICIE;  // enable the I2C interface
  EnableInterrupts;
}

// This is the I2C interrupt servicing function
void interrupt VectorNumber_Viic iic_isr(void)
{
  // verify the source of I2C interrupt:
  if (IICS_TCF)            // a transfer was completed
  {
    i2c_fsm(0);            // call I2C state machine
  }
```

```
     if (IICS_ARBL)                // the master lost the arbitration
     {
       IICS_ARBL = 1;              // clear ARBL
       i2c_fsm(I2C_IDLE);          // set the I2C state machine to idle state
     }
     IICS_IICIF = 1;              // clear the interrupt flag
}

//*************************************************************************
// I2C finite state machine
// Usage:
//    i2c_fsm(0) to read the current state
//    i2c_fsm(x) to change the current state
// I2C buffer structure:
// i2c_buffer[0] : slave address
// i2c_buffer[1] : EEPROM higher address byte
// i2c_buffer[2] : EEPROM lower address byte
// i2c_buffer[3] : data to be written to the EEPROM (or read from it)
enum ei2c_states i2c_fsm(char new_state)
{
  static enum ei2c_states i2c_state;
  static char current_command;
  if (new_state == I2C_EEPROM_RD)
  {
    current_command = RD;          // current command is read
    i2c_state = I2C_START;         // go to I2C_START state
  }
  if (new_state == I2C_EEPROM_WR)
  {
    current_command = WR;          // current command is write
    i2c_state = I2C_START;         // go to I2C_START state
  }

  // if the TCF flag is not set, do not process the state machine
  // and return now with the current state
  if (!IICS_TCF) return (i2c_state);
  IICS_TCF = 1;                    // clear TCF flag
  switch (i2c_state)
  {
    //*********************************************************************
    // I2C in idle state
    case I2C_IDLE:
      break;
    //*********************************************************************
    // send a start condition
    case I2C_START:
      IICC = bIICEN | bIICIE | bMST | bTX;
      IICD = i2c_buffer[0];        // send the address field
      i2c_state = I2C_WRITE_HIGHER_ADDRESS; // next state
      break;
    //*********************************************************************
    // send the higher address byte
    case I2C_WRITE_HIGHER_ADDRESS:
      IICD = i2c_buffer[1];        // send the higher address byte
      i2c_state = I2C_WRITE_LOWER_ADDRESS;  // next state
      break;
    //*********************************************************************
    // send the lower address byte
    case I2C_WRITE_LOWER_ADDRESS:
      IICD = i2c_buffer[2];        // send the lower address byte
      // now, depending on the current command being executed, the state
      // machine can take one of two possible paths
      if (current_command==WR)
        i2c_state = I2C_WRITE_BYTE; // if it is a write command
      else
        i2c_state = I2C_REPEATED_START; // if it is a read command
      break;
    //*********************************************************************
    // write one byte of data in the EEPROM
    case I2C_WRITE_BYTE:
      IICD = i2c_buffer[3];        // write data to the EEPROM
      i2c_state = I2C_STOP;        // next state
      break;
```

```
//*******************************************************************
// send a repeated start (to change data direction on the EEPROM)
case I2C_REPEATED_START:
  IICC_RSTA = 1;                    // repeated start
  IICD = i2c_buffer[0] | RD;        // send the address field (R/W = RD)
  i2c_state = I2C_DUMMY_READ;       // next state
  break;
//*******************************************************************
// dummy read (start the memory read operation)
case I2C_DUMMY_READ:
  IICC_TX = 0;                      // set I2C to receive mode
  i2c_buffer[3] = IICD;             // dummy read
  i2c_state = I2C_READ_BYTE;        // next state
  break;
//*******************************************************************
// read one byte of data from the EEPROM
case I2C_READ_BYTE:
  IICC_TXAK = 1;                    // send NACK on the next read
  i2c_buffer[3] = IICD;             // read data from the EEPROM
  i2c_state = I2C_STOP;             // go to I2C_STOP state
  break;
//*******************************************************************
// send a stop and go to slave mode
case I2C_STOP:
  IICC_MST = 0;                     // send a stop (go to slave mode)
  i2c_state = I2C_IDLE;             // next state
  command_ready = 1;                // set the command_read flag
  break;
  }
  return (i2c_state);
}
// Read a byte from the EEPROM
char eeprom_byteread(unsigned int address)
{
  while (i2c_fsm(0)!=I2C_IDLE);     // wait for the I2C interface to be idle
  i2c_buffer[0] = EEPROM_ID | EEPROM_ADR;
  i2c_buffer[1] = address >> 8;
  i2c_buffer[2] = address;
  command_ready = 0;                // clear the command_ready flag
  i2c_fsm(I2C_EEPROM_RD);           // set the I2C state machine
  while (!command_ready);           // wait until the command is completed
  return (i2c_buffer[3]);           // return the data read
}
// Write a byte into the EEPROM
void eeprom_bytewrite(unsigned int address, char data)
{
  while (i2c_fsm(0)!=I2C_IDLE);     // wait for the I2C interface to be idle
  i2c_buffer[0] = EEPROM_ID | EEPROM_ADR;
  i2c_buffer[1] = address >> 8;
  i2c_buffer[2] = address;
  i2c_buffer[3] = data;
  i2c_fsm(I2C_EEPROM_WR);           // set the I2C state machine
}
void main(void)
{
  mcu_init();                       // initialize the MCU
  eeprom_address = 0;
  eeprom_data = 0;
  operation = 0;
  while(1)
  {
    if (operation==1)         // write operation
    {
      eeprom_bytewrite(eeprom_address, eeprom_data);
      operation = 0;
    }
    if (operation==2)         // read operation
    {
      eeprom_data = eeprom_byteread(eeprom_address);
      operation = 0;
    }
  }
}
```

Example 11.5 – Communicating with an I^{2}C EEPROM (using interrupts)

11.3. SCI

The serial communication interface (SCI) is an asynchronous serial interface which allows communication with computers, printers and a number of peripheral devices.

The asynchronous interface is based on a known baud rate. The communication is initiated by a start bit (low level), followed by some data bits (typically eight), an optional parity bit and one (or two) stop bits (high level).

| Start | D0 | D1 | D2 | D3 | D4 | D5 | D6 | D7* | D8* | Stop |

* Bits D7 and D8 are optional and depend on the SCI configuration

Figure 11.10

The falling edge of the start bit synchronizes the transmitter and the receiver and considering their baud-rate is equal (or almost equal), the receiver will be able to successfully read the information transmitted.

The SCI operation principles are very simple: there are two shift registers, one (the transmitter) converts parallel data to serial data and the other (the receiver) converts serial data to parallel data. Both shift registers are clocked by a unique clock signal derived from the bus clock (the baud rate generator).

To better understand the operation of the SCI module, we will discuss about the transmitter and the receiver separately.

11.3.1. SCI Transmitter

The transmitter comprises a parallel-in/serial-out shift register (TSR – Transmit Shift Register) and the control circuitry that commands the transmitter operation and the TxD output pin.

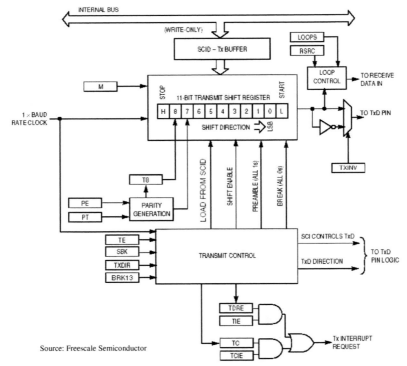

Figure 11.11 – Transmitter section

To transmit a byte of data, the transmitter section must be enabled (SCIC2:TE=1) and properly configured.

The width of the transmission is configured through the SCIC1:M bit. When M=0, eight bits of data are transmitted (plus a start and a stop bit). When M=1, nine bits of data are transmitted (plus a start and a stop bit).

The transmitter can also be configured to automatically generate a parity bit when SCIC1:PE is set. The parity calculation can be set to even or odd, depending on the state of the SCIC1:PT bit. When PT=0, even parity is selected. When PT=1, odd parity is selected.

Prior to transmitting a byte of data, the application needs to check the "transmit data register empty" flag (SCIS1:TDRE). This flag signals when the transmit data buffer is free to receive new data. When TDRE=1, any write to the SCID register starts a transmission.

When the SCI transmitter is operating in nine-bit mode (SCIC1:M=1), the ninth bit to be transmitted reflects the state of bit SCIC3:T8. This bit must be written prior to the writing into SCID (or the transmitted character will use the previous content of T8).

```
SCIC3_T8 = 1; // set the ninth bit
SCID = 0x50; // transmit the 0x50 data (0x150 in nine bit mode)
```

Following a write to the SCID register, the content of SCID is transferred to the transmit shift register (TSR) and the transmission is started (from the LSB to the MSB). The data is output through the TxD pin (which is automatically configured as an output when the transmit section is enabled) at a rate equal to the baud rate programmed into the SCIBD registers.

Note that immediately after the TSR is loaded with the SCID content, the TDRE flag is set again. This allows another byte of data to be written into SCID. This byte is transferred to the TSR as soon as the TSR finishes the current transmission.

Once the TSR finishes shifting out the data, the "transmission completed" flag (SCIS1:TC) is set.

To demonstrate the simplicity of use of the SCI, let us see a small example. The following application shows how to set up the SCI and how to transmit the "H" character through the serial port. A serial terminal application (such as HyperTerminal, Teraterm, etc.) can be used to visualize the data sent by the DEMO9S08QG8 board (do not forget to connect the DEMO9S08QG8 and the host PC through a serial cable!).

```
// DEMO9S08QG8 - SCI TX example 1
#include <hidef.h>           /* for EnableInterrupts macro */
#include "derivative.h"      /* include peripheral declarations */
#include "hcs08.h"           // This is our header file!

void main(void)
{
  SOPT1 = bBKGDPE;           // enable the debug pin
  // Following a reset, BUSCLK = 4MHz
  SCIBD = 26;                // SCI baud rate = 4MHz/(16*26)=9615 bps
  SCIC2 = bTE;               // enable the transmitter section
  SCID = 'H';                // send 'H' through the SCI
  while(1);
}
```

Example 11.6 – Transmitting a character

 Remove the COM_EN jumper before running this example! That jumper connects the FORCEON and FORCEOFF pins of the MAX3218 to the PTA4 pin of the MC9S08QG8 MCU. This pin is used for BDM communication, causing the MAX3218 to switch on and off repeatedly and making the host PC receive garbage data.

The next example shows how to implement a simple function to send a string (null terminated) through the SCI. This function is not based on interrupts!

```
// DEMO9S08QG8 - SCI TX example 2
#include <hidef.h>        /* for EnableInterrupts macro */
#include "derivative.h"   /* include peripheral declarations */
#include "hcs08.h"        // This is our header file!

// This is a simple string to serial function
void SCI_send_string(char *string)
{
  while (*string)         // while the current char of the string is not null
  {
    while (!SCIS1_TDRE);  // wait for the transmit buffer to be empty
    SCID = *string;       // write the current char into the transmit buffer
    string++;             // increment the current char position within the string
  }
}

void main(void)
{
  SOPT1 = bBKGDPE;                      // enable the debug pin
  // Following a reset, BUSCLK = 4MHz
  SCIBD = 26;                           // SCI baud rate = 4MHz/(16*26)=9615 bps
  SCIC2 = bTE;                          // enable the transmitter section
  SCI_send_string("Hello World!");      // write into the serial port ...
  while(1);
}
```

Example 11.7 – Transmitting strings

Figure 11.12 – HyperTerminal screen after running example 11.7

TxD Pin Polarity Inversion

An interesting feature of the SCI module (only on SCIv2, SCIv3 and SCIv4 modules) is the TxD pin polarity inversion. This feature is controlled by the SCIC3:TXINV bit. When TXINV=1, the TxD pin polarity is inverted (a zero data bit outputs a high level on TxD pin). When TXINV=0, the TxD polarity follows the polarity of the data bits (a zero data bit outputs a low level on TxD pin).

Break Character Transmission

The SCI interface is able to automatically generate a break character (a character with all bits in low state, including the stop bit). This feature is enabled by the SCIC2:SBK bit. While SBK=1, the SCI keeps sending break characters until SBK is cleared.

The standard length of the break character is the same for standard characters: 8 bits (when SCIC1:M=0) or 9 bits (when M=1). Some devices (those with SCIv2, SCIv3 or SCIv4 modules) allow extending the break character in three bits. This results in break characters with a total width of 13 or 14 bits. This extended time is obtained by setting the SCIS2:BRK bit.

11.3.2. SCI Receiver

The SCI receiver comprises a serial-in/parallel-out shift register (RSR – Receive Shift Register) and the control circuitry which commands the receiver operation and the RxD input pin.

To be able to receive data, the receiver must be enabled (SCIC2:RE=1) and properly configured. Most configurations already seen for the SCI transmitter section are also valid for the receiver section.

Data input through the RxD pin (automatically configured as the SCI input when SCIC2:RE=1 and SCIC1:LOOPS=0) or through the TxD pin (when the SCI is operating in single wire mode and SCIC3:TXDIR=0) is sampled at a rate sixteen times higher than the SCI baud rate (set in the SCIBD registers). This sampling technique allows the detection of noise in the received data (which is signaled through the appropriate flag, as we will se later).

The data read by the sampling circuit is applied into the receive shift register (RSR) and shifted until the programmed number of bits is received (including the stop bit). After that, the received data is transferred to the SCID register and the "receive data register full flag" (SCIS1:RDRF) is set.

When the receiver is operating in 9-bit mode (SCIC1:M=1), the ninth bit (bit 8) is stored into the SCIC3:R8 bit. The R8 bit must be read before reading SCID (where the remaining eight bits are stored) to ensure proper operation of the SCI.

Figure 11.13 shows the simplified block diagram of the SCI receiver section.

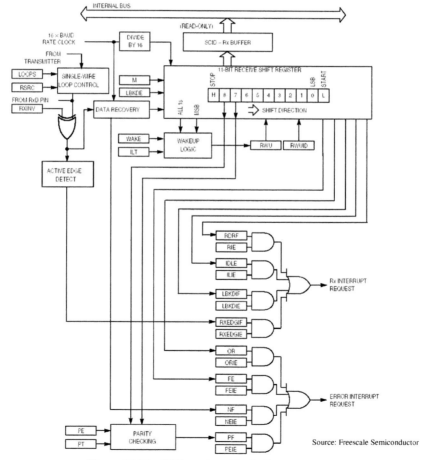

Figure 11.13 – SCI receiver block diagram

The receiver also includes an error detection circuitry, which can detect the following errors:

- **Receiver overrun** (SCIS1:OR flag) – this error signals that a new character was received before the previous one was read from the SCID register. In this case, the newly received character is lost (as well as all error information related to it) and the OR flag is set.

- **Noise flag** (SCIS1:NF flag) – this flag is related to the over sampling feature of the SCI receiver and is set when one sample differs from the other samples within the same bit time. This condition indicates the presence of noise in the RxD line (NF=1).

- **Framing error** (SCIS1:FE flag) – this flag is set when the SCI receiver can not detect the correct stop bit as expected.

- **Parity error** (SCIS1:PF flag) – this flag is set when the parity generation/checking is enabled (PE=1) and the parity calculated differs from the one received.

OR, NF, FE and PF flags are cleared by reading SCIS1 followed by reading SCID.

 The SCI receiver cannot receive new data characters while FE is set. Once it is set, the application must read the SCIS1 register and then read the SCID register. This clears the FE flag and enables the reception of new data.

The following example shows how to configure the SCI transmitter and the SCI receiver to echo the received character plus one (i.e.: the next ASCII character) back to the host PC (pressing A on the host terminal application echoes B on the terminal and so on). Each received character also toggles LED2 on the DEMO9S08QG8 board.

```
// DEMO9S08QG8 - SCI TX and RX example 1

#include <hidef.h>          /* for EnableInterrupts macro */
#include "derivative.h"     /* include peripheral declarations */
#include "hcs08.h"          // This is our header file!

#define LED2  PTBD_PTBD7

void main(void)
{
  char rxchar;
  SOPT1 = bBKGDPE;          // enable the debug pin
  PTBDD_PTBDD7 = 1;         // PTB7 as an output
  // Following a reset, BUSCLK = 4MHz
  SCIBD = 26;               // SCI baud rate = 4MHz/(16*26) = 9615 bps
  SCIC2 = bTE | bRE;        // enable the transmitter section
  while(1)
  {
    while (!SCIS1_RDRF);    // wait for a new character to be received
    while (!SCIS1_TDRE);    // wait for the transmit buffer to be empty
    rxchar = SCID;          // read the received char
    SCID = rxchar+1;        // send char+1 through the SCI
    LED2 = !LED2;           // toggle the led
  }
}
```

Example 11.8 – Echoing characters

The next example echoes all characters typed on the terminal program (and toggles the LED on each character typed). When the user presses the ENTER key, a message is displayed.

```
// DEMO9S08QG8 - SCI TX and RX example 2

#include <hidef.h>          /* for EnableInterrupts macro */
#include "derivative.h"     /* include peripheral declarations */
#include "hcs08.h"          // This is our header file!

#define LED2  PTBD_PTBD7
```

```
// This is a simple string to serial function
void SCI_send_string(char *string)
{
  while (*string)          // while the current char of the string is not null
  {
    while (!SCIS1_TDRE);   // wait for the transmit buffer to be empty
    SCID = *string;        // write the current char into the transmit buffer
    string++;              // increment the current char position within the string
  }
}

void main(void)
{
  char rxchar;
  SOPT1 = bBKGDPE;         // enable the debug pin
  PTBDD_PTBDD7 = 1;        // PTB7 as an output
  // Following a reset, BUSCLK = 4MHz
  SCIBD = 26;              // SCI baud rate = 4MHz/(16*26)=9615 bps
  SCIC2 = bTE | bRE;       // enable the transmitter section
  SCI_send_string("Hello World!");  // write into the serial port ...
  while(1)
  {
    while (!SCIS1_RDRF);   // wait for a new character to be received
    while (!SCIS1_TDRE);   // wait for the transmit buffer to be empty
    rxchar = SCID;         // read the received char
    SCID = rxchar;         // send the char back to the host PC
    LED2 = !LED2;          // toggle the led
    if (rxchar==13) SCI_send_string("HCS08 rules!\r\n");
  }
}
```

Example 11.9 – Echoing characters/string

The SCI receiver includes some interesting additional features such as:

RxD Pin Polarity Inversion

On the most recent devices (those with SCIv4 modules) the polarity of the RxD pin can be selected between the standard (not inverted) mode or inverted mode. This feature is controlled by the SCIS2:RXINV bit.

Idle Line Detection

The SCI receiver includes a special circuitry to detect when the RxD line is idle (high level) for full character time after any activity. In this case, the SCIS1:IDLE flag is set and remains set until it is cleared by the application.

The IDLE flag is cleared by a read of SCIS1 followed by a read of SCID.

There are two IDLE detection modes selected by the SCIC1:ILT bit: when ILT=0, the idle timeout starts counting when a start bit is received; when ILT=1, the idle timeout starts counting when a stop bit is received.

 Once IDLE is cleared by the application, it is only set again after a character is received (SCIS1:RDRF=1).

Edge Detection

SCI version four includes an edge detection circuitry which allows waking up the CPU from a low power mode (wait, LPrun, LPwait or stop3). This feature is enabled by the SCIxBDH:RXEDGIE bit.

Once enabled, a rising edge (if RXINV=1) or a falling edge (if RXINV=0) on the RX input wakes up the CPU and the SCI is able to receive the incoming character. The SCIxS2:RXEDGIF bit is also set upon an active edge is detected on the RX pin.

11.3.2.1. Address Detection

The SCI receiver can be configured to stay in standby mode until a wakeup event occurs. There are two selectable wakeup events: idle-line wakeup and address-mark wakeup.

These two wakeup modes can be used to enable an addressing system (multiple devices that share the same communication bus, each device with a unique address).

In the idle-line addressing mode, the addressing event is signaled by an idle line condition. This means that the first character following an idle line condition is an address character. All slave devices must receive it and compare it against their own addresses. If the addresses do not match, the device must ignore the next characters until a new addressing event occurs.

Figure 11.14 shows a typical idle-line communication. The master device is sending three data blocks for three different slave devices. The "A" fields correspond to the address byte; the "D" fields are the data bytes.

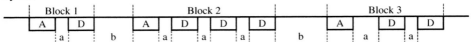

Figure 11.14 – Idle-line addressing

Each block comprises an address field (A) and one or more data fields (D). The space between each character ("a" on figure 11.14) cannot be higher than the break character length (typically 10 bit times). An idle time longer than the break length signals to all slave receivers that the next character is an address (this is signaled by "b" on figure 11.14).

In the address-mark addressing mode, the addressing event is signaled by the most significant bit. When the MSB bit is set, the character is an address, when the MSB is clear, the character is data.

Figure 11.15 shows a typical address-mark communication. The master device is sending two data blocks to two different slave devices. The "s" field is a start bit, the "S" field is a stop bit and the "A" field is the address field (the most significant bit of the character).

Figure 11.15 – Address-mark addressing

It is easy to notice that an address character has the A field set, while a data character has the A bit clear. This enables a slave receiver to distinguish which blocks are being sent to it and which are not.

Each mode has its pros and cons: the idle-line mode has the advantage of using one bit less than the address-mark mode. On the other hand, it depends on very strict timings (especially the interval between two characters for the same slave device). The advantage of the address-mark mode relies on its independence on strict timing between characters.

The selection between the two wakeup modes is done through the SCIC1:WAKE bit.

Idle-Line Addressing Mode

When WAKE = 0, the idle-line wakeup is selected. In this mode, the SCI receiver enters in standby mode when the SCIC2:RWU bit is set and stays in standby until the RxD line is detected idle.

Once an idle condition is detected, the receiver is woken up (IDLE and RDRF are not set in this case) and waits for a new character to be received.

The first character received after the receiver is woken up is an address. The slave device must compare it with its own address and if they match, the next characters should be processed/stored. If they do not match, the slave receiver must return to standby mode (SCIC2:RWU must be set).

Address-Mark Addressing Mode

When WAKE=1, the address-mark wakeup is selected. In this mode, the SCI receiver enters in standby mode when SCIC2:RWU is set and stays in standby until a character with the MSB set is received.

The address bit position depends on the SCIC1:M bit: when M=0 (8-bit data), the address bit is the eighth bit (bit 7); when M=1 (9-bit data), the address bit is the ninth bit (bit 8).

11.3.3. Baud Rate Generator

The SCI includes a baud rate generator which is responsible for generating the clock signals for the transmitter and receiver sections.

The baud rate generator is nothing but a frequency divider which divides the bus clock signal by a programmable factor between 1 and 8191. This factor (called BR) is set through the SCIBD registers (SCIBDH and SCIBDL).

The following formula is used to calculate the SCI baud rate:

$$\text{Baud rate} = \frac{\text{BUSCLK}}{\text{BR} * 16}$$

Tables 11.9 through 11.12 present the BR value for some BUSCLK frequencies and standard baud rates.

 The maximum baud rate difference allowed between the transmitter and the receiver is around 3%. Values higher than that can cause the receiver to misunderstand the last bits of the frame, thus leading to framing errors. For practical purposes, consider the oscillator frequency deviation and select speeds with a maximum 1% of error for each side.

Desired Speed (bps)	BUSCLK = 125kHz			BUSCLK = 500kHz			BUSCLK = 1MHz		
	Real Speed (bps)	BR	Error	Real Speed (bps)	BR	Error	Real Speed (bps)	BR	Error
1,200	1,302.08	6	8.51%	1,201.92	26	0.16%	1,201.92	52	0.16%
2,400	2,604.17	3	8.51%	2,403.85	13	0.16%	2,403.85	26	0.16%
4,800	-	-	-	5,208.33	6	8.51%	4,807.69	13	0.16%
9,600	-	-	-	10,416.67	3	8.51%	10,416.67	6	8.51%
19,200	-	-	-	-	-	-	20,833.33	3	8.51%
38,400	-	-	-	-	-	-	-	-	-
57,600	-	-	-	-	-	-	62,500.00	1	8.51%
115,200	-	-	-	-	-	-	-	-	-
230,400	-	-	-	-	-	-	-	-	-
minimum	0.95	8191	-	3.82	8191	-	7.63	8191	-
maximum	7,812.50	1	-	31,250.00	1	-	62,500.00	1	-

Table 11.9

Desired Speed (bps)	BUSCLK = 2MHz			BUSCLK = 4MHz			BUSCLK = 5MHz		
	Real Speed (bps)	BR	Error	Real Speed (bps)	BR	Error	Real Speed (bps)	BR	Error
1,200	1,201.92	104	0.16%	1,201.92	208	0.16%	1,201.92	260	0.16%
2,400	2,403.85	52	0.16%	2,403.85	104	0.16%	2,403.85	130	0.16%
4,800	4,807.69	26	0.16%	4,807.69	52	0.16%	4,807.69	65	0.16%
9,600	9,615.38	13	0.16%	9,615.38	26	0.16%	9,765.63	32	1.73%
19,200	20,833.33	6	8.51%	19,230.77	13	0.16%	19,531.25	16	1.73%
38,400	41,666.67	3	8.51%	41,666.67	6	8.51%	39,062.50	8	1.73%
57,600	62,500.00	2	8.51%	62,500.00	4	8.51%	62,500.00	5	8.51%
115,200	125,000.00	1	8.51%	125,000.00	2	8.51%	156,250.00	2	35.63%
230,400	-	-	-	250,000.00	1	8.51%	312,500.00	1	35.63%
minimum	15.26	8191	-	30.52	8191	-	19.08	8191	-
maximum	125,000.00	1	-	250,000.00	1	-	156,250.00	1	-

Table 11.10

Desired Speed (bps)	BUSCLK = 6MHz			BUSCLK = 8MHz			BUSCLK = 10MHz		
	Real Speed (bps)	BR	Error	Real Speed (bps)	BR	Error	Real Speed (bps)	BR	Error
1,200	1,201.92	312	0.16%	1,201.92	416	0.16%	1,201.92	520	0.16%
2,400	2,403.85	156	0.16%	2,403.85	208	0.16%	2,403.85	260	0.16%
4,800	4,807.69	78	0.16%	4,807.69	104	0.16%	4,807.69	130	0.16%
9,600	9,615.38	39	0.16%	9,615.38	52	0.16%	9,615.38	65	0.16%
19,200	19,736.84	19	2.80%	19,230.77	26	0.16%	19,531.25	32	1.73%
38,400	41,666.67	9	8.51%	38,461.54	13	0.16%	39,062.50	16	1.73%
57,600	62,500.00	6	8.51%	62,500.00	8	8.51%	62,500.00	10	8.51%
115,200	125,000.00	3	8.51%	125,000.00	4	8.51%	125,000.00	5	8.51%
230,400	375,000.00	1	-	250,000.00	2	8.51%	-	-	-
minimum	22.89	8191	-	30.52	8191	-	38.15	8191	-
maximum	187,500.00	1	-	250,000.00	1	-	312,500.00	1	-

Table 11.11

Desired Speed (bps)	BUSCLK = 16MHz			BUSCLK = 20MHz			BUSCLK = 25MHz		
	Real Speed (bps)	BR	Error	Real Speed (bps)	BR	Error	Real Speed (bps)	BR	Error
1,200	1,200.48	833	0.04%	1,200.77	1041	0.06%	1,200.08	1302	0.01%
2,400	2,403.85	416	0.16%	2,403.85	520	0.16%	2,400.15	651	0.01%
4,800	4,807.69	208	0.16%	4,807.69	260	0.16%	4,807.69	325	0.16%
9,600	9,615.38	104	0.16%	9,615.38	130	0.16%	9,645.06	162	0.47%
19,200	19,230.77	52	0.16%	19,230.77	65	0.16%	19,290.12	81	0.47%
38,400	38,461.54	26	0.16%	39,062.50	32	1.73%	39,062.50	40	1.73%
57,600	58,823.53	17	2.12%	59,523.81	21	3.34%	57,870.37	27	0.47%
115,200	125,000.00	8	8.51%	125,000.00	10	8.51%	120,192.31	13	4.33%
230,400	250,000.00	4	8.51%	250,000.00	5	8.51%	260,416.67	6	13.03%
minimum	61.04	8191	-	76.30	8191	-	95.38	8191	-
maximum	500,000.00	1	-	625,000.00	1	-	781,250.00	1	-

Table 11.12

11.3.4. Single Wire Operation

The SCI can operate in single wire mode when SCIC1:LOOPS and SCIC1:RSRC are set. While in this mode, the TxD pin operates as an output when SCIC3:TXDIR=1 or as input when TXDIR=0. The RxD pin is not used and reverts to general purpose I/O mode.

11.3.5. SCI Interrupts

The SCI is able to generate three interrupt requests (with three different vectors):

1. Transmit interrupt vector – there are two SCI interrupt sources able to generate a transmit interrupt request:

    ♦ **TDRE** (transmit buffer empty) – the SCIS1:TDRE flag is set when the transmit buffer is empty and a new character can be written into it. This flag generates an interrupt request if SCIC2:TIE=1.

    ♦ **TC** (transmission completed) – the SCIS1:TC flag is set immediately after the transmit shift register (TSR) has completed shifting out the current character. This flag generates an interrupt request if SCIC2:TCIE=1.

2. Receive interrupt vector – there are up to four SCI interrupt sources able to generate a receive interrupt request:

    ♦ **RDRF** (receive buffer full) – the SCIS1:RDRF flag is set when the receive shift register (RSR) stores a new character into the receive buffer (SCID). This flag generates an interrupt request if SCIC2:RIE=1.

    ♦ **IDLE** (RxD idle) – the SCIS1:IDLE flag is set when the RxD line is idle for at least a full character time. This flag generates an interrupt request if SCIC2:ILIE=1.

    ♦ **LBKDIF** (LIN break detect) – the SCIS2:LBKDIF flag is set when the LIN break detect circuitry is enabled (SCIS2:LBKDE=1) and a LIN break is detected (a break character longer than the standard one). This flag generates an interrupt request if SCIBDH:LBKDIE=1. LBKDIF is cleared by writing "1" into it. **This feature is only available on SCIv4!**

    ♦ **RXEDGIF** (RxD active edge detected) – the SCIS2:RXEDGIF flag is set when the receiver detects an edge (falling edge if RXINV=0, rising edge if RXINV=1). This flag generates an interrupt request if SCIBDH:RXEDGIE=1 and can be used for waking up the CPU from a low power stop3 mode. RXEDGIF is cleared by writing "1" into it. **This feature is only available on SCIv4!**

3. SCI error interrupt vector – there are four SCI interrupt sources able to generate an SCI error interrupt request:

    ♦ **OR** (receiver overrun) – the SCIS1:OR flag is set when the RSR is about to store a new character into SCID before the previous one is read (the new character is lost). This flag generates an interrupt request if SCIC3:ORIE=1.

    ♦ **NF** (noise error) – the SCIS1:NF flag is set when noise is detected on any bit of the received character. This flag generates an interrupt request if SCIC3:NEIE=1.

    ♦ **FE** (framing error) – the SCIS1:FE flag is set when the stop bit is not detected as expected. This flag generates an interrupt request if SCIC3:FEIE=1.

    ♦ **PF** (parity error) – the SCIS1:PF flag is set when the parity checking is enabled (SCIC1:PE=1) and a parity error is found on the received character. This flag generates an interrupt request if SCIC3:PEIE=1.

11.3.6. SCI External Connections

The SCI modules are connected to general purpose I/O pins. Table 11.13 shows the SCI connections for some HCS08 devices.

SCI Signal	AC/AW	Dx	EL/SL	EN	GB/GT	JM	LC/LL	QE*	QG	Rx	SG/SH
TxD	-	-	PTB1	-	-	-	PTC1	-	PTB1	-	PTB1
RxD	-	-	PTB0	-	-	-	PTC0	-	PTB0	-	PTB0
TxD1	PTE0	PTE0	-	PTE0	PTE0	PTE0	-	PTB1	-	PTB0	-
RxD1	PTE1	PTE1	-	PTE1	PTE1	PTE1	-	PTB0	-	PTB1	-
TxD2	PTC3	PTF0	-	-	PTC0	PTC3	-	PTC7	-	-	-
RxD2	PTC5	PTF1	-	-	PTC1	PTC5	-	PTC6	-	-	-

* QE4 and QE8 devices do not include SCI2.

Table 11.13 – SCI pins

Some devices allow alternate pins for the SCI interface. This selection is made through the bits SOPT1:SCIPS. Refer to topic 5.5 for further information on the SOPT1 register.

SCI Signal	Dx	EL/SL
TxD	-	PTA3
RxD	-	PTA2
TxD2	PTE6	-
RxD2	PTE7	-

Table 11.14 – Alternate SCI pins

 Some SCI operating modes do not use the RxD pin. In these modes, the RxD pin operates in general purpose I/O mode.

11.3.7. SCI Module Versions

The following table shows the evolution of SCI modules since version 1:

SCI Version	Devices	Features added/modified
1	GB, GT, Rx	First release.
2	AW	- TxD polarity inversion. - Extended break generation (3 bit times).
3	LC, QG	Minor changes since version 2.
4	AC, Dx, EL, EN, JM, LL, QE, SG, SH, SL	- LIN break detection. - RxD polarity inversion. - RxD edge detection.

Table 11.15 – SCI module history

11.3.8. SCI Registers

The SCI modules use eight registers for controlling its operation:

- SCI Control 1 (SCIC1) register.
- SCI Control 2 (SCIC2) register.
- SCI Control 3 (SCIC3) register.
- SCI Status 1 (SCIS1) register.
- SCI Status 2 (SCIS2) register.
- SCI Data (SCID) register.
- SCI Baud Rate High and Low (SCIBDH and SCIBDL) registers.

11.3.8.1. SCIC1 Register

Name	Model		BIT 7	BIT 6	BIT 5	BIT 4	BIT 3	BIT 2	BIT 1	BIT 0
SCIC1	EL, LC, LL, QG*, Sx	Read	LOOPS	SCISWAI	RSRC	M	WAKE	ILT	PE	PT
		Write								
		Reset	0	0	0	0	0	0	0	0
SCI1C1	Ax, Dx, EN, Gx, JM, QE, Rx	Read	LOOPS	SCISWAI	RSRC	M	WAKE	ILT	PE	PT
		Write								
		Reset	0	0	0	0	0	0	0	0
SCI2C1	Ax, Dx, Gx, JM, QE	Read	LOOPS	SCISWAI	RSRC	M	WAKE	ILT	PE	PT
		Write								
		Reset	0	0	0	0	0	0	0	0

* Except 8-pin devices.

Bit Name   Description   C Symbol

LOOPS
Loop mode select bit. This bit selects between the loop back mode and the standard full-duplex mode.
0 –   standard operation (full-duplex mode through TxD and RxD).
1 –   loop back mode or single wire mode (depending on the RSRC bit). The RxD pin is not used by the SCI and reverts to general purpose I/O mode.
bLOOPS

SCISWAI
SCI stops in wait mode. This bit controls whether the SCI stops its operation in wait mode.
0 – the SCI continue to operate in wait mode (it can generate an interrupt to wakeup the CPU).
1 – the SCI is stopped upon entering wait mode.
bSCISWAI

RSRC
Receiver source selection bit. This bit selects the source for the SCI RxD pin (provided LOOPS=1).
0 –   internal loop back mode (RxD internally connected to TxD). RxD pin reverts to general purpose I/O mode.
1 –   single wire mode (transmission and reception through the TxD pin. The direction is controlled by SCIC3:TXDIR).
bRSRC

M
Mode selection bit. This bit selects between 8-bit or 9-bit character widths.
0 – 8-bit mode (1 start + 8 data bits + 1 stop).
1 – 9-bit mode (1 start + 9 data bits + 1 stop).
bM

WAKE
Receiver wakeup mode selection bit.
0 – idle-line wakeup mode.
1 – address-mark wakeup mode.
bWAKE

ILT
Idle line type selection bit. This bit selects the detection mode for the idle condition.
0 – idle timeout starts after the start bit.
1 – idle timeout starts after the stop bit.
bILT

PE
Parity enable bit. This bit selects whether the parity generation and checking is enabled/disabled.
0 – parity disabled.
1 – parity enabled.
bPE

PT
Parity type selection bit. This bit selects the type of parity checking: even or odd. This bit is only valid when parity checking is enabled (PE=1).
0 – even parity.
1 – odd parity.
bPT

11.3.8.2. SCIC2 Register

Name	Model		BIT 7	BIT 6	BIT 5	BIT 4	BIT 3	BIT 2	BIT 1	BIT 0
SCIC2	EL, LC, LL, QG*, Sx	Read	TIE	TCIE	RIE	ILIE	TE	RE	RWU	SBK
		Write								
		Reset	0	0	0	0	0	0	0	0
SCI1C2	Ax, Dx, EN, Gx, JM, QE, Rx	Read	TIE	TCIE	RIE	ILIE	TE	RE	RWU	SBK
		Write								
		Reset	0	0	0	0	0	0	0	0
SCI2C2	Ax, Dx, Gx, JM, QE	Read	TIE	TCIE	RIE	ILIE	TE	RE	RWU	SBK
		Write								
		Reset	0	0	0	0	0	0	0	0

* Except 8-pin devices.

Bit Name	Description	C Symbol
TIE	Transmit interrupt enable bit. This bit controls whether the TDRE flag generates an interrupt request when it is set. 0 – TDRE interrupt disabled. When the flag is set it does not generate an interrupt. 1 – TDRE interrupt enabled.	bTIE
TCIE	Transmission complete interrupt enable bit. This bit controls whether the TC flag generates an interrupt request when it is set. 0 – TC interrupt disabled. When the flag is set it does not generate an interrupt. 1 – TC interrupt enabled.	bTCIE
RIE	Receiver interrupt enable bit. This bit controls whether the RDRF flag generates an interrupt request when it is set. 0 – RDRF interrupt disabled. When the flag is set it does not generate an interrupt. 1 – RDRF interrupt enabled.	bRIE
ILIE	Idle line interrupt enable bit. This bit controls whether the IDLE flag generates an interrupt request when it is set. 0 – IDLE interrupt disabled. When the flag is set it does not generate an interrupt. 1 – IDLE interrupt enabled.	bILIE
TE	Transmitter enable bit. This bit controls whether the SCI transmitter section is enabled or disabled. 0 – transmitter disabled (TxD pin operates in general purpose I/O). 1 – transmitter enabled (TxD pin is controlled by the SCI). **Note that the transition of TE from 1 to 0 may not happen instantly. If a transmission is in progress, it is first finished before the transmitter is disabled.**	bTE
RE	Receiver enable bit. This bit controls whether the SCI receiver section is enabled or disabled. 0 – receiver disabled. 1 – receiver enabled. Note that the RxD pin is used by the SCI only when SCIC1:LOOPS=0.	bRE
RWU	Receiver wakeup control bit. This bit controls whether the SCI receiver is operating or in standby mode. 0 – SCI receiver operating normally. 1 – SCI receiver in standby mode. A wakeup event (selected by SCIC1:WAKE) wakes up the SCI receiver and clears the RWU bit.	bRWU
SBK	Send break bit. This bit is used to instruct the SCI transmitter to send break characters instead of data. If this bit is set while a transmission is in progress, the break character is sent after the transmission is completed. 0 – standard transmitter operation (sends data). 1 – transmitter sends break characters.	bSBK

11.3.8.3. SCIC3 Register

Name	Model		BIT 7	BIT 6	BIT 5	BIT 4	BIT 3	BIT 2	BIT 1	BIT 0
SCIC3	EL, LC, LL, QG*, Sx	Read	R8	T8	TXDIR	TXINV	ORIE	NEIE	FEIE	PEIE
		Write	-							
		Reset	0	0	0	0	0	0	0	0
SCI1C3	GB, GT, Rx	Read	R8	T8	TXDIR	0	ORIE	NEIE	FEIE	PEIE
		Write	-			-				
		Reset	0	0	0	0	0	0	0	0
SCI2C3	GB, GT	Read	R8	T8	TXDIR	0	ORIE	NEIE	FEIE	PEIE
		Write	-			-				
		Reset	0	0	0	0	0	0	0	0
SCI1C3	Ax, Dx, EN, JM, QE	Read	R8	T8	TXDIR	TXINV	ORIE	NEIE	FEIE	PEIE
		Write	-							
		Reset	0	0	0	0	0	0	0	0
SCI2C3	Ax, Dx, JM, QE	Read	R8	T8	TXDIR	TXINV	ORIE	NEIE	FEIE	PEIE
		Write	-							
		Reset	0	0	0	0	0	0	0	0

* Except 8-pin QG devices.

Bit Name	Description	C Symbol
R8	Bit 8 of the received character (when the SCI is operating in nine bits mode). This is a read-only bit and must be read after RDRF is set and before reading SCID.	bR8
T8	Bit 8 of the character to be transmitted (when the SCI is operating in nine bits mode). This bit controls the state of the ninth bit to be transmitted by the SCI. If must be written before the character is written into SCID.	bT8
TXDIR	TxD pin direction control bit. This bit controls the direction of the TxD pin when the SCI is operating in single wire mode (SCIC1:LOOPS=1 and SCIC1:RSRC=1). 0 – TxD pin is an input (receive mode). 1 – TxD pin is an output (transmit mode).	bTXDIR
TXINV	TxD polarity inversion bit. This bit controls the polarity inversion of the TxD line. 0 – TxD is not inverted. 1 – TxD is inverted.	bTXINV
ORIE	Receive overrun interrupt enable bit. 0 – OR interrupt disabled. When the flag is set it does not generate an interrupt. 1 – OR interrupt enabled.	bORIE
NEIE	Noise error interrupt enable bit. 0 – NF interrupt disabled. When the flag is set it does not generate an interrupt. 1 – NF interrupt enabled.	bNEIE
FEIE	Framing error interrupt enable bit. 0 – FE interrupt disabled. When the flag is set it does not generate an interrupt. 1 – FE interrupt enabled.	bFEIE
PEIE	Parity error interrupt enable bit. 0 – PE interrupt disabled. When the flag is set it does not generate an interrupt. 1 – PE interrupt enabled.	bPEIE

11.3.8.4. SCIS1 Register

Name	Model		BIT 7	BIT 6	BIT 5	BIT 4	BIT 3	BIT 2	BIT 1	BIT 0
SCIS1	EL, LC, LL, QG*, Sx	Read	TDRE	TC	RDRF	IDLE	OR	NF	FE	PF
		Write	-	-	-	-	-	-	-	-
		Reset	0	0	0	0	0	0	0	0
SCI1S1	Ax, Dx, EN, Gx, JM, QE, Rx	Read	TDRE	TC	RDRF	IDLE	OR	NF	FE	PF
		Write	-	-	-	-	-	-	-	-
		Reset	0	0	0	0	0	0	0	0
SCI2S1	Ax, Dx, Gx, JM, QE	Read	TDRE	TC	RDRF	IDLE	OR	NF	FE	PF
		Write	-	-	-	-	-	-	-	-
		Reset	0	0	0	0	0	0	0	0

* Except 8-pin QG devices.

Bit Name	Description	C Symbol
TDRE	Transmit data register empty flag. This bit signals when the transmit buffer register (SCID) is free to receive another character. TDRE is cleared by reading SCIS1 and then writing into SCID. 0 – transmit buffer full. 1 – transmit buffer empty.	bTDRE
TC	Transmission complete flag. This bit signals when the transmit shift register (TSR) has finished shifting out a character. TC is cleared by reading SCIS1 and then writing into SCID or setting SCIC2:SBK. 0 – the transmitter is busy (a transmission is in progress). 1 – the transmitter is idle (no transmission is in progress).	bTC
RDRF	Receive data register full flag. This bit signals when the receive buffer (SCID) is full and cannot receive new characters. The application must read the SCID before new characters are received. RDRF is cleared by reading SCIS1 and then reading SCID. 0 – receive data register empty. 1 – receive data register full.	bRDRF
IDLE	Idle line flag. This flag indicates when the RxD line is found idle after a character has been received. IDLE is cleared by reading SCIS1 followed by a read of SCID. 0 – no idle line condition detected. 1 – Idle line condition detected.	bIDLE
OR	Receiver overrun error flag. This flag indicates when a new character is received before the previous character in SCID is read. In this case the new character is discarded. OR is cleared by reading SCIS1 and then reading SCID. 0 – No overrun error. 1 – Receive overrun.	bOR
NF	Noise error flag. This flag indicates noise was detected in the RxD line sampling. NF is cleared by reading SCIS1 and then reading SCID. 0 – No noise detected. 1 – Noise detected in the received character.	bNF
FE	Framing error flag. This flag indicates a failure in detecting the stop bit of the last character. FE is cleared by reading SCIS1 and then reading SCID. 0 – No framing error detected. 1 – Framing error detected in the received character. **The receiver cannot receive new data characters while FE is set!**	bFE
PF	Parity error flag. This flag indicates an error in the parity detection, providing the parity checking is enabled (PE=1). PF is cleared by reading SCIS1 and then reading SCID. 0 – No parity error. 1 – Parity error detected in the received character.	bPF

11.3.8.5. SCIS2 Register

Name	Model		BIT 7	BIT 6	BIT 5	BIT 4	BIT 3	BIT 2	BIT 1	BIT 0
SCI1S2 SCI2S2	Gx, Rx Gx	Read	0	0	0	0	0	0	0	RAF
		Write	-	-	-	-	-	-	-	-
		Reset	0	0	0	0	0	0	0	0
SCIS2	LC, QG*	Read	0	0	0	0	0	BRK13	0	RAF
		Write	-	-	-	-	-		-	-
		Reset	0	0	0	0	0	0	0	0
SCI1S2 SCI2S2	AW	Read	0	0	0	0	0	BRK13	0	RAF
		Write	-	-	-	-	-		-	-
		Reset	0	0	0	0	0	0	0	0
SCIS2 SCI1S2	EL, LL, Sx EN	Read	LBKDIF	RXEDGIF	0	RXINV	RWUID	BRK13	LBKDE	RAF
		Write	LBKDIF	RXEDGIF	-	RXINV	RWUID	BRK13	LBKDE	-
		Reset	0	0	0	0	0	0	0	0
SCI1S2 SCI2S2	AC, Dx, JM, QE	Read	LBKDIF	RXEDGIF	0	RXINV	RWUID	BRK13	LBKDE	RAF
		Write	LBKDIF	RXEDGIF	-	RXINV	RWUID	BRK13	LBKDE	-
		Reset	0	0	0	0	0	0	0	0

* Except 8-pin QG devices.

Bit Name	Description	C Symbol
RAF	Receiver active flag. This bit indicates whether a reception is in progress. 0 – receiver is idle (no reception in progress). 1 – receiver is busy (reception in progress).	bRAF
BRK13	Break character length. This bit controls the width of the break character sent when SCIC2:SBK=1. 0 – break character is 10 bit times (SCIC1:M=0) or 11 bit times (SCIC1:M=1). 1 – break character is 13 bit times (SCIC1:M=0) or 14 bit times (SCIC1:M=1). **This feature in not present in GB/GT/Rx devices!**	bBRK13
LBKDIF	LIN break detect interrupt flag. This flag indicates whether a LIN break character is detected by the SCI receiver (when LBKDE=1). LBKDIF is cleared by writing "1" into it. 0 – no LIN break character received. 1 – LIN break character detected.	bLBKDIF
RXEDGIF	RxD pin active edge interrupt flag. This flag indicates whether an active edge (rising edge when RXINV=1, falling edge when RXINV=0) was detected in the RxD line. RXEDGIF is cleared by writing "1" into it. 0 – no active edge detected. 1 – an active edge was detected.	bRXEDGIF
RXINV	Received data polarity inversion. This bit controls the polarity inversion of the received data. 0 – received data polarity not inverted. 1 – received data polarity inverted.	bRXINV
RWUID	Receive wakeup idle detect control bit. This bit controls whether an idle event sets the IDLE flag when the SCI receiver is in standby mode. 0 – IDLE is not set when an idle condition is detected and the receiver is in standby. 1 – IDLE is set when idle condition is detected and the receiver is in standby.	bRWUID
LBKDE	LIN break detection enable bit. This bit controls the break detection length. While LBKDE is set, FE and RDRF are prevented from setting. 0 – break character is 10 bit times (11 if SCIC1:M=1). 1 – break character is 11 bit times (12 if SCIC1:M=1).	bLBKDE

11.3.8.6. SCID Register

Name	Model		BIT 7	BIT 6	BIT 5	BIT 4	BIT 3	BIT 2	BIT 1	BIT 0
SCID	EL, LC, LL, QG*, Sx	Read	Received data							
		Write	Transmit data							
		Reset	0	0	0	0	0	0	0	0
SCI1D	Ax, Dx, EN, Gx, JM, QE, Rx	Read	Received data							
		Write	Transmit data							
		Reset	0	0	0	0	0	0	0	0
SCI2D	Ax, Dx, Gx, JM, QE	Read	Received data							
		Write	Transmit data							
		Reset	0	0	0	0	0	0	0	0

* Except 8-pin devices.

11.3.8.7. SCIBD Registers

Name	Model		BIT 7	BIT 6	BIT 5	BIT 4	BIT 3	BIT 2	BIT 1	BIT 0
SCIBDH	LC, QG*	Read	0	0	0	SBR12	SBR11	SBR10	SBR9	SBR8
		Write	-	-	-					
		Reset	0	0	0	0	0	0	0	0
SCIBDH	EL, LL, Sx	Read	LBKDIE	RXEDGIE	0	SBR12	SBR11	SBR10	SBR9	SBR8
		Write			-					
		Reset	0	0	0	0	0	0	0	0
SCIBDL	EL, LC, LL, QG*, Sx	Read	SBR7	SBR6	SBR5	SBR4	SBR3	SBR2	SBR1	SBR0
		Write								
		Reset	0	0	0	0	0	0	0	0
SCI1BDH SCI2BDH	AW, Gx, Rx AW, Gx	Read	0	0	0	SBR12	SBR11	SBR10	SBR9	SBR8
		Write	-	-	-					
		Reset	0	0	0	0	0	0	0	0
SCI1BDH SCI2BDH	AC, Dx, EN, JM, QE AC, Dx, JM, QE	Read	LBKDIE	RXEDGIE	0	SBR12	SBR11	SBR10	SBR9	SBR8
		Write			-					
		Reset	0	0	0	0	0	0	0	0
SCI1BDL SCI2BDL	Ax, Dx, EN, JM, QE Ax, Dx, JM, QE	Read	SBR7	SBR6	SBR5	SBR4	SBR3	SBR2	SBR1	SBR0
		Write								
		Reset	0	0	0	0	0	0	0	0

* Except 8-pin devices.

Bit Name	Description	C Symbol
SBR12 ... SBR0	SCI baud rate dividing factor (BR).	-
LBKDIE	LIN break detect interrupt enable bit. This bit controls whether the SCIS2:LBKDIF flag generates an interrupt request when it is set. 0 – LBKDIF interrupt disabled. 1 – LBKDIF interrupt enabled.	bLBKDIE
RXEDGIE	RxD input active edge interrupt enable bit. This bit controls whether the SCIS2:RXEDGIF flag generates an interrupt request when it is set. 0 – RXEDGIF interrupt enabled. 1 – RXEDGIF interrupt disabled.	bRXEDGIE

11.3.9. A Circular Buffer

The next example shows how to implement a circular buffer to store data to be transmitted or received. The application creates two buffers: txbuffer and rxbuffer. Two `#define` control the size of the buffers.

Each buffer has two pointers: the read pointer is dedicated to reading data from the buffer and the write pointer is dedicated to writing data into the buffer.

The txbuffer is used for transmitting data through the SCI. The application writes into the buffer (using the `write_string_sci_buffer()` and `write_char_sci_buffer()` functions) and the TX interrupt servicing function reads the buffer and transmits the data.

Figure 11.16 shows the TX buffer right after executing the `write_string_sci_buffer("Hello World!");` function. The write pointer (tx_buf_write_pointer) is pointing to the last character in the buffer (the null character).

Position	0	1	2	3	4	5	6	7	8	9	10	11	12	13	14	15	16	17	18	19	20	21	22	23	24	25	26	27	28	29	30	31
Content	H	E	L	L	O		W	O	R	L	D	!																				
Content (hex)	46	65	6C	6C	7F	20	57	6F	72	6C	64	21	0																			

Read pointer ⬆ (position 0)

Write pointer ⬆ (position 12)

Figure 11.16 – TX buffer is not empty

After writing into the TX buffer, the `write_string_sci_buffer()` also enables the TX interrupt. This interrupt is called when the SCI TX buffer is empty (TDRE flag is set). As this is the case at this time, an interrupt is requested and the `isr_sciltx()` function is called.

Each TX interrupt reads one character from the buffer (pointed to by the read pointer), writes it into SCID and increments the read pointer. This will repeat until the read pointer (tx_buf_read_pointer) is equal to the write pointer (tx_buf_write_pointer). This indicates there are no more characters to be transmitted, the TX interrupt is disabled and the TX buffer empty flag is set (indicating there is no more data to be transmitted in the buffer).

Disabling the TX interrupt is necessary because while the SCI TX buffer is empty, the TDRE flag is set and generates an interrupt request if TIE is set. If there is no more data to be transmitted, TIE must be cleared to prevent new TX interrupts.

Position	0	1	2	3	4	5	6	7	8	9	10	11	12	13	14	15	16	17	18	19	20	21	22	23	24	25	26	27	28	29	30	31
Content	H	E	L	L	O		W	O	R	L	D	!																				
Content (hex)	46	65	6C	6C	7F	20	57	6F	72	6C	64	21	0																			

Read pointer ⬆ (position 12)

Write pointer ⬆ (position 12)

Figure 11.17 – TX buffer is empty

```
// DEMO9S08QE128 - SCI example (circular buffer)
// RxD connected to PTB0
// TxD connected to PTB1
// COM_EN connected to PTC5

#include <hidef.h>              /* for EnableInterrupts macro */
#include "derivative.h"         /* include peripheral declarations */
#include "hcs08.h"              // This is our header file!

#define LED2     PTCD_PTCD0
#define COM_EN   PTCD_PTCD5

#define TX_BUF_SIZE   32        // TX buffer size = 32 characters
#define RX_BUF_SIZE   32        // RX buffer size = 32 characters

volatile char txbuffer[TX_BUF_SIZE], rxbuffer[RX_BUF_SIZE];
volatile char tx_buf_read_pointer, tx_buf_write_pointer;
volatile char rx_buf_read_pointer, rx_buf_write_pointer;
```

```
#pragma DATA_SEG __DIRECT_SEG MY_ZEROPAGE
struct
{
  char rx_flag      : 1;    // there are characters in the rx buffer
  char tx_buf_empty : 1;    // the tx buffer is empty
} __near flags;
#pragma DATA_SEG DEFAULT

/*
  SCI receive interrupt servicing function
  The received characters are stored into the receive buffer
*/
void interrupt VectorNumber_Vsci1rx isr_sci1rx(void)
{
  char temp;
  temp = SCI1S1;           // read SCI1S1 to check for receive errors
  if (temp & 7)            // is there any rx errors?
  { // yes, make a dummy read of the rx data and return
    temp = SCI1D;          // dummy read of rx data to clear the interrupt flag
    return;
  }
  // read and store the received character, increment the rx write pointer
  rxbuffer[rx_buf_write_pointer++] = SCI1D;
  if (rx_buf_write_pointer>=RX_BUF_SIZE) rx_buf_write_pointer = 0;
  flags.rx_flag = 1;       // set the rx_flag (there is data into the rx buffer)
}

/*
  SCI transmit interrupt servicing function
  Get characters from the transmit buffer and transmit them through the SCI
  If there are no more characters to be transmitted, disable the TX interrupt
*/
void interrupt VectorNumber_Vsci1tx isr_scitx(void)
{
  if (tx_buf_read_pointer==tx_buf_write_pointer)
  {
    // if read and write pointers are equal, there is no more data to transmit
    // disable tx interrupt and set the tx buffer empty flag
    SCI1C2_TIE = 0;        // disable the TX interrupt
    flags.tx_buf_empty = 1; // tx buffer is empty
  } else
  {
    if (SCI1S1_TDRE)
    {
      // if there is any data in the tx buffer, transmit the next char and
      // advance the tx buffer read pointer
      SCI1D = txbuffer[tx_buf_read_pointer++];
      // if the tx read buffer is bigger than the buffer size, set the pointer to zero
      if (tx_buf_read_pointer>=TX_BUF_SIZE) tx_buf_read_pointer = 0;
    }
  }
}

// Function for writing a string into the sci tx buffer
void write_string_sci_buffer(unsigned char *data)
{
  flags.tx_buf_empty = 0;
  while (*data)
  {
    txbuffer[tx_buf_write_pointer++] = *data++;
    // if the tx write buffer is bigger than the buffer size, set the pointer to zero
    if (tx_buf_write_pointer>=TX_BUF_SIZE) tx_buf_write_pointer = 0;
    SCI1C2_TIE = 1;        // enable TX interrupt
  }
}

// Function for writing a char into the sci tx buffer
void write_char_sci_buffer(unsigned char data)
{
  flags.tx_buf_empty = 0;  // the tx buffer is not empty anymore
  // write the data into the tx buffer and increment the tx write buffer
  txbuffer[tx_buf_write_pointer++] = data;
  // if the tx write buffer is bigger than the buffer size, set the pointer to zero
  if (tx_buf_write_pointer>=TX_BUF_SIZE) tx_buf_write_pointer = 0;
  SCI1C2_TIE = 1;          // enable TX interrupt
}
```

```
// Function for reading a character from the rx buffer
char read_char_sci_buffer(void)
{
    char temp;
    temp = rxbuffer[rx_buf_read_pointer++];
    // if the read pointer is bigger than the buffer size, set the pointer to zero
    if (rx_buf_read_pointer>=RX_BUF_SIZE) rx_buf_read_pointer = 0;
    // if the read pointer reached the write pointer, clear the rx_flag
    SCI1C2_RIE = 0;              // disable the receive interrupt
    if (rx_buf_read_pointer==rx_buf_write_pointer) flags.rx_flag = 0;
    SCI1C2_RIE = 1;              // re-enable the receive interrupt
    return (temp);               // return the character
}

void main(void)
{
    char rxchar;
    SOPT1 = bBKGDPE;             // enable the debug pin
    PTCDD = BIT_0 | BIT_5;       // PTC0 and PTC5 as outputs
    // Following a reset, BUSCLK = 4MHz
    SCI1BD = 26;                 // SCI baud rate = 4MHz/(16*26)=9615 bps
    // enable TX and RX section, RX and TX interrupts
    SCI1C2 = bTIE | bRIE | bTE | bRE;
    COM_EN = 1;                  // enable MAX3218
    write_string_sci_buffer("Hello World!");  // write into the tx buffer
    EnableInterrupts;
    while(1)
    {
        while (!flags.rx_flag);  // wait for a new character to be received
        rxchar = read_char_sci_buffer();
        write_char_sci_buffer(rxchar);
        LED2 = !LED2;            // toggle the led
        if (rxchar==13) write_string_sci_buffer("HCS08 rules!\r\n");
    }
}
```

A more efficient approach is the use of a logical AND operation to mask the read pointer (this is valid only for power-of-2 buffer sizes):
rx_buf_read_pointer &= (RX_BUF_SIZE-1);

Disabling and re-enabling the RIE flag prevents that the **if** statement disables the rx_flag if a new character is received during the executing of the statement!

Example 11.10

12

FLASH Memory Controller

The HCS08 microcontrollers use FLASH memories as their program memories. This memory technology allows non-volatile storage and also enables the erasing and writing of the memory.

To handle erase and write operations, the HCS08 microcontrollers include a special circuitry: the FLASH memory controller.

12.1. The FLASH Controller

The FLASH memory controller handles the major elements related to the erasing/programming process: the internal charge pump voltage source (used to generate the voltage levels necessary for the erasing and writing operations), the timings necessary for erasing and writing into the FLASH cells and the selection of which cells are programmed/erased.

A clock divider is responsible for reducing the bus clock frequency down to a range between 150 and 200 kHz, which is the range specified by the manufacturer. Of course, the higher the controller operating-frequency, the faster the programming and erasing times, but working outside the specified range is not recommended by the manufacturer.

The controller automates the erasing and the programming processes through the use of hardware state machines. There are five commands possible: blank check, byte program, byte program (burst mode), sector erase and mass erase. Table 12.1 shows the total timings for a FLASH clock (FCLK) of 150 and 200kHz.

Command	Cycles (FCLK)	Total Time	
		FCLK=150kHz	FCLK=200kHz
single byte program	9	60μs	45μs
multiple bytes program (burst mode)*	4	26.6μs	20μs
sector erase	4,000	26.6ms	20ms
mass erase	20,000	133.3ms	100ms

* a single byte programmed in burst mode.

Table 12.1 – Flash commands timing

The FLASH controller clock frequency (FCLK) is calculated using the following formula:

$$FCLK = \frac{BUSCLK}{(8*PRDIV8)*(DIV+1)}$$

In which: PRDIV8 = pre-dividing by 8 bit (FCDIV register).

DIV = clock dividing factor (FCDIV register).

The FLASH controller operation is very simple: once the FCDIV register is correctly programmed, the application writes the desired data in the target address (the address to be programmed or erased), writes the desired command into the FCMD register and writes "1" into the FSTAT:FCBEF bit. This operation starts the command execution. Once the command is finished, the controller automatically sets the FSTAT:FCCF bit.

Notice that prior to starting a new command, it is necessary to verify the state of the FSTAT:FCBEF bit. If FCBEF is clear, the controller is busy and a new command cannot be started. Starting a new command while another is in progress causes a FLASH access error to be raised (FSTAT:FACCERR=1).

Note that a program running from the FLASH memory cannot erase or program a FLASH address within the same FLASH block where it is located. This is only possible on LC and QE devices and only when the program is running from a different FLASH block.

On most devices, the programming and erasing of the FLASH memory can only be executed by a program running from the RAM memory.

12.1.1. FLASH Controller Commands

As we said, the FLASH controller can handle five different commands: blank check, single-byte program, multiple-byte program, page erase and mass erase.

12.1.1.1. Blank Check

This command reads all FLASH memory addresses and verifies if they are all erased (an erase bit reads "1"). In case all bits within the FLASH memory are found set (erased), the FSTAT:FBLANK bit is set. If one or more bits are found cleared (programmed) FBLANK is cleared.

This command also causes the memory to be unsecured if it is found completely erased (FBLANK=1). In this case, the FOPT:SEC01 bit is set and FOPT:SEC00 is cleared.

12.1.1.2. Single-Byte Program

This command allows a single byte to be programmed into a FLASH address. The command usage is very simple, as long as the following steps are watched:

1. The FCDIV register must be properly configured. Note that this register can be written only once after a reset.

2. The application verifies the FLASH access error (FSTAT:FACCERR) bit. If FACCERR is set, all commands are inhibited. This bit must be cleared before issuing a new command to the FLASH controller. Writing "1" into FACCERR clears it.

3. The application verifies the FLASH command buffer empty flag (FSTAT:FCBEF). If it is clear, the FLASH controller is busy and the application need to wait until it completes the current command (this can be verified by reading the FSTAT:FCCF bit, it is set once the command is completed).

4. The data to be programmed must be written into the desired address. Remember to erase the FLASH memory before attempting to program it.

5. The byte program command (0x20) must be written into the FCMD register.

6. The application needs to write "1" into the FCBEF flag. This clears FCBEF and starts the command execution.

7. The application must verify the FLASH access error (FSTAT:FACCERR) flag and the FLASH protection violation (FSTAT:FPVIOL) flag. If any of them is set, the command was

not completed due to an error. Note that the FLASH protection violation occurs when the controller tries to program a protected address (as defined by the FPROT register).

8. If no error is found, the application needs to wait until the command is finished. This is indicated by the FSTAT:FCCF flag's being set. The FCCF flag is automatically cleared upon the start of a command and set when the command is finished. The manufacturer states that the application needs to wait for at least four BUSCLK cycles before reading FCCF or FCBEF.

 The manufacturer recommends writing only once into each address after an erase operation. New writes should only be attempted after a new sector erase (or mass erase) operation. Trying to program the same address two or more times can lead to a corruption of the FLASH memory.

12.1.1.3. Multiple-Byte Program

This mode, called "burst program" by the manufacturer, is characterized by not turning off the charge pump between sequential writes (which occurs in the single byte mode). This leads to faster programming times.

For optimal speed, the addresses of the first and the last byte to be programmed must be located into the same FLASH row. Each FLASH row comprises 64 addresses in which the bits A15 through A6 are equal, as shown in figure 12.1.

| | Address | Internal Address Bus Bits | | | | | | | | | | | | | | | |
---	---	A15	A14	A13	A12	A11	A10	A9	A8	A7	A6	A5	A4	A3	A2	A1	A0
FLASH row	0xFFFF	1	1	1	1	1	1	1	1	1	1	1	1	1	1	1	1
	to																
	0xFFC0	1	1	1	1	1	1	1	1	1	1	0	0	0	0	0	0
FLASH row	0xFFBF	1	1	1	1	1	1	1	1	1	0	1	1	1	1	1	1
	to																
	0xFF80	1	1	1	1	1	1	1	1	1	0	0	0	0	0	0	0
FLASH row	0xFF7F	1	1	1	1	1	1	1	1	0	1	1	1	1	1	1	1
	to																
	0xFF40	1	1	1	1	1	1	1	1	0	1	0	0	0	0	0	0

Figure 12.1

The following steps should be followed for using the burst program command:

1. The FCDIV register must be properly configured. Note that this register can be written only once after a reset.

2. The application verifies the FLASH access error (FSTAT:FACCERR) bit. If FACCERR is set, all commands are inhibited. This bit must be cleared before issuing a new command to the FLASH controller. Writing "1" into FACCERR clears it.

3. The application verifies the FLASH command buffer empty flag (FSTAT:FCBEF). If it is clear, the FLASH controller is busy and the application needs to wait until it completes the current command.

4. The first byte of data to be programmed must be written into the starting address. Remember to erase the FLASH memory before attempting to program it.

5. The burst program command (0x25) must be written into the FCMD register.

6. The application needs to write "1" into the FCBEF flag. This clears FCBEF and starts the command execution.

7. The application must verify the FLASH access error (FSTAT:FACCERR) flag and the FLASH protection violation (FSTAT:FPVIOL) flag. If any of them is set, the command was not completed due to an error. Note that the FLASH protection violation occurs when the controller tries to program a protected address (as defined by the FPROT register).

8. If no error is found, and there are remaining bytes to be programmed, the application should verify the FLASH command buffer empty flag (FSTAT:FCBEF). If it is clear, the FLASH controller is busy and the application needs to wait until it completes the current command. Once it is set, the next byte to be programmed should be written into the desired address and the application should go to step 5.

9. If there are no remaining bytes to be programmed, the application needs to wait until the command is finished. This is indicated by the FSTAT:FCCF flag's being set. The manufacturer states that the application needs to wait for at least four BUSCLK cycles before reading FCCF or FCBEF.

Note that the first byte programmed in burst mode takes 9 FCLK cycles to be programmed. The next bytes within the same FLASH row are programmed in 4 FCLK cycles. If a FLASH row boundary is crossed, the first byte on the next FLASH row takes 9 cycles to be programmed.

12.1.1.4. Sector Erase

Erasing the FLASH memory can be done only on a sector basis. The size of each sector is typically 512 bytes, but some devices (such as the DN, DV, DZ and EN) present a sector size of 768 bytes.

This means that erasing a single byte of FLASH is not possible. To erase a byte, the whole sector to which it belongs must be erased.

The sector erase command is executed in the same way as a single-byte program. The address written in step 4 selects the sector to be erased (the sector to which the address belongs). Note that this write operation is a dummy write (the data written is not important). The only difference is that the command to be written into FCMD (step 5) must be 0x40.

Once the erase operation is completed (FSTAT:FCCF=1), the content of all addresses in the sector is equal to 0xFF (all bits in high level).

 Erasing a FLASH sector which is protected from writing (through the FPROT register) is not possible. Attempts to erase a protected sector result in a FPVIOL error.

Aborting a Sector Erase

Some devices (DN, DV, DZ, EL and EN) include a special command to abort a sector erase already in progress.

The procedure to abort a sector erase operation is the following:

1. The FSTAT:FCCF is verified. If it is found set, the previous command (the sector erase) has completed and no abort is necessary.

2. If FCCF=0, a dummy write operation must be done. The target address and the data written are not important.

3. Write the abort command (0x47) into the FCMD register.

4. Write "1" into the FSTAT:FCBEF bit. This starts the sector abort command.

5. The application needs to wait until the FSTAT:FCCF bit is set indicating the abort operation is completed.

6. Once the abort operation is completed, the application needs to check the FSTAT:FACCERR flag. If FACCERR=1, the sector erase was aborted successfully and the sector content is unknown. If FACCERR=0, the sector erase was completed (despite the abort command) and the sector content is all "1s" (0xFF).

12.1.1.5. Mass Erase

This erase command is intended for erasing the whole FLASH memory content. The command flow is similar to the sector erase, but the command for mass erase is 0x41.

Notice that in step 4, the mass erase command requires a dummy write into any valid FLASH address.

12.1.2. FLASH Controller Registers

There are three register related to the FLASH controller write/erase operations:

♦ FLASH Clock Divider (FCDIV) register.

♦ FLASH Status (FSTAT) register.

♦ FLASH Command (FCMD) register.

12.1.2.1. FCDIV Register

Name	Model		BIT 7	BIT 6	BIT 5	BIT 4	BIT 3	BIT 2	BIT 1	BIT 0
FCDIV	All	Read	DIVLD	PRDIV8	DIV					
		Write	-							
		Reset	0	0	0	0	0	0	0	0

* This a write-once register. It can be written only once after a reset.

Bit Name	Description	C Symbol
DIVLD	Divider loaded indicator bit. This bit is set after the DIV bits are written for the first time after a reset. No write or erase operations are allowed when DIVLD=0. 0 – DIV bits not written. 1 – DIV bits were written after the last reset.	bDIVLD
PRDIV8	Divide by eight prescaler selection bit. This bit controls the clock selection for the FLASH controller. 0 – BUSCLK. 1 – BUSCLK/8.	bPRDIV8
DIV	These bits select the dividing factor applied to the clock source selected by PRDIV8. The resulting clock must stay between 150 and 200kHz. $$FCLK = \frac{BUSCLK}{(8*PRDIV8)*(DIV+1)}$$	-

12.1.2.2. FSTAT Register

Name	Model		BIT 7	BIT 6	BIT 5	BIT 4	BIT 3	BIT 2	BIT 1	BIT 0
FSTAT	All	Read	FCBEF	FCCF	FPVIOL	FACCERR	0	FBLANK	0	0
		Write		-			-	-	-	-
		Reset	1	1	0	0	0	0	0	0

Bit Name	Description	C Symbol
FCBEF	FLASH command buffer empty flag. This flag indicates whether the FLASH controller can receive a new command in FCMD. 0 – command buffer is full. No more commands allowed. 1 – command buffer is empty.	bFCBEF
FCCF	FLASH command complete flag. This flag indicates whether a command was successfully completed. FCCF is automatically cleared when a new command is loaded into FCMD. 0 – command in progress. 1 – command completed.	bFCCF
FPVIOL	FLASH protection violation flag. This flag indicates a violation of a write protected address. FPVIOL is cleared by writing "1" into it. 0 – no violation. 1 – the controller tried to program or erase a write protected address.	bFPVIOL
FACCERR	FLASH access error flag. This flag indicates an error in the FLASH controller command sequence. FACCERR is cleared by writing "1" into it. 0 – no access error. 1 – an access error occurred in the last operation of the FLASH controller.	bFACCERR
FBLANK	FLASH blank flag. This bit is set after a blank check command. 0 – FLASH array is not erased. 1 – FLASH array is fully erased.	bFBLANK

12.1.2.3. FCMD Register

Name	Model		BIT 7	BIT 6	BIT 5	BIT 4	BIT 3	BIT 2	BIT 1	BIT 0
FCMD	All	Read	0	0	0	0	0	0	0	0
		Write				FCMD				
		Reset	0	0	0	0	0	0	0	0

Bit Name	Description	C Symbol
FCMD	FLASH controller command:	
	0x05 – blank check.	CMD_BLANK
	0x20 – single byte program.	CMD_BYTEPROG
	0x25 – burst program.	CMD_BURSTPROG
	0x40 – sector erase.	CMD_SECTORERASE
	0x41 – mass erase.	CMD_MASSERASE
	0x47 – sector erase abort. This command is not available on all devices!	CMD_ERASEABORT

12.2. FLASH Write Protection

The FLASH controller includes a protection mechanism to prevent writing and erasing of specific memory areas. This mechanism is controlled by the FPROT register operates differently on each HCS08 device.

In all cases the FPROT register cannot be directly written by the application code. Following a reset, FPROT is loaded with the content of the NVPROT register (located in the FLASH memory).

The main application of the memory protection is to avoid that a program erases itself due to a malfunction. One possible case is the use of FLASH memory for nonvolatile data storage.

Another possible application is to protect a bootloader code. A bootloader is a small application used to reprogram the FLASH memory using a communication port (such as the SCI, I²C or USB interface). A bootloader enables the firmware update directly in the field or even remotely.

By using the write protection feature, the bootloader can protect itself from being erased due to a failure in the loaded application.

12.2.1. Write Protection on Ax, JM, LC, QA, QD, QG, SG and SH Devices

In the AW, AC8/16/32/48/60, JM, LC, LL, QA, QD, QG, SG and SH devices, the write protection mechanism comprises from FPS1 through FPS7 bits of the FPROT register.

These bits store the upper seven bits of the last unprotected memory address. All addresses above that are write protected.

Let us suppose an application needs to protect the last 512 bytes of memory (0xFE00 to 0xFFFF). The last unprotected memory address is 0xFDFF. All you have to do is copy bit 15 to FPS7, bit 14 to FPS6 and so on. The last bit is bit 9, which is copied to FPS1. FPDIS must be cleared to enable the write protection. Figure 12.2 shows graphical representation of the operation.

Address	A15	A14	A13	A12	A11	A10	A9	A8	A7	A6	A5	A4	A3	A2	A1	A0
0xFDFF	1	1	1	1	1	1	0	1	1	1	1	1	1	1	1	1

FPROT	FPS7	FPS6	FPS5	FPS4	FPS3	FPS2	FPS1	FPDIS
0xFC	1	1	1	1	1	1	0	0

Figure 12.2 – FPROT configuration

The structure of the FPROT and NVPROT registers is represented below:

Name		BIT 7	BIT 6	BIT 5	BIT 4	BIT 3	BIT 2	BIT 1	BIT 0
FPROT	Read	FPS7	FPS6	FPS5	FPS4	FPS3	FPS2	FPS1	FPDIS
	Write	-	-	-	-	-	-	-	-
	Reset	Loaded with the content of NVPROT							
NVPROT	Read	FPS7	FPS6	FPS5	FPS4	FPS3	FPS2	FPS1	FPDIS
	Write	-	-	-	-	-	-	-	-
	Reset	1	1	1	1	1	1	1	1

The FPDIS bit, when set, disables the protection mechanism (regardless of the state of the FPS bits).

In order to use the protection mechanism, the application needs to program the NVPROT register (located in the FLASH memory) during the programming of the FLASH memory of the microcontroller. This can be done by inserting the following line of code at the beginning of the application:

```
// Write-protects addresses ranging from 0xFE00 to 0xFFFF
const byte NVPROT_INIT @0xFFBD = 0xFC;
```

12.2.2. Write Protection on GB, GT and Rx Devices

On GB, GT, RC, RD, RE and RG devices, the write protection mechanism comprises five bits of the FPROT register: FPOEN, FPDIS, FPS2, FPS1 and FPS0.

When FPOEN=1 and FPDIS=0, the memory area specified by the three FPS bits is protected. Table 12.2 shows the corresponding protected addresses.

FPS2	FPS1	FPS0	Range of Protected Addresses	Total Protected Block Size	Redirected Vector Addresses
0	0	0	0xFE00 a 0xFFFF	512 bytes	0xFDC0 a 0xFDFD
0	0	1	0xFC00 a 0xFFFF	1024 bytes	0xFBC0 a 0xFBFD
0	1	0	0xF800 a 0xFFFF	2048 bytes	0xF7C0 a 0xF7FD
0	1	1	0xF000 a 0xFFFF	4096 bytes	0xEFC0 a 0xEFFD
1	0	0	0xE000 a 0xFFFF	8192 bytes	0xDFC0 a 0xDFFD
1	0	1	0xC000 a 0xFFFF	16384 bytes	0xBFC0 a 0xBFFD
1	1	0	0x8000 a 0xFFFF	32768 bytes	0x7FC0 a 0x7FFD
1	1	1	0x8000 a 0xFFFF	32768 bytes	0x7FC0 a 0x7FFD

Table 12.2 – Protected ranges by the FPS bits

The structure of the FPROT and NVPROT registers is shown below.

Name		BIT 7	BIT 6	BIT 5	BIT 4	BIT 3	BIT 2	BIT 1	BIT 0
FPROT	Read	FPOEN	FPDIS	FPS2	FPS1	FPS0	0	0	0
	Write	-	-	-	-	-	-	-	-
	Reset	Loaded with the content of NVPROT							
NVPROT	Read	FPOEN	FPDIS	FPS2	FPS1	FPS0	0	0	0
	Write	-	-	-	-	-	-	-	-
	Reset	1	1	1	1	1	1	1	1

12.2.3. Write Protection on AC96/128 and QE32/64/96/128 Devices

On the AC96/128 and QE32/64/96/128 devices, the protection scheme works in the following way: the seven most significant bits of the FPROT register indicate the number of KiB (kilobytes) of the protected area.

To calculate the total size of the protected area all you have to do is subtract the FPS value from 0x7F.

Some examples:

If FPS = 0x00, the total protected memory size is equal to 0x7F – 0x00 = 0x7F (127 decimal) = 127 KiB (127 kilobytes).

If FPS = 0x10, the total protected memory size is equal to 0x7F – 0x10 = 0x6F (111 decimal) = 111 KiB (111 kilobytes).

If FPS = 0x3F, the total protected memory size is equal to 0x7F – 0x3F = 0x40 (64 decimal) = 64 KiB (64 kilobytes).

If FPS = 0x7E, the total protected memory size is equal to 0x7F – 0x7E = 0x01 (1 decimal) = 1 KIB (1 kilobyte).

Name	BIT 7	BIT 6	BIT 5	BIT 4	BIT 3	BIT 2	BIT 1	BIT 0
FPROT Read	FPS							FPOPEN
FPROT Write								-
FPROT Reset	Loaded with the content of NVPROT							
NVPROT Read	FPS							FPOPEN
NVPROT Write	-	-	-	-	-	-	-	-
NVPROT Reset	1	1	1	1	1	1	1	1

The FPOPEN bit, when clear, protects the whole FLASH memory from writes/erases (regardless of the state of the FPS bits). When FPOPEN is set, the FPS bits specify the amount of protected memory (when FPS = 0x7F and FPOPEN = 1, the memory is not protected).

In order to use the protection mechanism, the application needs to program the NVPROT register (located in the FLASH memory) during the programming of the FLASH memory of the microcontroller. This is done by inserting the following line of code at the beginning of the application:

```
// Write-protect addresses ranging from 0x0F000 to 0xFFFF (4 kilobytes)
// FPS = 0x7F - 4 = 0x7B
const byte NVPROT_INIT @0xFFBD = (0x7B<<1) | bFPOPEN;
```

 Remember to shift the FPS value one bit to the left before writing it into NVPROT!

12.3. FLASH Security

One interesting feature of the HCS08 microcontrollers is the ability to secure the FLASH memory. This feature prevents any external reading of the FLASH contents, allowing the application code to stay away from "curious eyes".

Once the device is secured, the FLASH and the RAM area are considered secure resources and cannot be read by a program running from an unsecured resource (such as the direct page registers, high page registers and the background debug controller).

A secured resource can read and write into any other resources (either secured or unsecured). An unsecured resource cannot read or write into a secured resource (writes are ignored and reads return all zeros).

The security mechanism is controlled by the FOPT:SEC bits. The FOPT register is a read-only register and its bits cannot be changed by the application. Following a reset, the FOPT register is loaded with the content of the NVOPT register (located in the FLASH memory).

In order to enable the security mechanism, the application needs to program the NVOPT register (located in the FLASH memory) during the programming of the FLASH memory of the microcontroller. This is done by inserting the following line of code at the beginning of the application:

```
// Enable the security mechanism, notice that vector redirection is also enabled (if
// available and the memory is protected through the FPROT/NVPROT register)
const byte NVOPT_INIT @0xFFBF = MEM_SECURE;
```

 No protection scheme is 100% effective. Reading the application code can be done by using advanced (and highly expensive) techniques (such as etching, micro probing, etc.).

12.3.1.1. FOPT and NVOPT Registers

Name	Model		BIT 7	BIT 6	BIT 5	BIT 4	BIT 3	BIT 2	BIT 1	BIT 0
FOPT	AW, AC*, EN, GB, GT, JM, LC, LL, QA, QB, QD, QE*, QG, Rx, SG, SH	Read	KEYEN	FNORED	0	0	0	0	SEC01	SEC00
		Write	-	-	-	-	-	-	-	-
		Reset	Loaded with NVOPT content following a reset							
NVOPT	AW, AC, EN, GB, GT, JM, LC, LL, QA, QB, QD, QE*, QG, Rx, SG, SH	Read	KEYEN	FNORED	0	0	0	0	SEC01	SEC00
		Write	-	-	-	-	-	-	-	-
		Reset	1	1	1	1	1	1	1	1
FOPT	DN, DV, DZ, EL, SL	Read	KEYEN	FNORED	EPGMOD	0	0	0	SEC01	SEC00
		Write	-	-	-	-	-	-	-	-
		Reset	Loaded with NVOPT content following a reset							
NVOPT	DN, DV, DZ, EL, SL	Read	KEYEN	FNORED	EPGMOD	0	0	0	SEC01	SEC00
		Write	-	-	-	-	-	-	-	-
		Reset	1	1	1	1	1	1	1	1

* AC8, AC16, AC32, AC48, AC60, QE4, QE8, QE16 and QE32 only.

Bit Name Description   C Symbol

KEYEN	Backdoor key mechanism enable bit. This bit controls whether the backdoor mechanism can be used to disengage the security. 0 – no backdoor access allowed. 1 – backdoor access enabled. Security is disabled if the firmware writes the correct 8-byte value into the NVBACKKEY.	**bKEYEN**
FNORED	Vector redirection disable bit. 0 – interrupt vectors are redirected to other addresses when any memory address is write protected. 1 - interrupt vectors are not redirected when memory is write protected.	**bFNORED**

SEC01
SEC00   Security state code. These bits control the MCU security state.

00 – secure.	**MEM_SECURE**
01 – secure.	**MEM_SECURE**
10 – unsecured.	**MEM_UNSECURE**
11 – secure.	**MEM_SECURE**

Name	Model		BIT 7	BIT 6	BIT 5	BIT 4	BIT 3	BIT 2	BIT 1	BIT 0
FOPT	AC96/128, QE64/96/128	Read	KEYEN		0	0	0	0	SEC01	SEC00
		Write	-		-	-	-	-	-	-
		Reset	Loaded with NVOPT content following a reset							
NVOPT	AC96/128, QE64/96/128	Read	KEYEN		0	0	0	0	SEC01	SEC00
		Write	-		-	-	-	-	-	-
		Reset	1	1	1	1	1	1	1	1

Bit Name Description   C Symbol

KEYEN	Backdoor key mechanism enable bit. This bit controls whether the backdoor mechanism can be used to disengage the security.

00 – disabled.	**KEY_DISABLED**
01 – disabled.	**KEY_DISABLED**
10 – enabled.	**KEY_ENABLED**
11 – disabled.	**KEY_DISABLED**

SEC01
SEC00   Security state code. These bits control the MCU security state.

00 – secure.	**MEM_SECURE**
01 – secure.	**MEM_SECURE**
10 – unsecured.	**MEM_UNSECURE**
11 – secure.	**MEM_SECURE**

12.3.2. Storing Non-volatile Data into the FLASH Memory

The next example shows how to write, erase and read data from the FLASH memory. As we said earlier in this chapter, for most HCS08 devices, directly writing and erasing the FLASH memory is not possible for a program running from the FLASH memory itself.

In such cases, a small routine must be loaded in RAM to perform the desired operations. Freescale provides a set of assembly routines in a single file called "doonstack.asm". This file must be included in the project as shown in figure 12.3.

```
;****************************************************************
;* This stationery is meant to serve as the framework for a    *
;* user application. For a more comprehensive program that     *
;* demonstrates the more advanced functionality of this        *
;* processor, please see the demonstration applications        *
;* located in the examples subdirectory of the                 *
;* Metrowerks Codewarrior for the HC08 Program directory       *
;****************************************************************
; export symbols
            XDEF DoOnStack
            XDEF FlashErase
            XDEF FlashProg
            ; we use export 'Entry' as symbol. This allows us to
            ; reference 'Entry' either in the linker .prm file
            ; or from C/C++ later on

; include derivative specific macros
            Include 'MC9S08GB60.inc'

;mPageErase equ $40         ◄──────────────      Comment these two lines!
;mByteProg equ $20
mFACCERR equ $10
mFPVIOL equ $20
mFCBEF equ $80


; variable/data section
MY_ZEROPAGE: SECTION  SHORT
; Insert here your data definition. For demonstration, temp_byte is used.
; temp_byte ds.b 1


; code section
MyCode:     SECTION
;****************************************************************
; this assembly routine is called the C/C++ application
DoOnStack:  pshx
            pshh
            psha            ;save command on stack     ;save pointer to flash
            ldhx #SpSubEnd  ;point at last byte to move to stack;
SpMoveLoop: lda ,x          ;read from flash
            psha            ;move onto stack
            aix #-1         ;next byte to move
            cphx #SpSub-1   ;past end?
            bne SpMoveLoop  ;loop till whole sub on stack
            tsx             ;point to sub on stack
            tpa             ;move CCR to A for testing
            and #$08        ;check the I mask
            bne I_set       ;skip if I already set
            sei             ;block interrupts while FLASH busy
            lda SpSubSize+6,sp ;preload data for command
            jsr ,x          ;execute the sub on the stack
            cli             ;ok to clear I mask now
            bra I_cont      ;continue to stack de-allocation
    I_set:  lda SpSubSize+6,sp ;preload data for command
            jsr ,x          ;execute the sub on the stack
    I_cont: ais #SpSubSize+3 ;deallocate sub body + H:X + command ;H:X flash pointer OK
from SpSub
            lsla            ;A=00 & Z=1 unless PVIOL or ACCERR
            rts             ;to flash where DoOnStack was called
;****************************************************************
SpSub:      ldhx LOW(SpSubSize+4),sp ;get flash address from stack
            sta 0,x ;write to flash; latch addr and data
            lda SpSubSize+3,sp ;get flash command
            sta FCMD        ;write the flash command
            lda #mFCBEF     ;mask to initiate command
```

```
                    sta FSTAT        ;[pwpp] register command
                    nop              ;[p] want min 4~ from w cycle to r
ChkDone:            lda FSTAT        ;[prpp] so FCCF is valid
                    lsla             ;FCCF now in MSB
                    bpl ChkDone      ;loop if FCCF = 0
SpSubEnd:           rts              ;back into DoOnStack in flash
SpSubSize:          equ (*-SpSub)
;****************************************************************
FlashErase:         psha             ;adjust sp for DoOnStack entry
                    lda #(mFPVIOL+mFACCERR) ;mask
                    sta FSTAT        ;abort any command and clear errors
                    lda #mPageErase  ;mask pattern for page erase command
                    bsr DoOnStack    ;finish command from stack-based sub
                    ais #1           ;deallocate data location from stack
                    rts
;****************************************************************
FlashProg:          psha             ;temporarily save entry data
                    lda #(mFPVIOL+mFACCERR) ;mask
                    sta FSTAT        ;abort any command and clear errors
                    lda #mByteProg   ;mask pattern for byte prog command
                    bsr DoOnStack    ;execute prog code from stack RAM
                    ais #1           ;deallocate data location from stack
                    rts
;****************************************************************
```

Listing 12.1 – Doonstack.asm listing

Figure 12.3

Another important file is the "doonstack.h" header. It must also be included in the C application (this time, a simple #include directive can be used), as shown in the example 12.1.

```
/****************************************************************/
/*                                                            */
/* Project Name:  doonstack.h                                 */
/* Last modified: 04/11/2004                                  */
/* By:            r60817                                       */
/*                                                            */
/****************************************************************/
/*                                                            */
/* Description: MC9S08GB60_FLASH_DOONSTACK - demo             */
/*                                                            */
/* Documentation: MC9S08GB60/D Rev. 2.2                       */
/*                HCS08RMv1/D  Rev. 1(4.8FLASH Application Examples)  */
/*                                                            */
/* This software is classified as Engineering Sample Software.  */
/*                                                            */
/****************************************************************/
/*                                                            */
/* Services performed by FREESCALE in this matter are performed AS IS */
/* and without any warranty. CUSTOMER retains the final decision  */
/* relative to the total design and functionality of the end product. */
/* FREESCALE neither guarantees nor will be held liable by CUSTOMER   */
/* for the success of this project. FREESCALE DISCLAIMS ALL       */
/* WARRANTIES, EXPRESSED, IMPLIED OR STATUTORY INCLUDING, BUT NOT   */
/* LIMITED TO, IMPLIED WARRANTY OF MERCHANTABILITY OR  FITNESS FOR A  */
```

```
/* PARTICULAR PURPOSE ON ANY HARDWARE, SOFTWARE ARE ADVISE SUPPLIED    */
/* TO THE PROJECT BY FREESCALE, AND OR ANY PRODUCT RESULTING FROM       */
/* FREESCALE SERVICES. IN NO EVENT SHALL FREESCALE BE LIABLE FOR        */
/* INCIDENTAL OR CONSEQUENTIAL DAMAGES ARISING OUT OF THIS AGREEMENT.   */
/*                                                                      */
/* CUSTOMER agrees to hold FREESCALE harmless against any and all       */
/* claims demands or actions by anyone on account of any damage, or     */
/* injury, whether commercial, contractual, or tortuous, rising         */
/* directly or indirectly as a result of the advise or assistance       */
/* supplied CUSTOMER in connection with product, services or goods      */
/* supplied under this Agreement.                                       */
/*                                                                      */
/**********************************************************************/

/*- this file API between main.c and doonstack.asm */

#ifndef _doonstack
#define _doonstack

#ifdef __cplusplus
  extern "C" {                    /* our assembly functions have C calling convention */
#endif

void DoOnStack(void);             /* prototype for DoOnStack routine */
void FlashErase(unsigned char *); /* prototype for FlashErase routine */
                                  /* Page Erase command */
void FlashProg(unsigned char *, unsigned char);    /* prototype for FlashProg routine */
                                  /* Byte Program command */
#ifdef __cplusplus
  }
#endif
#endif /* _doonstack */
```

Listing 12.2 – Doonstack.h listing

Our example reserves a full FLASH sector (512 bytes) for non-volatile storage. This size was selected as this is the smallest erasable area on the MC9S08QG8.

The non-volatile area starts at the beginning of the FLASH area (0xE000) and extends for 512 bytes up to 0xE1FF. The program area starts at 0xE200 and extends until the end of the FLASH area (0xFFFF).

It is necessary to instruct the linker not to store application code or constant data into the first 512 bytes of the FLASH memory. This is done by changing ROM area in the linker parameter file (.PRM). From:

```
ROM              = READ_ONLY          0xE000 TO 0xFFAF;
```

To:

```
ROM              = READ_ONLY          0xE200 TO 0xFFAF;
```

The example also write-protects the FLASH area from 0xE200 to 0xFFFF, preventing it from being modified by the FLASH erasing and writing functions, which could lead to a potentially catastrophic application failure.

Figure 12.4 shows the calculation of the value to be stored into the FPROT register to enable the write-protection of the memory area between 0xE200 and 0xFFFF (the last unprotect address is 0xE1FF).

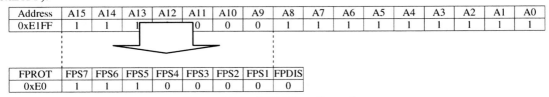

Address	A15	A14	A13	A12	A11	A10	A9	A8	A7	A6	A5	A4	A3	A2	A1	A0
0xE1FF	1	1	1	0	0	0	0	1	1	1	1	1	1	1	1	1

FPROT	FPS7	FPS6	FPS5	FPS4	FPS3	FPS2	FPS1	FPDIS
0xE0	1	1	1	0	0	0	0	0

Figure 12.4 – FPROT configuration

```
// DEMO9S08QG8 - FLASH writing/erasing/reading example

#include <hidef.h>              /* for EnableInterrupts macro */
#include "derivative.h"         /* include peripheral declarations */
#include "hcs08.h"              // This is our definition file!
#include "doonstack.h"

#define BUSCLK   8000000
#define vFCDIV   (BUSCLK/200000-1)

char fdata, operation;
unsigned int faddress;

// Write-protect addresses ranging from 0xE200 up to 0xFFFF
const byte NVPROT_INIT @0xFFBD = 0xE0;

// MCU innitialization
void mcu_init(void)
{
  SOPT1 = bBKGDPE;             // Enable debug pin
  ICSSC = NV_FTRIM;            // configure FTRIM value
  ICSTRM = NV_ICSTRM;          // configure TRIM value
  ICSC2 = 0;                   // ICSOUT = DCOOUT / 1
  // BUSCLK is now 8 MHz
  FCDIV = vFCDIV;              // set FCLK divider (FCLK = 200kHz)
}

#pragma inline
// Read the content of a memory address
char flash_read(unsigned int address)
{
  unsigned char *pointer;
  pointer = (char*) address;
  return (*pointer);
}

// Write a byte into the specified FLASH address
char flash_write(unsigned int address, unsigned char data)
{
  unsigned char *pointer;
  pointer = (char*) address;
  FlashProg(pointer,data);     // call the FLASH programming function
  if (FSTAT_FACCERR) data=1; else data=0;
  if (FSTAT_FPVIOL) data|=2;
  return(data);
}

// Erase a sector of the FLASH memory
unsigned char flash_sector_erase(unsigned int address)
{
  unsigned char *pointer, res;
  pointer = (char*) address;
  FlashErase(pointer);
  if (FSTAT_FACCERR) res=1; else res=0;
  if (FSTAT_FPVIOL) res|=2;
  return(res);
}

void main(void)
{
  mcu_init();
  fdata = 0;
  faddress = 0xE000;
  operation = 0;
  while (1)
  {
    switch (operation)
    {
      case 1: // sector erase
        fdata = flash_sector_erase(faddress);
        operation = 0;
        break;
      case 2: // write operation
        fdata = flash_write(faddress,fdata);
        operation = 0;
```

```
        break;
    case 3: // read operation
        fdata = flash_read(faddress);
        operation = 0;
        break;
    }
  }
}
```

Example 12.1

Testing the application is very simple: by using the real-time debug, all you have to do is add three variables ("faddress", "fdata" and "operation") to data watch window and set the window to update periodically (200ms for example).

By running the application, set the desired address (in the "faddress" variable) and the value to be written (in the "fdata" variable) and write "2" into the "operation" variable. This starts the writing of the byte into the specified FLASH memory address (the address must be erased (0xFF) prior to attempting to write on it).

To erase the page from 0xE000 to 0xE1FF, just set the "faddress" variable to a value between 0xE000 and 0xE1FF and set the "operation" variable to 1.

Reading from the FLASH memory is also very simple: set the desired address (in the "faddress" variable) and write "3" into the "operation" variable. The content of that FLASH address is stored into the "fdata" variable.

Note that the `flash_write()` and `flash_sector_erase()` functions return a **char** value with the error status of the operation: 0 for no error, bit 1 set for an access error and bit 2 set for a violation error.

 The flash_write() and flash_sector_erase() functions require up to 35 bytes of stack to operate correctly. If the application stack is sub dimensioned, the application is very likely to crash!

In order to view the changes in the debugger memory window, it is necessary to modify the configuration of the debugging tool. By default, the debugger reads the FLASH area only once, at the beginning of the debug session. To modify this configuration, select the "MultilinkCyclonePro>Debug Memory Map..." option in the main menu. This opens the window shown in figure 12.5A. Select the memory block 3 and click on Modify/Details button. On the window shown in figure 12.5B, select the option highlighted in the figure. This enables the debugger to update (even periodically) the memory window contents.

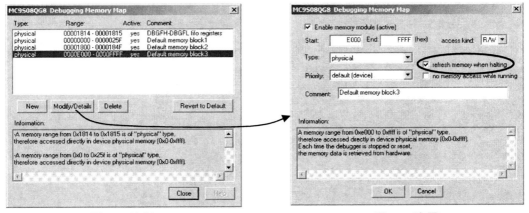

Figure 12.5A   **Figure 12.5B**

13

Project Examples

13.1. Application Design Checklist

In this topic we will review some important designing principles that can make the difference between a successful project and a failing one.

13.1.1. Hardware

Microcontrolled circuits designing checklist:

1. Unused pins must be configured as outputs (and left unconnected) or configured as inputs and tied to V_{DD} or V_{SS}. Unconnected input pins can draw current in excess and pick up noise. The advantage of configuring the pin to output mode and leaving it unconnected is the possibility of future usage (which is not so simple when the pin is connected to V_{DD} or V_{SS}).

2. External reset: some applications may need to be synchronized with external elements through a common reset signal. On some HCS08 devices, the reset pin does not operate as a reset input and this function must be enabled by the application code. Some HCS08 devices may also output the reset signal, enabling them to reset other devices. Refer to topic 5.1 for further information.

3. The slew rate feature can be enabled to aid reducing EMI. This is especially important when generating high frequency signals such as PWM.

4. Drive strength and ganged outputs (when available) can be used to increase the current capacity of I/O ports.

5. The BKGD/MS pin can be used as a general purpose output, but remember that the load on this pin must be very low (the external load capacitance must be very low and the impedance must be high). An impedance too low forces the pin to low state, making the chip enter in active debug mode upon a reset (the application code is not run). A capacitance too high may interfere with BDM communication, preventing programming and debugging.

13.1.2. Software

The following items must always be checked when designing microcontroller software:

1. Initialize all important registers. On HCS08 devices, it is important to initialize the option registers and all write-once registers/bits. This prevents their modification by a faulty code.

    Four registers must be written by the application in the initialization phase:

- ◆ SOPT/SOPT1 (to configure the COP options, STOP instruction, RESET pin and BKGD pin).
- ◆ SOPT2 (to configure other MCU options related to the peripherals and COP).
- ◆ SPMSC1 (to configure the low-voltage detect module).
- ◆ SPMSC2 (to configure the desired stop mode).

2. Set the desired clock rate and clock option. Most microcontrolled applications need precise clocks, especially when using timers and asynchronous communication peripherals.

3. Use the COP/watchdog carefully: select key points for placing COP-clearing instructions. Avoid placing COP-clearing instructions in ISRs (in several cases, the main code may be locked but the interrupts can still be serviced).

4. Use the interrupts wisely: in any application, the smaller is the ISR code, the better. A big ISR increases interrupt latency (a new interrupt can only be serviced after the previous one has been serviced). The best ISR code would simply set or reset a flag and the main application would process the event later on. Of course, it is up to the programmer to decide which code must be embedded into the ISR and which code can run outside the ISR (triggered by it).

   Avoid using high interrupt rates (such as high frequency timer interrupts) as they can lead to high CPU usage and delay other process (such as other interrupts) running on the system. A typical scenario is the performing of high rate time bases: if the system needs a 1ms time base, avoid setting up the timer to generate sub-multiples of the time (such as ten 100µs interrupts, for example), as this only leads to increased interrupt processing time and can steal precious CPU time that could be dedicated to other important processes.

   Notice that, in some cases, a high interrupt rate with a slow ISR code can lead to a system lock up (a new interrupt event arises while the ISR is running, preventing the ISR code from returning to the main task).

5. Stack sizing: this is an important topic that is often forgotten by most programmers. When programming small devices with limited memory resources, any byte of memory can be useful. By default, the HCS08 stack size is set to 0x50 (80) bytes. While most applications work comfortably with such stack size, for small applications it may be too large, on the other hand, it may be too small for complex applications. There is no simple way to calculate the ideal stack size and the effects of a sub dimensioned stack can be catastrophic: the common symptom is the unexpected changes of variables (especially those located near the bottom end of the stack) and in most cases the application crashes unexpectedly. A good measure to help debug such applications is to read the SRS register at the beginning of the application code. Two bits (ILAD and ILOP) indicate whether the reset source was the execution of an illegal opcode (ILOP) or an access to an illegal or unimplemented address (ILAD). Both conditions can be a symptom of a stack overflow.

13.2. Controlling an Alphanumeric LCD Module

A microcontrolled application frequently needs to interface with human beings either for data input or for data output. Data input is done by using keys, switches, keyboards and many other input devices. Data output is frequently done by using LEDs, buzzers and displays. This topic shows how to interface an HCS08 device to an alphanumeric LCD module.

An alphanumeric LCD module comprises a liquid crystal display (LCD) and an integrated controller circuit (HD44780 or KS0066). On these modules, the display is divided into columns and rows (with small spaces between them). Each display position can show any character (alphanumeric or graphic), but it is not possible to address a single pixel within the display (however, it is possible to create special graphic characters with a single pixel turned on).

To interface the microcontroller to the display controller (either HD44780 or KS0066, as they are compatible) it is necessary to know how the controller operates and which lines and signals are used to interface to it.

The controller uses eleven lines to control the operating mode and display data: eight bidirectional data lines (DB0 to DB7), an enable input, a read/write selection input and the R/S input (to select between command and data). There are also other lines: V_{SS} (ground reference), V_{DD} (positive supply) and V_O (contrast voltage). Some modules also include a LED backlight and the LED pins are available as A (anode) and K (cathode).

Note that most alphanumeric LCD modules on the market operate on 5V. 3.3V devices are not so easy to find (especially in Brazil). The module used in this example is an SDEC LMC-SSC2E16-01 (a 5V LCD module based on the KS0066 LCD controller). The module pinout is described on table 13.1.

Pin	Name	Function
1	V_{SS}	Ground
2	V_{DD}	Supply (+5V)
3	V_O	Contrast
4	RS	Register select: command (0) or data (1)
5	R/$\overline{W}$	Read (1) or write (0) operation
6	E	Enable pin (when E=1 the display can receive commands or data)
7	DB0	Command/data bit 0. Not used in 4-bit mode
8	DB1	Command/data bit 1. Not used in 4-bit mode
9	DB2	Command/data bit 2. Not used in 4-bit mode
10	DB3	Command/data bit 3. Not used in 4-bit mode
11	DB4	Command/data bit 4
12	DB5	Command/data bit 5
13	DB6	Command/data bit 6
14	DB7	Command/data bit 7
15	A	LED backlight (anode)
16	K	LED backlight (cathode)

Table 13.1 – LCD module pins

Communicating with an LCD module is a matter of configuring its operating mode and writing into the display memory area.

It is important to notice that the controller can operate with two different bus widths: the standard 8-bit bus and an "economic" 4-bit mode.

The 4-bit mode is intended for using with low pin count microcontrollers. Using this mode, it is possible to interface a microcontroller to an LCD controller by using only six lines: four data lines (DB4, DB5, DB6 and DB7), plus RS and E.

The manufacturer recommends the use of the following startup sequence for configuring the controller for operation in 4-bit mode:

1. Wait for 15ms after the V_{DD} supply voltage rises above 90% of V_{DD}.

2. Send data 0x3 (DB4 and DB5 set) with RS=R/W=0.

3. Wait for at least 5ms.

4. Send data 0x3 (DB4 and DB5 set) with RS=R/W=0.

5. Wait for at least 100µs.

6. Send data 0x3 (DB4 and DB5 set) with RS=R/W=0.

7. Wait for at least 100µs (or until the busy flag is cleared).

8. Send data 0x3 (DB4 and DB5 set) with RS=R/W=0.

Now the controller is configured to operate in 4-bit mode. The microcontroller completes the display configuration by sending the following commands: function set, display off, display clear and entry mode set.

Once properly configured, the microcontroller can start writing data into the display memory.

Table 13.2 shows all commands recognized by the HD44780/KS0066 controller. The table also shows the approximate execution time for each command.

Command	RS	R/W	D7	D6	D5	D4	D3	D2	D1	D0	Description	Execution Time
Clear display	0	0	0	0	0	0	0	0	0	1	Clear the display and return the cursor to the first column of the first line.	1.64ms
Return home	0	0	0	0	0	0	0	0	1	x	Return the cursor to the first column of the first line and cancel any message shifting.	1.64ms
Entry mode set	0	0	0	0	0	0	0	1	I/D	S	I/D – set the display shifting direction: 1 – shift the cursor to the right on each character written. 0 – shift the cursor to the left on each character written. S – turn on (S=1) or off (S=0) the display shifting on each reading of the DDRAM.	40µs
Display on/off	0	0	0	0	0	0	1	D	C	B	Turn the display on (D=1) or off (D=0), turn the cursor on (C=1) or off (C=0) and enable (B=1) or disable (B=0) the cursor blinking.	40µs
Cursor or display shift	0	0	0	0	0	1	S/C	R/L	x	x	S/C – select between the cursor shifting (S/C=0) and the cursor+message shifting (S/C=1). R/L – Shifting direction: right (R/L=1) or left (R/L=0).	40µs
Function set	0	0	0	0	1	DL	N	F	x	x	DL – data bus width: 8-bit (DL=1) or 4-bit (DL=0). N – number of lines: one line (N=0) or two lines (N=1). F – character size: 8x5 (F=0) or 10x5 (F=1).	40µs
Set CGRAM address	0	0	0	1	A	A	A	A	A	A	Specify the current address (6-bit) of the CGRAM.	40µs
Set DDRAM address	0	0	1	A	A	A	A	A	A	A	Specify the current address (7-bit) of the DDRAM.	40µs
Read the address counter (AC) and the busy flag	0	1	BF	AC6	AC5	AC4	AC3	AC2	AC1	AC0	Read the content of the address counter (AC) and the state of the busy flag: BF = 1, controller busy. BF = 0, controller idle.	40µs
Write data into the DDRAM or CGRAM	1	0	D7	D6	D5	D4	D3	D2	D1	D0	Write data in the current address of the CGRAM or DDRAM pointed to by the AC register.	40µs
Read data from the DDRAM or CGRAM	1	1	D7	D6	D5	D4	D3	D2	D1	D0	Read data from the current address of the CGRAM or DDRAM pointed to by the AC register.	40µs

Table 13.2 – LCD module commands

The display controller has two different memory areas: the display data RAM (DDRAM) and the character generator RAM (CGRAM).

The DDRAM stores the data that is shown in the display. Each DDRAM address refers to a specific line and column on the display.

Figure 13.1 shows a DDRAM map for a 20x4 display (20 columns by 4 lines). The first line comprises address 0 up to 19. Addresses 20 up to 39 are mapped into the third line. The second line is mapped into addresses 64 up to 83 and the fourth line is mapped into addresses 84 up to 103.

Line	Address																			
1	0	1	2	3	4	5	6	7	8	9	10	11	12	13	14	15	16	17	18	19
2	64	65	66	67	68	69	70	71	72	73	74	75	76	77	78	79	80	81	82	83
3	20	21	22	23	24	25	26	27	28	29	30	31	32	33	34	35	36	37	38	39
4	84	85	86	87	88	89	90	91	92	93	94	95	96	97	98	99	100	101	102	103

Figure 13.1 – LCD module addresses

To write SCTEC into the first line and fourth column, the microcontroller must issue a "set DDRAM address" command with an address equal to 3 (the fourth column). If we wanted to write into the same column, but on the second line, the address should be equal to 67.

Once the address is set, a "write data" command (with data equal to "S") makes the character "S" appear into the desired display position.

If the controller is configured for automatically shift the cursor to the left on each character written, the microcontroller can send new "write data" commands with the remaining characters (CTEC). Doing so makes the word SCTEC to appear in the display.

Note that each LCD module has a specific character set stored into its ROM memory. Most modules have a character set similar (or identical) to the one shown in table 13.3.

Lower 4 Bits \ Upper 4 Bits	0000	0001	0010	0011	0100	0101	0110	0111	1000	1001	1010	1011	1100	1101	1110	1111
xxxx0000	CG RAM (1)			0	@	P	`	p				―	タ	ミ	α	p
xxxx0001	(2)		!	1	A	Q	a	q			。	ア	チ	ム	ä	q
xxxx0010	(3)		"	2	B	R	b	r			「	イ	ツ	メ	β	θ
xxxx0011	(4)		#	3	C	S	c	s			」	ウ	テ	モ	ε	∞
xxxx0100	(5)		$	4	D	T	d	t			、	エ	ト	ヤ	μ	Ω
xxxx0101	(6)		%	5	E	U	e	u			・	オ	ナ	ユ	σ	Ü
xxxx0110	(7)		&	6	F	V	f	v			ヲ	カ	ニ	ヨ	ρ	Σ
xxxx0111	(8)		'	7	G	W	g	w			ア	キ	ヌ	ラ	g	π
xxxx1000	(1)		(	8	H	X	h	x			ィ	ク	ネ	リ	√	x̄
xxxx1001	(2)		)	9	I	Y	i	y			ゥ	ケ	ノ	ル	⁻¹	y
xxxx1010	(3)		*	:	J	Z	j	z			エ	コ	ハ	レ	j	千
xxxx1011	(4)		+	;	K	[	k	{			オ	サ	ヒ	ロ	ˣ	万
xxxx1100	(5)		,	<	L	¥	l	|			ャ	シ	フ	ワ	¢	円
xxxx1101	(6)		-	=	M	]	m	}			ュ	ス	ヘ	ン	Ł	÷
xxxx1110	(7)		.	>	N	^	n	→			ョ	セ	ホ	゛	ñ	
xxxx1111	(8)		/	?	O	_	o	←			ッ	ソ	マ	゜	ö	█

Table 13.3 – LCD module ROM character set

Notice the presence of sixteen special characters on the first column. These are the user defined characters which can be stored into the CGRAM memory. The controller allows up to eight characters to be defined by the user application and each character comprises eight bytes.

Each user-defined character on the character set (table 13.3) is mapped into eight CGRAM addresses. Table 13.4 shows the mapping for the eight user-defined characters.

Character code	User character	
	First CGRAM address	Last CGRAM address
0 or 8	0	7
1 or 9	8	15
2 or 10	16	23
3 or 11	24	31
4 or 12	32	39
5 or 13	40	47
6 or 14	48	55
7 or 15	56	63

Table 13.4 – LCD module CGRAM ranges

To draw your own characters, all you need to do is store the character pattern into the desired CGRAM addresses. Figure 13.2 shows how the patterns are obtained (an up arrow is drawn).

Data	Bits							
	7	6	5	4	3	2	1	0
0x04	0	0	0	0	0	1	0	0
0x0E	0	0	0	0	1	1	1	0
0x15	0	0	0	1	0	1	0	1
0x04	0	0	0	0	0	1	0	0
0x04	0	0	0	0	0	1	0	0
0x04	0	0	0	0	0	1	0	0
0x04	0	0	0	0	0	1	0	0
0x00	0	0	0	0	0	0	0	0

Figure 13.2 – Designing a custom character

To store our character into the CGRAM, all we need to do is to store the data shown in figure 13.2 (data column) into the CGRAM area. If we want our character to be accessed by character code 2, we must store the pattern into addresses 16 up to 23. Example 13.1 shows how to use user-defined characters to create a nice bar graph on the display.

The listing below shows the implementation of some functions to deal with most low level communication between the microcontroller and the LCD module. The initialization function configures the LCD controller to operate in 4-bit mode. To minimize the number of connections, no read operation is allowed (the R/W line is connected to the ground).

In order to run the examples 13.1 and 13.2, the following listing must be saved as filed named "my_lcd.c" in the same folder of the application (or in a folder in the compiler search path). The file is also available for download at www.sctec.com.br/hcs08.

```
//************************************************************************
//* Basic Character LCD functions
//* By Fábio Pereira
//* 01/15/08
//************************************************************************

// The following defines set the default pins for the LCD display
#ifndef LCD_ENABLE                          // if lcd_enable is not defined
#define LCD_ENABLE        PTBD_PTBD1         // LCD enable pin on PTB1
#define LCD_ENABLE_DIR    PTBDD_PTBDD1       // LCD enable pin direction
#define LCD_RS            PTBD_PTBD0         // LCD R/S pin on PTB0
#define LCD_RS_DIR        PTBDD_PTBDD0       // LCD R/S pin direction
#define LCD_D4            PTBD_PTBD4         // LCD data D4 pin
#define LCD_D4_DIR        PTBDD_PTBDD4       // LCD data D4 pin direction
#define LCD_D5            PTBD_PTBD5         // LCD data D5 pin
#define LCD_D5_DIR        PTBDD_PTBDD5       // LCD data D5 pin direction
#define LCD_D6            PTBD_PTBD6         // LCD data D6 pin
#define LCD_D6_DIR        PTBDD_PTBDD6       // LCD data D6 pin direction
#define LCD_D7            PTBD_PTBD7         // LCD data D7 pin
#define LCD_D7_DIR        PTBDD_PTBDD7       // LCD data D7 pin direction
#endif
#define LCD_SEC_LINE      0x40               // Address of the second line of the LCD
```

```
// LCD configuration constants
#define CURSOR_ON        2
#define CURSOR_OFF       0
#define CURSOR_BLINK     1
#define CURSOR_NOBLINK   0
#define DISPLAY_ON       4
#define DISPLAY_OFF      0
#define DISPLAY_8X5      0
#define DISPLAY_10X5     4
#define _2_LINES         8
#define _1_LINE          0

// Display configuration global variable
static char lcd_mode;

union ubyte
{
  char _byte;
  struct
  {
    char b0 : 1;
    char b1 : 1;
    char b2 : 1;
    char b3 : 1;
    char b4 : 1;
    char b5 : 1;
    char b6 : 1;
    char b7 : 1;
  } bit;
};

//******************************************************************************
//* Prototypes
//******************************************************************************
void LCD_delay_ms (unsigned char time);
void LCD_send_nibble(char data);
void LCD_send_byte(char address, char data);
void LCD_init(char mode1, char mode2);
void LCD_pos_xy(char x, char y);
void LCD_write_char(char c);
void LCD_write_string (char *c);
void LCD_display_on(void);
void LCD_display_off(void);
void LCD_cursor_on(void);
void LCD_cursor_off(void);
void LCD_cursor_blink_on(void);
void LCD_cursor_blink_off(void);

//******************************************************************************
//* A simple delay function (used by LCD functions)
//******************************************************************************
//* Calling arguments
//* unsigned char time: aproximate delay time in miliseconds
//******************************************************************************
void LCD_delay_ms (unsigned char time)
{
  unsigned int temp;
  for(;time;time--) for(temp=(BUS_CLOCK/23);temp;temp--);
}

//******************************************************************************
//* Send a nibble to the LCD
//******************************************************************************
//* Calling arguments
//* char data : data to be sent to the display
//******************************************************************************
void LCD_send_nibble(char data)
{
  union ubyte my_union;
  my_union._byte = data;
  // Output the four data bits
  LCD_D4 = my_union.bit.b0;
  LCD_D5 = my_union.bit.b1;
  LCD_D6 = my_union.bit.b2;
  LCD_D7 = my_union.bit.b3;
```

```
  // pulse the LCD enable line
  LCD_ENABLE = 1;
  for (data=20; data; data--);
  LCD_ENABLE = 0;
}

//*********************************************************************
//* Write a byte into the LCD
//*********************************************************************
//* Calling arguments:
//* char address : 0 for instructions, 1 for data
//* char data : command or data to be written
//*********************************************************************
void LCD_send_byte(char address, char data)
{
  unsigned int temp;
  LCD_RS = address;                    // configure the R/S line
  LCD_ENABLE = 0;                      // set LCD enable line to 0
  LCD_send_nibble(data >> 4);          // send the higher nibble
  LCD_send_nibble(data & 0x0f);        // send the lower nibble
  for (temp=1000; temp; temp--);
}


//*********************************************************************
//* LCD initialization
//*********************************************************************
//* Calling arguments:
//* char mode1 : display mode (number of lines and character size)
//* char mode2 : display mode (cursor and display state)
//*********************************************************************
void LCD_init(char mode1, char mode2)
{
  char aux;
  // Configure the pins as outputs
  LCD_ENABLE_DIR = 1;
  LCD_RS_DIR = 1;
  LCD_D4_DIR = 1;
  LCD_D5_DIR = 1;
  LCD_D6_DIR = 1;
  LCD_D7_DIR = 1;
  // Set the LCD data pins to zero
  LCD_D4 = 0;
  LCD_D5 = 0;
  LCD_D6 = 0;
  LCD_D7 = 0;
  LCD_RS = 0;
  LCD_ENABLE = 0;
  LCD_delay_ms(50);
  // LCD 4-bit mode initialization sequence
  // send three times 0x03 and then 0x02 to finish configuring the LCD
  for(aux=0;aux<3;++aux)
  {
    LCD_send_nibble(3);
    LCD_delay_ms(5);
  }
  LCD_send_nibble(2);
  // Now send the LCD configuration data
  LCD_send_byte(0,0x20 | mode1);
  LCD_send_byte(0,0x08 | mode2);
  lcd_mode = 0x08 | mode2;
  LCD_send_byte(0,1);
  LCD_send_byte(0,6);
}

//*********************************************************************
//* LCD cursor position set
//*********************************************************************
//* Calling arguments:
//* char x : column (starting by 0)
//* char y : line (0 or 1)
//*********************************************************************
void LCD_pos_xy(char x, char y)
{
  char address;
  if (y) address = LCD_SEC_LINE; else address = 0;
```

```c
    address += x;
    LCD_send_byte(0,0x80|address);
  }

//*****************************************************************************
//* Write a character on the display
//*****************************************************************************
//* Calling arguments:
//* char c : character to be written
//*****************************************************************************
//* Notes :
//* \f clear the display
//* \n and \r return the cursor to line 1 column 0
//*****************************************************************************
void LCD_write_char(char c)
{
  switch (c)
  {
    case '\f' :
      LCD_send_byte(0,1);
      LCD_delay_ms(5);
      break;
    case '\n' :
    case '\r' :
      LCD_pos_xy(0,1);
      break;
    default:
      LCD_send_byte(1,c);
  }
}

//*****************************************************************************
//* Write a string on the display
//*****************************************************************************
//* Calling arguments:
//* char *c : pointer to the string
//*****************************************************************************
void LCD_write_string (char *c)
{
  while (*c)
  {
    LCD_write_char(*c);
    c++;
  }
}


//*****************************************************************************
//* Turn the display on
//*****************************************************************************
void LCD_display_on(void)
{
  lcd_mode |= 4;
  LCD_send_byte (0,lcd_mode);
}


//*****************************************************************************
//* Turn the display off
//*****************************************************************************
void LCD_display_off(void)
{
  lcd_mode &= 0b11111011;
  LCD_send_byte (0,lcd_mode);
}


//*****************************************************************************
//* Turn the cursor on
//*****************************************************************************
void LCD_cursor_on(void)
{
  lcd_mode |= 2;
  LCD_send_byte (0,lcd_mode);
}
```

```
//************************************************************
//* Turn the cursor off
//************************************************************
void LCD_cursor_off(void)
{
  lcd_mode &= 0b11111101;
  LCD_send_byte (0,lcd_mode);
}


//************************************************************
//* Turn on the cursor blink function
//************************************************************
void LCD_cursor_blink_on(void)
{
  lcd_mode |= 1;
  LCD_send_byte (0,lcd_mode);
}


//************************************************************
//* Turn off the cursor blink function
//************************************************************
void LCD_cursor_blink_off(void)
{
  lcd_mode &= 0b11111110;
  LCD_send_byte (0,lcd_mode);
}
```

Listing 13.1 – Character LCD functions

The next example shows how to implement an "analog-like" bar graph display using an MC9S08QG8 and a 16x2 LCD module.

The application defines five custom characters representing one, two, three, four and five column bars (a five column bar appears as a black block on the display). The ADC converter reads the voltage from the trimpot and the LCD_bargraph function is called periodically to draw the bar graph according to the converted voltage read from the ADC.

This application also uses the ADC as a simple timer, to control the update of the display (one update at every 50ms).

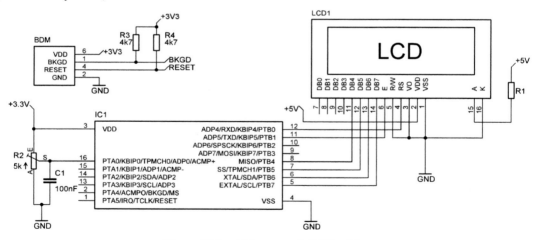

Figure 13.3 – Alphanumeric LCD wiring

Note that the direct interfacing between the 3.3V MCU and the 5V LCD module is possible due to the LCD module's ability to detect voltages higher than 2.2V as logic level "1". Additionally, as the LCD data bus is always an input (R/W pin = 0), no other circuitry is necessary to connect the devices together. R1 should be calculated according to the current drained by the module backlight.

```
/*
   LCD demonstration - MC9S08QG8 analog bar graph

   By Fábio Pereira (01/21/08)
   Joinville - SC - Brasil
   www.sctec.com.br

   LCD display connected to:
   DB0 - not connected
   DB1 - not connected
   DB2 - not connected
   DB3 - not connected
   DB4 - PTB4
   DB5 - PTB5
   DB6 - PTB6
   DB7 - PTB7
   E   - PTB1
   RS  - PTB0
   Trimpot connected to PTA0

*/

#include <hidef.h>           /* for EnableInterrupts macro */
#include "derivative.h"      /* include peripheral declarations */
#include "hcs08.h"           // this is our definition file

#define BUS_CLOCK 8000
#include "my_lcd.c"

unsigned int average;
unsigned int sum;
unsigned int timeout;

const char custom_char[] =
{
  0x10, 0x10, 0x10, 0x10, 0x10, 0x10, 0x10, 0x10, // one column
  0x18, 0x18, 0x18, 0x18, 0x18, 0x18, 0x18, 0x18, // two columns
  0x1C, 0x1C, 0x1C, 0x1C, 0x1C, 0x1C, 0x1C, 0x1C, // three columns
  0x1E, 0x1E, 0x1E, 0x1E, 0x1E, 0x1E, 0x1E, 0x1E, // four columns
  0x1F, 0x1F, 0x1F, 0x1F, 0x1F, 0x1F, 0x1F, 0x1F  // five columns
};

// ADC ISR
void interrupt VectorNumber_Vadc adc_isr(void)
{
  sum = sum + ADCR - average;
  if (timeout) timeout--; // ADC embedded timer
}

// LCD bar graph function
// "pix" = number of pixels to plot (0 up to 80)
void LCD_bargraph(char pix)
{
  char counter = 16;  // number of display columns
  while (pix)          // while pix>0
  {
    if (pix>=5)        // if there are more than 5 pixels
    {
      LCD_write_char(4); // draw a black block
      counter--;         // one column less
      pix -= 5;          // five pixels less
    }
    else               // if there is less than 5 pixels
    {
      LCD_write_char(pix-1); // draw a partial block
      pix = 0;           // no more pixels remaining
      counter--;         // one column less
    }
  }
  // fill the remaining columns with white spaces
  while (counter--) LCD_write_char(' ');
}
```

```
void main(void)
{
  char aux;
  SOPT1 = bBKGDPE;            // configure SOPT1 register, enable pin BKGD
  ICSSC = NV_FTRIM;           // configure FTRIM value
  ICSTRM = NV_ICSTRM;         // configure TRIM value
  ICSC2 = 0;                  // ICSOUT = DCOOUT / 1
  PTAD = PTBD = 0;            // force output latches to "0"
  PTADD = 0xFE;               // port A pins as outputs (except PTA0)
  PTBDD = 0xFF;               // port B pins as outputs
  // enable long sampling, 10-bit mode, ADICLK = 11b, ADCK = BUSCLK/16
  // ADC clock = 500 kHz - conversion time = 80us
  ADCCFG = bADLSMP | ADC_10BITS | ADC_BUSCLK_DIV2 | ADIV_8;
  // ADC in continuous mode, interrupt enabled, internal temperature sensor
  ADCSC1 = bAIEN | bADCO | ADC_CH0;
  APCTL1 = bADPC0;            // PTA0 as an analog input
  LCD_init(DISPLAY_8X5|_2_LINES,DISPLAY_ON|CURSOR_OFF|CURSOR_NOBLINK);
  // Now we set up the user custom characters
  LCD_send_byte(0,0x40);      // point AC to the CGRAM area
  for (aux=0; aux<sizeof(custom_char); aux++)
  {
    LCD_send_byte(1,custom_char[aux]);  // send the data to the LCD module
  }
  LCD_write_char('\f');       // clear the display
  LCD_pos_xy(0,0);            // set cursor to the first column of the first line
  LCD_write_string("HCS08 Unleashed!"); // write on the display
  LCD_pos_xy(0,1);            // set cursor to the second column of the second line
  sum = 0;
  average = 0;
  EnableInterrupts;
  while(1)
  {
    average = sum / 8;       // update the average
    if (!timeout)            // if the timeout expired
    {
      LCD_pos_xy(0,1);       // set cursor to the first column of the second line
      LCD_bargraph(average/13); // draw the bar graph
      timeout = 625;         // wait for 50ms (625 * 80us)
    }
  }
}
```

Example 13.1 – LCD bar graph

The next example demonstrates how to use the internal temperature sensor. This small application reads the average temperature (using the internal temperature sensor) and shows it on the LCD display (using the same schematic shown in figure 13.3).

To calculate the current temperature we used the formula shown in topic 10.2.3:

$$\text{Temp}(^{o}C) = 25 - \frac{V_{sensor} - V_{25degrees}}{m}$$

To avoid using floating point math, we performed all calculations using integer math. Prior to calculating the temperature itself, it is important to measure the actual voltage supply of the MCU (and the internal ADC), as it affects all ADC measurements.

To measure the actual supply voltage we can use the internal bandgap reference. As its voltage is well known (1.2V), measuring it enables us to calculate the actual ADC supply voltage:

$$V_{DD} = \frac{1.2V * 1023}{ADCR_{BANDGAP}}$$

For improved accuracy, we are going to multiply the bandgap voltage by a factor of ten, resulting in the following formula (V_{DD_CALC} is also multiplied by ten):

$$V_{DD\_CALC} = \frac{12 * 1023}{ADCR_{BANDGAP}} = \frac{12276}{ADCR_{BANDGAP}}$$

Now, we calculate the equivalent ADC result for a 701.2mV input voltage (sensor output at 25°C):

$$ADCR_{25degrees} = \frac{701.2mV * 1023}{V_{DD}}$$

Note that our V_{DD_CALC} is ten times the actual ADC voltage supply (V_{DD}). In order to calculate the correct ADCR value, we need to multiply the numerator by a factor of ten. Actually, multiplying the numerator by ten also improves the overall accuracy:

$$ADCR_{25degrees} = \frac{701.2mV * 1023 * 10}{V_{DD\_CALC}} \cong \frac{7173}{V_{DD\_CALC}}$$

In order to correctly calculate the temperature of the internal sensor, we must also calculate the correct slope for the sensor. The slope is represented in a mV/°C relation, which must be converted to an ADC count/°C relation. A 1.646mV/°C slope (cold slope) can be represented as the following hypothetical ADC result:

$$ADCR_{m\_cold} = \frac{1.646mV * 1023}{V_{DD}}$$

Unfortunately, the result is going to be lower than one. For a 3.3V supply, the $ADCR_{m\_cold}$ is going to be 0.51! In order to use such value in our temperature calculation, we are going to need to convert it to an integer number.

A little bit of basic math and "voilà": fractions! As you may remember, 0.51 can be represented by the fraction (51/100). This fraction (51/100) can be easily used in our calculations!

Now, in order to obtain the numerator of such fraction, our m_cold formula must be changed to:

$$ADCR_{m\_cold} = \frac{1.646mV * 1023 * 1000}{V_{DD\_CALC}} \cong \frac{1684}{V_{DD\_CALC}}$$

Applying the same reasoning to the hot slope we find the hot slope equivalent:

$$ADCR_{m\_hot} = \frac{1.769mV * 1023 * 1000}{V_{DD}} \cong \frac{1809}{V_{DD\_CALC}}$$

Using the formulas above, it is easy to calculate the current temperature of the chip (actually the temperature of the die).

Our example application also includes a function (bin_to_string) to convert an integer value into its ASCII representation (in order to be displayed). The application code is below.

```
/*
   LCD demonstration - MC9S08QG8 thermometer (internal temperature sensor)
   By Fábio Pereira (01/22/08)
   Joinville - SC - Brasil
   www.sctec.com.br

   LCD display module connected to:
   DB0 - not connected
   DB1 - not connected
   DB2 - not connected
   DB3 - not connected
   DB4 - PTB4
   DB5 - PTB5
   DB6 - PTB6
   DB7 - PTB7
   E   - PTB1
   RS  - PTB0
*/
```

```c
#include <hidef.h>            /* for EnableInterrupts macro */
#include "derivative.h"       /* include peripheral declarations */
#include "hcs08.h"            // this is our definition file

#define BUS_CLOCK 8000
#include "my_lcd.c"

unsigned int average_temp, average_bandgap;
unsigned long int sum_temp, sum_bandgap;
unsigned int temperature;
unsigned int reference_temp;
unsigned int supply_voltage;
unsigned int timeout;
unsigned int m_slope;

void interrupt VectorNumber_Vtpmovf tpm_isr(void)
{
  TPMSC_TOF = 0;
  if (timeout) timeout--;
}

// ADC ISR
void interrupt VectorNumber_Vadc adc_isr(void)
{
  static char adc_seq;
  if (!adc_seq)
  {
    sum_temp = sum_temp + ADCR - average_temp;
    ADCSC1_ADCH = ADCH_BANDGAP;
    adc_seq++;
  }
  else
  {
    sum_bandgap = sum_bandgap + ADCR - average_bandgap;
    ADCSC1_ADCH = ADCH_TEMP_SENSOR;
    adc_seq = 0;
  }

}

// Converts a 16-bit value into a ASCII string
// "value" is the 16-bit value
// "*str" is a pointer to the string
// This is a very simple way to convert a number to decimal format
void bin_to_string(unsigned int value, char *str)
{
  char aux;
  unsigned int unit;
  aux = 0;
  str[0] = str[1] = str[2] = str[3] = str[4] = '0';
  str[5] = 0;
  unit = 10000;
  while (value)
  {
    if (value>=unit)
    {
      value -= unit;
      str[aux]++;
    } else
    {
      aux++;
      unit /= 10;
    }
  }
  while (str[0]=='0')
  {
    for (aux=0; aux<=4; aux++) str[aux]=str[aux+1];
  }
}

void main(void)
{
  char string[10];
  SOPT1 = bBKGDPE;                // configure SOPT1 register, enable pin BKGD
  ICSSC = NV_FTRIM;               // configure FTRIM value
  ICSTRM = NV_ICSTRM;             // configure TRIM value
```

```
      ICSC2 = 0;                      // ICSOUT = DCOOUT / 1
      // BUSCLK = 16MHz/2 = 8MHz
      SPMSC1 = bBGBE;                 // enable bandgap reference
      PTAD = PTBD = 0;                // force output latches to "0"
      PTADD = 0xFE;                   // port A pins as outputs (except PTA0)
      PTBDD = 0xFF;                   // port B pins as outputs
      // enable long sampling, 10-bit mode, ADICLK = 11b, ADCK = BUSCLK/16
      // ADC clock = 500 kHz - conversion time = 80us
      ADCCFG = bADLSMP | ADC_10BITS | ADC_BUSCLK_DIV2 | ADIV_8;
      // ADC: interrupt enabled, bandgap reference
      ADCSC1 = bAIEN | ADCH_BANDGAP;
      // Configure TPM for 1ms periodic interrupt
      // TPMCK = 1MHz, modulo = 1000 (1MHz/1000 = 1000Hz = 1ms)
      TPMSC = bTOIE | TPM_BUSCLK | TPM_DIV8;
      TPMMOD = 999;
      LCD_init(DISPLAY_8X5|_2_LINES,DISPLAY_ON|CURSOR_OFF|CURSOR_NOBLINK);
      LCD_write_char('\f');           // clear the display
      LCD_pos_xy(0,0);                // set cursor to the first column of the first line
      LCD_write_string("HCS08 Unleashed!"); // write on the display
      LCD_pos_xy(1,1);                // set cursor to the second column of the second line
      LCD_write_string("Int temp = ");        // write on the display
      timeout = 300;                  // wait for 300ms
      sum_temp = 0;
      sum_bandgap = 0;
      average_temp = 0;
      average_bandgap = 0;
      EnableInterrupts;
      while(1)
      {
        average_temp = sum_temp / 64;        // update the average temperature
        average_bandgap = sum_bandgap / 64; // update the average bandgap value
        if (average_bandgap) supply_voltage = 12276/average_bandgap;
        if (supply_voltage) reference_temp = 7173/supply_voltage;
        if (!timeout)                 // if the timeout expired
        {
          // read the averaged temperature, if it is higher than or equal to
          // the reference temperature (700mv approximately), then we use the
          // cold slope formula
          if (average_temp>=reference_temp)     // cold slope
          {
            if (supply_voltage)
            {
              m_slope = 1684/supply_voltage;
              temperature = 25 - ((average_temp - reference_temp)*100)/m_slope;
            }
          }
          else
          { // if average is less than the reference temperature, then we
            // use the hot slope formula
            if (supply_voltage)
            {
              m_slope = 1809/supply_voltage;
              temperature = 25 + ((reference_temp - average_temp)*100)/m_slope;
            }
          }
          bin_to_string(temperature,string);  // convert the temperature to string
          LCD_pos_xy(12,1);           // set cursor position
          LCD_write_string(string);   // write the converted temperature on the LCD
          LCD_write_string("   ");    // clear the next characters after the temperature
          timeout = 300;              // wait for 300ms
        }
      }
    }
```

These **ifs** prevent divide-by-zero operations!

Example 13.2 – Thermometer example

13.3. RC and Robots Interfacing

Controlling small robots and radio-controlled vehicles (such as cars, airplanes, etc.) is a fascinating hobby. In this topic, we will learn how to build many circuits and applications that can be used to control or assist in controlling of such devices.

13.3.1. Understanding Small Servos

Most radio-controlled vehicles and robots are built around servomechanisms. These servomechanisms use a small DC motor to drive a gear that can be used to drive a robot arm, a steering wheel or any other mechanical device.

The servomechanism also needs a feedback sensor, so that it can track the current position of the driving gear (and actuate on the motor to reach the desired position). This is generally accomplished by a small potentiometer.

RC-type servos are usually driven by a pulse-position modulation (PPM) signal, a very simple system where the pulse width dictates the servo position. A typical PPM signal has a minimum width of 1ms and a maximum width of 2ms, with an interval of about 30ms between pulses. The servomechanism decodes those pulses and actuates on the motor so that it reaches the desired position (1ms = servo fully turned to one side, 1.5ms = servo in central position and 2ms = servo fully turned to the other side).

The following example applications demonstrate how to generate and decode such signals.

13.3.2. Generating PPM Signals

Generating PPM signals is a simple task and can be done with the help of a simple timer module. A typical use for a PPM generator is as a servo controller, which can control one or more servos according to a specific program. Another application could be a servo tester.

The following application example shows how to implement a serial servo controller able to generate 10 simultaneous and independent signals. This controller is compatible with standard MINI-SSC controllers and can be used with software such as the Visual Show Automation[45] or the ServoMotion software[44].

The serial protocol is very simple and consists of 3-byte frames:

	Header	Servo	Value
data:	255	0 to 254	0 to 254

Table 13.5 – Serial servo controller protocol

The application uses the serial communications interface (SCI) and the two TPM channels. Channel 0 and channel 1 are configured to operate in compare mode generating an interrupt each time the TPM counter (TPMCNT) reaches the compare value stored into the compare register (TPMC0V/TPMC1V).

The signal generation is quite straightforward: channel 1 is used to cycle between all servos (a sequence). For each servo, the channel 1 ISR takes care of the following tasks:

- Activate the corresponding output pin (high level).

- Update channel 0 compare register with the pulse width value (channel 0 ISR takes care of turning off the pin state).

- Update channel 1 compare register with the time for the next servo in the sequence (this time is actually calculated using the MAXIMUM_CYCLE_uS value defined at the beginning of the program).

- After all servo signals are generated, the "seq_complete" flag is set so that the new servo cycle values can be calculated (within the main function).

Channel 0 is used only to turn off the output pins (it ends the current pulse for the active servo).

The SCI module is configured to receive mode and operates at 9600 bps. Each received character generates an interrupt. The ISR then stores the received char into a circular buffer (16 bytes long) and sets a flag (rx_flag).

The "rx_flag" is tested within the main function. If it is found set, a small finite state machine (FSM) called "com_state_machine" is then executed. This FSM processes all received chars and searches for a valid frame sequence. Once found, the corresponding servo position (in the array "servo_pos") is updated with the new received value.

As always, we tried to keep all ISRs as short as possible. In this application, the time wasted on ISRs can produce wrong signal timings.

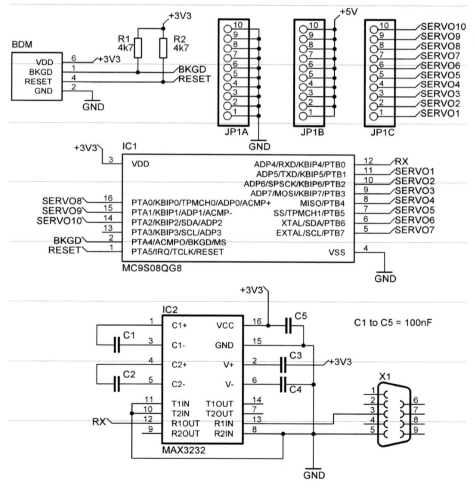

Figure 13.4 – Serial servo controller

```c
/*
   Serial servo controller with the MC9S08QG8
   By Fábio Pereira (12/04/07)
   Joinville - SC - Brasil
   www.sctec.com.br

   PTB0 - serial rx data
   PTB1 - servo 1 output
   PTB2 - servo 2 output
   PTB3 - servo 3 output
   PTB4 - servo 4 output
   PTB5 - servo 5 output
   PTB6 - servo 6 output
   PTB7 - servo 7 output
   PTA0 - servo 8 output
   PTA1 - servo 9 output
   PTA2 - servo 10 output

   ICS operating at 16 MHz (8 MHz BUSCLK)
   TPM operating from BUSCLK divided by 8 (1us counts)
   SCI operating at 9600 bps (RX only)
*/

#include <hidef.h>            /* for EnableInterrupts macro */
#include "derivative.h"       /* include peripheral declarations */
#include "hcs08.h"            // this is our definition file

#define SERVO1   PTBD_PTBD1
#define SERVO2   PTBD_PTBD2
#define SERVO3   PTBD_PTBD3
#define SERVO4   PTBD_PTBD4
#define SERVO5   PTBD_PTBD5
#define SERVO6   PTBD_PTBD6
#define SERVO7   PTBD_PTBD7
#define SERVO8   PTAD_PTAD0
#define SERVO9   PTAD_PTAD1
#define SERVO10  PTAD_PTAD2

#define MINIMUM_CYCLE_uS  1000        /* this is the minimum pulse width in us    */
#define MAXIMUM_CYCLE_uS  2000        /* this is the maximum pulse width in us    */
#define INTERVAL_uS       30000       /* this is the interval between pulses in us */
#define MAX_SERVO_POS     254         /* this is the maximum servo position       */

#define MULT_VALUE        (((MAXIMUM_CYCLE_uS-MINIMUM_CYCLE_uS)*10)/MAX_SERVO_POS)

char servo_pos[10];              // array for each servo position
unsigned int new_cycle[10];      // array for each servo cycle (us)
char servo_index;                // current active servo

#define RX_BUF_SIZE   16         // total rx buffer size in bytes

volatile char rx_buffer[RX_BUF_SIZE];
volatile char rx_buf_read_pointer, rx_buf_write_pointer;
volatile char rx_com_state;

#pragma DATA_SEG __DIRECT_SEG MY_ZEROPAGE
struct
{
  char rx_flag      : 1;    // new data received
  char seq_complete : 1;    // servo sequence is completed
} __near flags;
#pragma DATA_SEG DEFAULT

/*** Function prototypes ****************************************************/
void com_state_machine(void);
void init(void);

// TPM channel 0 ISR - clears all outputs after a specified delay
void interrupt VectorNumber_Vtpmch0 ISR_channel0(void)
{
  TPMC0SC_CH0F = 0;          // clear interrupt (CH0F)
  // Now we clear all output pins (all servo signals go low)
  PTBD = 0;
  PTAD = 0;
}
```

```c
/*****************************************************************************
   TPM channel 1 ISR:
   - Controls the servo sequencing
   - Activates each servo output in the sequence
   - Updates the channel 0 compare register (current servo pulse width)
   - Updates the channel 1 compare register (next servo)
   - Flags the start of global servo width calculation
   - Controls the delay between pulse sequences
 *****************************************************************************/
void interrupt VectorNumber_Vtpmch1 ISR_channel1(void)
{
  TPMC1SC_CH1F = 0;               // clear interrupt (CH1F)
  // the following switch cycles through all servos,
  // enabling the corresponding output pin
  switch (servo_index)
  {
    case 0:                    // Servo 1 (output high)
      SERVO1 = 1; break;
    case 1:                    // Servo 2 (output high)
      SERVO2 = 1; break;
    case 2:                    // Servo 3 (output high)
      SERVO3 = 1; break;
    case 3:                    // Servo 4 (output high)
      SERVO4 = 1; break;
    case 4:                    // Servo 5 (output high)
      SERVO5 = 1; break;
    case 5:                    // Servo 6 (output high)
      SERVO6 = 1; break;
    case 6:                    // Servo 7 (output high)
      SERVO7 = 1; break;
    case 7:                    // Servo 8 (output high)
      SERVO8 = 1; break;
    case 8:                    // Servo 9 (output high)
      SERVO9 = 1; break;
    case 9:                    // Servo 10 (output high)
      SERVO10 = 1; break;
  }
  if (servo_index<10)          // if the index refers to a valid servo
  {
    // then we update channel 0 compare register to the servo cycle value
    TPMC0V = TPMCNT + new_cycle[servo_index] - 19;
    // we also update channel 1 compare register to generate an interrupt
    // to initiate the next servo update
    TPMC1V = TPMCNT + MAXIMUM_CYCLE_uS + 10;
    servo_index++;             // next servo to be updated
  } else
  {
    // if the servo index does not point to a valid servo, then we
    // finished the current sequence
    flags.seq_complete = 1;
    // we update the channel 1 compare register to generate a delay
    // corresponding to the remaining interval time
    TPMC1V = TPMCNT + INTERVAL_uS - (MAXIMUM_CYCLE_uS * 10);
    servo_index = 0;           // restart the servo sequence
  }
}


/*****************************************************************************
   SCI reception interrupt servicing function
   Received data is stored into the receive buffer (rx_buffer)
 *****************************************************************************/
void interrupt VectorNumber_Vscirx ISR_scirx(void)
{
  char temp;
  temp = SCIS1;                // read sci status for the received char
  if (temp & 7)                // is there any error ?
  {
    // yes, read the char and return (no data is stored into the buffer)
    temp = SCID;               // dummy read of rx data for clearing interrupt flag
    return;
  }
  // an error free char was received, write it into the buffer
  rx_buffer[rx_buf_write_pointer++] = SCID;
  // verify if the buffer pointer exceeded the buffer size, if so, restart
  // pointer to zero (this is a circular buffer)
```

```c
    if (rx_buf_write_pointer>=RX_BUF_SIZE) rx_buf_write_pointer = 0;
    flags.rx_flag = 1;          // set rx flag
}

/*****************************************************************************
  State machine for controlling the communication with the host computer
  One state for each frame part:
  0 - header (0xFF)
  1 - servo number
  2 - cycle value
*****************************************************************************/
void com_state_machine(void)
{
  unsigned char data;
  static unsigned char servo;
  data = rx_buffer[rx_buf_read_pointer++];
  if (rx_buf_read_pointer>=RX_BUF_SIZE) rx_buf_read_pointer = 0;
  if (rx_buf_read_pointer==rx_buf_write_pointer) flags.rx_flag = 0;
  switch (rx_com_state)
  {
    case 0:               // wait for the header
      if (data==0xFF) rx_com_state++;  // if the header is ok, advance to the next state
      break;
    case 1:               // wait for the servo number
      if (data<10)        // if it is a valid servo
      {
        servo = data;
        rx_com_state++;   // advance to the next state
      } else rx_com_state = 0;
      break;
    case 2:               // wait for the cycle value
      servo_pos[servo] = data;       // store it into the array
      rx_com_state=0;     // restart rx state machine
      break;
  }
}


// MCU initialization function
void MCU_init(void)
{
  SOPT1 = bBKGDPE;        // enable debug pin
  SOPT2 = 0;             // initialize SOPT2
  // Trim the internal oscillator
  ICSTRM = NV_ICSTRM;
  ICSSC = NV_FTRIM;
  ICSC2 = 0;
  // CPU = 16MHz, BUSCLK = ICSCLK/2 = 8MHz
  PTAD = 0;
  PTADD = 0xFF;          // PTA0 through PTA4 as outputs (PTA5 is input only)
  PTBD = 0;
  PTBDD = 0xFE;          // PTB1 through PTB7 as outputs (PTB0 as an input)
  // TPM config
  TPMSC = TPM_BUSCLK + TPM_DIV8; // prescaler /8 (1us @ 8MHz)
  // Channel 0 config (interrupt on compare)
  TPMC0SC = bCHIE | TPM_COMPARE_INT;
  // Channel 1 config (interrupt on compare)
  TPMC1SC = bCHIE | TPM_COMPARE_INT;
  // Initialize channel 1 compare register
  TPMC1V = TPMCNT + 10;
  // SCI config (9600 bps, RX only)
  SCIBD = 52;            // 9600 bps
  SCIC2 = bRIE | bRE;    // enable RX module and RX interrupt
  // Variables initialization
  flags.rx_flag = 0;
  flags.seq_complete = 0;
  servo_index = 10;
  EnableInterrupts;      // enable interrupts (CCR:I = 0)
}


void main(void)
{
  char index;
  MCU_init();            // initialize the MCU
  while(1)
  {
    // if there is new data on rx buffer, call the state machine
```

```
          if (flags.rx_flag) com_state_machine();
          // if the servo sequence is completed, calculate new cycles
          if (flags.seq_complete)
          {
            for (index=0; index<10; index++)
              new_cycle[index]=MINIMUM_CYCLE_uS + (servo_pos[index]*MULT_VALUE)/10;
            flags.seq_complete = 0;        // clear seq_complete flag
          }
        }
      }
```

Example 13.3 – Serial servo controller

13.3.3. Reading PPM Signals

Reading PPM signals can be done by using a TPM capture channel; however, this example shows an alternate method for implementing capture without using a TPM channel. In this example, we used the TPM counter as the time base for a simple relay controller with the MC9S08QD4. The current TPM counting is stored on the PPM input rising edge and on the falling edge. The difference between the two values is the width of the captured pulse. The detection of rising and falling edges is done using the KBI module.

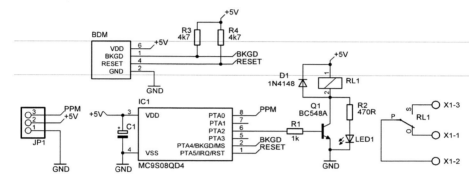

Figure 13.5 – PPM relay controller

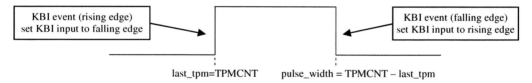

last_tpm=TPMCNT   pulse_width = TPMCNT – last_tpm

Figure 13.6 – Interrupt events

Note that the TPM modulo is set to 4096. This was only done because the measuring function was borrowed from the next example (PPM DC motor speed controller), which is based on an MC9S08QG8 device and uses the TPM for measuring PPM pulses and generating PWM pulses.

Due to the TPM modulo count is set to 4096 (TPMMOD=4095), if the TPM counter rolls from 4095 to 0 while a measurement is in progress, the application needs to subtract the last TPM count ("last_tpm") from 4096, before adding the value to the current TPM count. This is done by the following code ("temp" has the current TPM count):

```
if (last_tpm>temp)        // if the last TPM counting is higher the the current one
{                         // calculate the pulse width
  pulse_width = (4096 - last_tpm) + temp;
}
else                      // if the last TPM counting is lower than the current one
{                         // calculate the pulse width
  pulse_width = temp - last_tpm;
```

}

Each time a new capture is done, the "new_sample" flag is set. The main function monitors the flag and when it is found set, the new captured width is averaged (along with the last 8 samples), resulting in the "average_width" value.

The relay is turned on when the pulse width is found greater than 1.8ms and turned off when the pulse width is found lower than 1.3ms.

```c
/*
   PPM relay controller using the MC9S08QD4
   By Fábio Pereira (12/17/07)
   Joinville - SC - Brasil
   www.sctec.com.br
   PTA0 - PPM input   PTA2 - Relay output
   TPM:
   - pulse width measurements (detected with KBI interrupt)
*/
#define RELAY   PTAD_PTAD2

#include <hidef.h>
#include "derivative.h"
#include "hcs08.h"

volatile unsigned int pulse_width, average_width, sum_width;
volatile unsigned char new_sample;
#define MY_NVFTRIM  (*(const char * __far)0xFFAE)
#define MY_NVICSTRM (*(const char * __far)0xFFAF)

// KBI ISR
void interrupt VectorNumber_Vkeyboard1 kbi_isr (void)
{
  static unsigned int last_tpm;
  unsigned int temp;
  KBISC_KBACK = 1;          // clear interrupt flag
  if (KBIES)                // rising edge ?
  {
    last_tpm = TPMCNT;      // store the TPM counting
    KBIES = 0;              // next edge = falling
  } else                    // falling edge ?
  {
    temp = TPMCNT;          // read current TPM count
    if (last_tpm>temp)      // if the last TPM counting is higher the the current one
    {                       // calculate the pulse width
      pulse_width = (4096 - last_tpm) + temp;
    } else                  // if the last TPM counting is lower than the current one
    {                       // calculate the pulse width
      pulse_width = temp - last_tpm;
    }
    KBIES = BIT_0;          // next edge = rising
    new_sample = 1;         // set the flag
  }
}
// MCU initialization
void MCU_init(void)
{
  SOPT1 = bBKGDPE;          // enable debug pin
  PTAD = 0;
  PTADD = BIT_1 | BIT_2 | BIT_3 | BIT_4;  // PTA0 as input, PTA1 through PTA4 as outputs
  PTADS = BIT_0;            // enable PTA0 drive strength
  SOPT2 = 0;
  // Trim the internal oscillator and config ICS module
  ICSSC = MY_NVFTRIM;       // configure FTRIM value
  ICSTRM = MY_NVICSTRM;     // configure TRIM value
  ICSC2 = 0;                // BDIV = 00b
  // CPU = 16MHz, BUSCLK = ICSCLK/2 = 8MHz
  TPMSC = TPM_BUSCLK | TPM_DIV8;  // TPM clocked by BUSCLK/8 (1MHz)
  TPMMOD = 4095;            // TPM modulo = 4096
  //KBI config
  KBISC = bKBIE;            // KBI interrupt enabled
  KBIPE = BIT_0;            // PTA0 as KBI input
  KBIES = BIT_0;            // rising edge sensitivity
  KBISC_KBACK = 1;          // clear interrupt flag
  // Initialize global variables
  sum_width = new_sample = average_width = 0;
```

```
      EnableInterrupts;           // enable interrupts (CCR:I=0)
  }
  void main(void)
  {
    MCU_init();
    while (1)
    {
      if (new_sample)             // if a new sample is ready
      {
        // add the new sample to the sum of samples and subtract the average
        sum_width = sum_width + pulse_width - average_width;
        // calculate the new average (the sum divided by 8)
        average_width = sum_width>>3;
        new_sample = 0;           // clear the new_sample flag
      }
      // turn on the relay if the pulse width is greater than 1.8ms
      if (average_width>1800) RELAY = 1;
      // turn off the relay if the pulse width is smaller than 1.3ms
      if (average_width<1300) RELAY = 0;
    }
  }
```

Example 13.4 – PPM relay

13.3.4. A Simple DC Motor Speed Controller

The next example shows the implementation of a simple electronic speed controller (ESC). This circuit can control small brushed DC motors. The input signal is a pulse-position modulated (PPM) signal generated by the servo controller (such as the one we saw some pages ago) or by any RC receiver.

The circuit is based on an L298 integrated circuit. The L298 is a dual H-bridge driver which can drive loads with a voltage supply of up to 46V and maximum DC current of 2A (the maximum power dissipation of 25W must be respected). In this example we connected the two H-bridges in parallel for a maximum current of up to 4A (note that the example circuit is not intented for operating with current levels higher than 1A). The schematic is shown in figure 13.7.

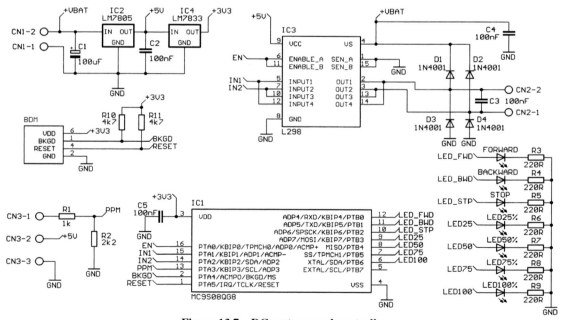

Figure 13.7 – DC motor speed controller

The L298 enable inputs (ENABLE_A and ENABLE_B) are connected together and a PWM signal is applied into them. The rotation direction is selected through the inputs IN1 and IN2 according to the following truth table:

IN1	IN2	Direction
0	0	brake (not used)
0	1	forward
1	0	backward
1	1	brake

Table 13.6 – L298 direction selection

The PWM signal is generated by the MCU proportionally to the PPM pulse width. Pulse widths higher than 1.6ms move the motor forward, the higher the pulse width, the higher is the PWM duty cycle:

```
new_pwm_cycle = (average_width-1600) * 10;
```

On the other hand, a pulse width lower than 1.4ms moves the motor backwards, the lower the pulse width, the higher is the PWM duty cycle:

```
new_pwm_cycle = (400-(average_width-1000)) * 10;
```

If the pulse width is between 1.4ms and 1.6ms, the motor is stopped. The two L298 inputs (IN1 and IN2) are set to level "1", making the motor be short-circuited. The PWM duty cycle still controls the amount of "braking force" applied to the motor. In this case, we set the software for full brake (PWM duty cycle = 100%).

The TPM counter and the KBI interrupt are used for reading the PPM pulse width. This is the same approach used in the last example.

The circuit uses seven LEDs to signal the current direction (forward, backward and stopped), and the approximate throttle (25%, 50%, 75%, 100%).

```
/*
A simple electronic speed controller (ESC) for DC motors using the MC9S08QG8

By Fábio Pereira (01/21/08)
Joinville - SC - Brasil
www.sctec.com.br

PTA0 - L298 enable pin (PWM)
PTA1 - L298 direction input 1 (IN1)
PTA2 - L298 direction input 2 (IN2)
PTA3 - PPM input
PTA4 - debug pin (BKGD)
PTA5 - reset (not used)
PTB0 - forward direction led
PTB1 - backward direction led
PTB2 - motor stopped led
PTB3 - <25% throttle
PTB4 - <50% throttle
PTB5 - <75% throttle
PTB6 - full throttle
PTB7 - not used

TPM:
 - PWM generation for the L298 (channel 0)
 - Pulse width measurement (KBI interrupt)
*/

#define EN       PTAD_PTAD0
#define IN1      PTAD_PTAD1
#define IN2      PTAD_PTAD2
#define LED_FWD  PTBD_PTBD0
#define LED_BWD  PTBD_PTBD1
#define LED_STP  PTBD_PTBD2
#define LED25    PTBD_PTBD3
#define LED50    PTBD_PTBD4
#define LED75    PTBD_PTBD5
#define LED100   PTBD_PTBD6
```

```c
#include <hidef.h>
#include "derivative.h"
#include "hcs08.h"

volatile unsigned int pulse_width, average_width;
volatile unsigned char new_sample;
volatile unsigned int new_pwm_cycle;

//const byte NVOPT_INIT @0xFFBF = MEM_SECURE;

// TPM overflow ISR
void interrupt VectorNumber_Vtpmovf tpm_overflow_isr(void)
{
  TPMSC_TOF = 0;            // clear interrupt flag
  TPMC0V = new_pwm_cycle;   // update PWM cycle
}


// KBI ISR
void interrupt VectorNumber_Vkeyboard kbi_isr (void)
{
  static unsigned int last_tpm;
  unsigned int temp;
  KBISC_KBACK = 1;          // clear interrupt flag
  if (KBIES)                // rising edge ?
  {
    last_tpm = TPMCNT;      // store the TPM counting
    KBIES = 0;              // next edge = falling
  }
  else                      // falling edge ?
  {
    temp = TPMCNT;
    if (last_tpm>temp)      // if the last TPM counting is higher than the the current one
    {                       // calculate the pulse width
      pulse_width = (4096 - last_tpm) + temp;
    }
    else                    // if the last TPM counting is lower than the current one
    {                       // calculate the pulse width
      pulse_width = temp - last_tpm;
    }
    KBIES = BIT_3;          // next edge = rising
    new_sample = 1;         // set the flag (a new sample is ready)
  }
}


// MCU initialization
void MCU_init(void)
{
  SOPT1 = bBKGDPE;          // enable debug pin
  SOPT2 = 0;                // initialize SOPT2
  PTAD = 0;
  PTBD = 0;
  PTADD = BIT_2 | BIT_1 | BIT_0;  // PTA0, PTA1 and PTA2 as outputs
  PTBDD = 0xFF;             // all port B pins as outputs
  // Set the ICS module
  // Trim the internal oscillator
  ICSTRM = NV_ICSTRM;
  ICSSC = NV_FTRIM;
  ICSC2 = 0;
  // CPU = 16MHz, BUSCLK = ICSCLK/2 = 8MHz
  // Configure TPM: TPMCK=BUSCLK/8, overflow interrupt enabled
  TPMSC = TPM_BUSCLK | TPM_DIV8 | bTOIE;
  TPMMOD = 4095;           // TPM modulo = 4096
  // Channel 0 as a PWM output (pin PTA0)
  TPMC0SC = TPM_PWM_HIGH;
  // KBI configuration:
  KBISC = bKBIE;           // KBI interrupt enabled
  KBIPE = BIT_3;           // input 3 (PTA3) as a KBI input
  KBIES = BIT_3;           // rising edge on PTA3 generates an interrupt
  KBISC_KBACK = 1;
  // Global variables initialization
  new_sample = 0;
  average_width = 0;
  EnableInterrupts;        // enable interrupts (CCR:I=0)
}
```

Uncomment this line to enable security (this prevents reading/copying the code in the device FLASH memory).

```
void main(void)
{
  volatile unsigned int sum_width;
  MCU_init();              // initialize the MCU
  sum_width = 12000;       // start the sum at 1.5ms (1500*8 = 12000)
  while (1)
  {
    if (new_sample)        // if a new sample (pulse width) is ready
    {
      // add the new sample to the sum of samples and subtract the average
      sum_width = sum_width + pulse_width - average_width;
      // calculate the new average (the sum divided by 8)
      average_width = sum_width>>3;
      new_sample = 0;      // clear the new_sample flag
    }
    if (average_width > 1600)
    {                      // if the pulse is higher than 1.6ms
      IN1 = 0;             // set the L298 direction
      IN2 = 1;
      LED_FWD = 1;         // turn on the FORWARD LED
      LED_BWD = 0;
      LED_STP = 0;
      //if the pulse width is higher than 1.98ms, set the PWM duty cycle to 100%
      if (average_width > 1980) new_pwm_cycle = 4096;
      // calculate the PWM duty cycle (proportional to the pulse width)
      else new_pwm_cycle = (average_width-1600) * 10;
    }
    else if (average_width < 1400)
    {                      // if the pulse is lower than 1.4ms
      IN1 = 1;             // set the L298 direction
      IN2 = 0;
      LED_FWD = 0;
      LED_BWD = 1;         // turn on the BACKWARD LED
      LED_STP = 0;
      //if the pulse width is lower than 1.02ms, set the PWM duty cycle to 100%
      if (average_width < 1020) new_pwm_cycle = 4096;
      // calculate the PWM duty cycle (proportional to the pulse width)
      else new_pwm_cycle = (400-(average_width-1000)) * 10;
    }
    else
    {                      // if the pulse is >= 1.4ms and <= 1.6ms
      IN1 = 1;             // stop the motor
      IN2 = 1;
      LED_FWD = 0;
      LED_BWD = 0;
      LED_STP = 1;         // turn on the STOP LED
      new_pwm_cycle = 4096; // set the PWM duty cycle to 100% (brake the motor)
    }
    if (!LED_STP)          // if the STOP LED is not lit
    {
      // turn on the LEDs according to the PWM duty cycle
      if (new_pwm_cycle>3072) LED100 = 1; else LED100 = 0;
      if (new_pwm_cycle>2048) LED75 = 1; else LED75 = 0;
      if (new_pwm_cycle>1024) LED50 = 1; else LED50 = 0;
      if (new_pwm_cycle>64) LED25 = 1; else LED25 = 0;
    }
  }
}
```

Example 13.5 – DC motor speed controller

13.4. I²C ADC and I/O Expander

This example application demonstrates how to use the I²C interface of the MC9S08QG8 to implement an I²C ADC and I/O expander.

The circuit shown in figure 13.8 comprises two microcontrollers: one operating as a master (IC1) and the other as a slave device (IC2).

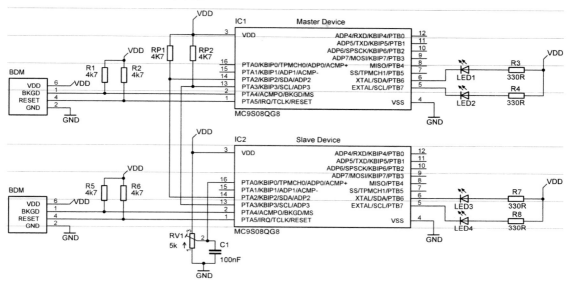

Figure 13.8 – Master/slave I²C ADC and I/O expander (V_DD=3.3V)

The slave device implements a simple protocol to allow full control of port B and reading the conversion result of the ADC (channels 0 and 1) through the I²C bus.

There are seven registers to know:

0. ADCH: allows selecting the ADC channel for reading through the ADCR_HIGH and ADCR_LOW registers.

    ADCH = 0, selects the ADC channel 0 (PTA0 input).

    ADCH > 0, selects the ADC channel 1 (PTA1 input).

1. ADCR_HIGH: allows reading the two most significant bits of the 10-bit conversion result for the selected ADC channel. Reading ADCR_HIGH latches the ADCR_LOW register until it is read.

2. ADCR_LOW: allows reading the eight lowest significant bits of the 10-bit conversion result for the selected ADC channel.

3. IO_DIR: allows configuring the direction for each pin of port B. This register directly changes the content of PTBDD.

4. IO_OUT: allows writing into the port B pins configured as outputs. This register directly changes the content of PTBD.

5. IO_IN: allows reading the state of the port B pins. This register reads the PTBD register.

6. IO_PUE: allows turning the internal pull up devices on port B on and off. This register directly modifies the PTBPE register.

Any register can be read or written through the I²C interface. A write operation comprises an address field (the slave device address is 1001000b and the R/W bit must be cleared) followed by a byte indicating the register to be written (REG) and a byte with the value to be written into the specified register (VALUE). Note that trying to write into an unimplemented register (REG>6) produces not effect. Figure 13.9 shows a typical I²C frame for a write register operation.

Figure 13.9 – Write register command

A read operation comprises an address field (the slave device address is 1001000b and the R/W bit must be cleared) followed by a byte indicating the register (REG) to be written in. After the Ack response from the slave, the master must send a repeated start followed by another address field (the same slave address, but the R/W bit must be set). Following the address field, the slave device sends the value of the specified register.

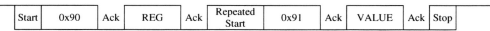

Figure 13.10 – Read register command

The listing below shows the code for the slave device (based on the MC9S08QG8 device).

```
/*
    DEMO9S08QG8 - I2C slave ADC converter and I/O expander
    SDA pin connected to PTA2/SDA
    SCL pin connected to PTA3/SCL
    Registers:
    0 - channel selection
    1 - higher byte of ADC conversion
    2 - lower byte of ADC conversion
    3 - I/O direction
    4 - output register
    5 - input register
*/

#include <hidef.h>          /* for EnableInterrupts macro */
#include "derivative.h"     /* include peripheral declarations */
#include "hcs08.h"          // This is our definition file!

#define ADC_ID      0x90    // I2C slave device ID
#define ADC_ADR     0       // I2C slave address

#define ADCH        0       // ADC channel selection register
#define ADCR_HIGH   1       // ADC result (high byte)
#define ADCR_LOW    2       // ADC result (low byte)
#define IO_DIR      3       // I/O direction selection register
#define IO_OUT      4       // I/O output register
#define IO_IN       5       // I/O input register
#define IO_PUE      6       // I/O pull up enable register

char registers[7], current_reg, block;

//const byte NVOPT_INIT @0xFFBF = MEM_SECURE;
```

> Uncomment this line to enable security (this prevents reading/copying the code in the device FLASH memory).

```
// MCU initialization
void MCU_init(void)
{
    SOPT1 = bBKGDPE;        // Enable debug pin
    ICSSC = NV_FTRIM;       // configure FTRIM value
    ICSTRM = NV_ICSTRM;     // configure TRIM value
    // BUSCLK is 4 MHz
    // Now we configure the I2C interface:
    // I2C clock is 4MHz*1/40 = 100kHz, SDA hold = 9/4MHz=2.25us
```

```
           IICF = I2C_M1 | 0x0B;
           IICA = ADC_ID | ADC_ADR;    // set the slave own address
           IICC = bIICEN | bIICIE;     // enable the I2C interface
           // Configure the ADC
           ADCCFG = ADIV_8 | bADLSMP | ADC_10BITS;
           ADCSC1 = bADCO | ADC_CH0;
           EnableInterrupts;
       }

       // This is the I2C interrupt servicing function
       void interrupt VectorNumber_Viic iic_isr(void)
       {
           char temp;
           static char rx_counter;
           // verify the source of I2C interrupt:
           if (IICS_IAAS)              // slave device addressed
           {
               if (IICS_SRW)
               {                         // slave transmits data to master
                   IICC_TX = 1;          // set transmit mode
                   if (current_reg<7) temp = registers[current_reg];
                   else temp = 0;
                   IICD = temp;
                   if (current_reg==ADCR_HIGH) block = 1;
                   if (current_reg==ADCR_LOW) block = 0;
               }
               else
               {                         // slave receives data from master
                   IICC_TX = 0;          // go to receive mode
                   temp = IICD;          // dummy read of IICD
                   rx_counter = 0;
               }
           }
           else if (IICS_TCF)          // a transfer was completed
           {
               if (IICC_TX)
               {                         // TX=1 -> transmit mode
                   if (IICS_RXAK)
                   {                     // NACK received from the master receiver
                       IICC_TX = 0;      // go to receive mode
                       temp = IICD;      // dummy read of IICD
                   }
                   else
                   {                     // ACK received from the master receiver
                       if (current_reg<7) temp = registers[current_reg];
                       else temp = 0;
                       // send the current register content (or 0x00 if no valid register is selected)
                       IICD = temp;
                   }
               }
               else
               {                         // TX=0 -> receive mode
                   if (!rx_counter)
                   {                     // if it is the first byte
                       IICC_TXAK = 0;    // send an ACK
                       current_reg = IICD; // select an internal register
                   }
                   else
                   {                     // if it is the second byte
                       IICC_TXAK = 1;    // send a NACK
                       temp = IICD;      // read data from the I2C bus
                       if (!current_reg)
                       { // if the current register is 0x00 (ADCH), config the ADC channel
                           if (registers[ADCH]) ADCSC1 = bADCO | ADC_CH1;
                           else ADCSC1 = bADCO | ADC_CH0;
                       }
                       // if the current register is valid (<7), update the register content
                       else if (current_reg<7) registers[current_reg] = temp;
                   }
                   rx_counter++;
               }
           }
           IICS_IICIF = 1;             // clear the interrupt flag
       }
```

> This is the blocking mechanism. It blocks ADCR_HIGH and ADCR_LOW updates when ADCR_HIGH is read and releases them when ADCR_LOW is read.

```
void main(void)
{
  MCU_init();                 // initialize the MCU
  for (current_reg=0; current_reg<7; current_reg++) registers[current_reg] = 0;
  current_reg = 0;            // set the current register to ADCH
  while(1)
  {
    if (ADCSC1_COCO && !block)
    { // if an ADC conversion is ready and the ADC is not blocked
      registers[ADCR_HIGH] = ADCRH;   // update the ADCR_HIGH register
      registers[ADCR_LOW] = ADCRL;    // update the ADCR_LOW register
    }
    PTBDD = registers[IO_DIR];        // update the PTBDD register with the IO_DIR content
    PTBPE = registers[IO_PUE];        // update the PTBPE register with the IO_PUE content
    registers[IO_IN] = PTBD;          // update the IO_IN register with the PTBD content
    PTBD = registers[IO_OUT];         // update the PTBD register with the IO_OUT content
  }
}
```

Example 13.6 – I^{2}C slave ADC and I/O expander

The listing below shows the code for the master device (based on the MC9S08QG8 device). The master and the slave devices were tested using two DEMO9S08QG8 boards. The external pull up devices were connected externally.

The master code configures the PTB6 and PTB7 pins on the slave device as outputs and selects the channel 0 (PTA0, connected to the trimpot RV1) as the ADC input.

The program then enters an infinite loop, reading the slave ADC channel 0 conversion result and toggling leds LED3 and LED4 on the slave device.

The conversion result read from the slave device is also compared to a fixed threshold (511). If it is lower than 511, LED2 (connected to the master PTB7 pin) lights. If the value read is higher than or equal to 511, LED2 is turned off.

```
/*

    DEMO9S08QG8 - I2C master reader for the I2C ADC and I/O expander
    SDA pin connected to PTA2/SDA
    SCL pin connected to PTA3/SCL
    Registers:
    0 - channel selection
    1 - higher byte of ADC conversion
    2 - lower byte of ADC conversion
    3 - I/O direction
    4 - output register
    5 - input register

*/

#include <hidef.h>          /* for EnableInterrupts macro */
#include "derivative.h"     /* include peripheral declarations */
#include "hcs08.h"          // This is our definition file!

#define ADC_ID      0x90
#define ADC_ADR     0

#define ADCH        0
#define ADCR_HIGH   1
#define ADCR_LOW    2
#define IO_DIR      3
#define IO_OUT      4
#define IO_IN       5
#define IO_PUE      6

#define RD          1
#define WR          0

#define LED         PTBD_PTBD7
```

```
enum ei2c_states
{
  I2C_IDLE,
  I2C_START, I2C_REPEATED_START,
  I2C_WRITE_REGISTER,
  I2C_WRITE_BYTE,
  I2C_DUMMY_READ, I2C_READ_BYTE,
  I2C_STOP,
  I2C_SLAVE_REG_WR, I2C_SLAVE_REG_RD
};

char i2c_buffer[3], command_ready;
unsigned int adc_result;

enum ei2c_states i2c_fsm(char new_state);

// MCU initialization
void MCU_init(void)
{
  SOPT1 = bBKGDPE;              // Enable debug pin
  ICSSC = NV_FTRIM;            // configure FTRIM value
  ICSTRM = NV_ICSTRM;          // configure TRIM value
  // BUSCLK is now 4 MHz
  // Now we configure the I2C interface:
  // I2C clock is 4MHz*1/40 = 100kHz, SDA hold = 9/4MHz=1.125us
  IICF = I2C_M1 | 0x0B;        // set the I2C bus speed (100kHz)
  IICC = bIICEN | bIICIE;      // enable the I2C interface
  // Configure the I/O pins
  PTBDD = BIT_7;               // PTB7 as an output
  EnableInterrupts;
}


// This is the I2C interrupt servicing function
void interrupt VectorNumber_Viic iic_isr(void)
{
  // verify the source of I2C interrupt:
  if (IICS_TCF)                // a transfer was completed
  {
    i2c_fsm(0);               // call I2C state machine
  }
  if (IICS_ARBL)              // the master lost the arbitration
  {
    IICS_ARBL = 1;           // clear ARBL
    i2c_fsm(I2C_IDLE);       // put the I2C state machine in idle
  }
  IICS_IICIF = 1;            // clear the interrupt flag
}

//**********************************************************************
// I2C finite state machine
// Usage:
//   i2c_fsm(0) to read the current state
//   i2c_fsm(x) to change the current state
// I2C buffer structure:
// i2c_buffer[0] : slave address
// i2c_buffer[1] : slave register
// i2c_buffer[2] : data to be written to the slave register (or read from it)
enum ei2c_states i2c_fsm(char new_state)
{
  static enum ei2c_states i2c_state;
  static char current_command;
  if (new_state == I2C_SLAVE_REG_RD)
  {
    current_command = RD;                 // current command is read
    i2c_state = I2C_START;                // go to I2C_START state
  }
  if (new_state == I2C_SLAVE_REG_WR)
  {
    current_command = WR;                 // current command is write
    i2c_state = I2C_START;                // go to I2C_START state
  }
  // if the TCF flag is not set, do not process the state machine
  // and return now with the current state
  if (!IICS_TCF) return (i2c_state);
  IICS_TCF = 1;                           // clear TCF flag
```

```c
    switch (i2c_state)
    {
    //*********************************************************************
    // I2C in idle state
    case I2C_IDLE:
        break;
    //*********************************************************************
    // send a start condition
    case I2C_START:
        IICC = bIICEN | bIICIE | bMST | bTX;
        IICD = i2c_buffer[0];              // send the address field
        i2c_state = I2C_WRITE_REGISTER;    // next state
        break;
    //*********************************************************************
    // send the register number
    case I2C_WRITE_REGISTER:
        IICD = i2c_buffer[1];              // send the register number
        // now, depending on the current command being executed, the state
        // machine can take one of two paths
        if (current_command==WR)
            i2c_state = I2C_WRITE_BYTE;    // if it is a write command
        else
            i2c_state = I2C_REPEATED_START; // if it is a read command
        break;
    //*********************************************************************
    // write one byte of data into the slave device
    case I2C_WRITE_BYTE:
        IICD = i2c_buffer[2];              // write data into the slave
        i2c_state = I2C_STOP;              // next state
        break;
    //*********************************************************************
    // send a repeated start (to change data direction on the slave)
    case I2C_REPEATED_START:
        IICC_RSTA = 1;                     // repeated start
        IICD = i2c_buffer[0] | RD;         // send the address field (R/W = RD)
        i2c_state = I2C_DUMMY_READ;        // next state
        break;
    //*********************************************************************
    // dummy read (start the memory read operation)
    case I2C_DUMMY_READ:
        IICC_TX = 0;                       // set I2C to receive mode
        i2c_buffer[2] = IICD;              // dummy read
        i2c_state = I2C_READ_BYTE;         // next state
        break;
    //*********************************************************************
    // read one byte of data from the slave device
    case I2C_READ_BYTE:
        IICC_TXAK = 1;                     // send NACK on the next read
        i2c_buffer[2] = IICD;              // read data from the slave device
        i2c_state = I2C_STOP;              // go to I2C_STOP state
        break;
    //*********************************************************************
    case I2C_STOP:
        IICC_MST = 0;                      // send a stop (go to slave mode)
        i2c_state = I2C_IDLE;              // next state
        command_ready = 1;                 // set the command_read flag
        break;
    }
    return (i2c_state);
}

// Read a slave register
char slave_register_read(char reg)
{
    char temp;
    while (i2c_fsm(0)!=I2C_IDLE);          // wait for the I2C interface to be idle
    // write the slave device ID and address into the buffer
    i2c_buffer[0] = ADC_ID | ADC_ADR;
    i2c_buffer[1] = reg;                   // write the register number into the buffer
    command_ready = 0;                     // clear the command_ready flag
    i2c_fsm(I2C_SLAVE_REG_RD);             // set the I2C state machine
    for (temp=250; temp; temp--);
    while (!command_ready);                // wait until the command is completed
    return (i2c_buffer[2]);                // return the data read
}
```

```
// Write into a slave register
void slave_register_write(char reg, char data)
{
  char temp;
  while (i2c_fsm(0)!=I2C_IDLE);        // wait for the I2C interface to be idle
  // write the slave device ID and address into the buffer
  i2c_buffer[0] = ADC_ID | ADC_ADR;
  i2c_buffer[1] = reg;                 // write the register number into the buffer
  i2c_buffer[2] = data;                // write the data into the buffer
  i2c_fsm(I2C_SLAVE_REG_WR);           // set the I2C state machine
  for (temp=250; temp; temp--);
}

void main(void)
{
  unsigned int temp;
  MCU_init();                          // initialize the MCU
  // set PTB6 and PTB7 as outputs on the slave device
  slave_register_write(IO_DIR,0xC0);
  slave_register_write(IO_OUT,0xFF);   // clear outputs on the slave device
  slave_register_write(ADCH,0);        // select slave analog channel 0  while(1)
  {
    slave_register_write(IO_OUT,BIT_7);
    adc_result = slave_register_read(ADCR_HIGH) << 8;
    adc_result |= slave_register_read(ADCR_LOW);
    if (adc_result<511) LED = 1; else LED = 0;
    for (temp = 15000; temp; temp--);
    slave_register_write(IO_OUT,BIT_6);
    for (temp = 15000; temp; temp--);
  }
}
```

Example 13.7 – I²C master controller

13.5. Playing Music

In this topic, we will learn about the RTTTL (Ringing Tones Text Transfer Language) standard and how to implement an RTTTL player on the HCS08 devices. The RTTTL standard was designed by John Mostelo and describes a monophonic music using ASCII text.

The RTTTL standard originated the RTX format, which is used by a number of cellphone manufacturers.

The RTTTL standard comprises three sections: the title of the ring tone, the default settings and the notes. Sections are separated by colons.

```
song_title:defaults:notes
```

The title section describes the name of the ring tone. The standard recommends the maximum length to be no longer than 10 characters (though many ringtones, including some presented in this book, have titles larger than 10 characters).

Following the title, we have the default settings section. This section is intended to specificy some global parameters to be used within the ring tone. There are three parameters defined by the RTTTL standard (five on RTX). The parameters are separated by commas.

- **Duration** – this parameter specifies the standard duration for all notes within the song. It can be one of the following values: 1, 2, 4, 8, 16, 32 or 64. On musical notation, 1 means a whole note (or Semibreve), 2 means a half note (or Minim), 4 means a quarter note (or Crotchet), 8 means an eighth note (or Quaver), 16 means a sixteenth note (Semiquaver or Demiquaver), 32 means a thirty-second note (Demisemiquaver) and 64 means a sixty-fourth note (or Hemidemisemiquaver). Note that all durations are relative to a whole note. That means that a half note has half the duration of a whole note. The duration specifier has the following format: **d**=duration. If no duration is specified, duration defaults to 4 (quarter note).

- **Octave** – this parameter specifies the default octave or scale. The RTTTL standard specifies four possible octaves: 4th (A4 or La = 440Hz), 5th (A5 or La = 880Hz), 6th (A6 or La = 1760Hz) and 7th (A7 or La = 3520Hz). The octave specifier has the following format: **o**=octave. If no octave is specified, octave defaults to 6.

- **Tempo** – this parameter specifies the number of notes in an interval of sixty seconds (also referred to as beats per second or BPM). This parameter is used to calculate the length of each note in seconds. For example: a 63 BPM means that each note plays for 60/63 = 0.952 seconds. A BPM of 180 means that the song plays 180 notes every 60 seconds and so, each note plays for 0.333 seconds. The tempo specifier has the following format: **b**=tempo. If no tempo is specified, tempo defaults to 63 BPM.

Following the defaults is the notes section. This section defines sequentially all notes that comprise the song. The RTTTL standard uses the English natural notation to represent each musical note: c (do), c#, d (re), d#, e (mi), f (fa), f#, g (sol), g#, a (la), a#, and b(si).

Each note can be prefixed by a duration value and suffixed by an octave. If no duration or octave is specified, the defaults are used.

A note can also be suffixed by a dot ".". A dotted note plays for 50% more time than the standard duration (or the duration specified in the note suffix).

Below we have a simple ring tone of the "Itchy and Scratchy" theme song:

```
itchy:d=8,o=6,b=160:c,a5,4p,c,a,4p,c,a5,c,a5,c,a,4p,p,c,d,e,p,e,f,g,4p,d,c,4d,f,4a#,4a,2c7
```

Separating each section:

Name: itchy.

Defaults: d=8, o=6, b=160 (an eighth duration, sixth octave and 160 BPM).

Notes: "c,a5,4p,c,a,4p,c,a5,c,a5,c,a,4p,p,c,d,e,p,e,f,g,4p,d,c,4d,f,4a#,4a,2c7".

Our RTTTL player is based on a simple parser that decodes each section of the ring tone, reads the note frequency from an array, calculates the duration of the note (based on the tempo and duration modifiers) and plays each note using the sound() function presented in topic 9.4.8.

Note that the duration calculation is performed on-the-fly, using the following formula:

$$\text{note\_duration} = \frac{\dfrac{60000}{\text{tempo(BPM)}}}{\text{duration}}$$

In which: note_duration is the total note duration in milliseconds.

tempo is the ring tone tempo in BPM.

duration is the note duration (1, 2, 4, 8, 16, 32 or 64).

The following example application was developed on the DEMO9S08QE128 board using the onboard piezoelectric buzzer. Porting the application to another HCS08 CPU is very simple as the hardware used is common to all available devices (including the tiny QA, QD and QG devices).

```
// DEMO9S08QE128 - RTTTL player
// By Fábio Pereira (01/09/08)
// Joinville - SC - Brasil
// www.sctec.com.br
// TPM1 channel 1 pin (PTB5) connected to a piezoelectric buzzer (mounted in the
// demonstration board)
```

```c
#include <hidef.h>        /* for EnableInterrupts macro */
#include "derivative.h"   /* include peripheral declarations */
#include "hcs08.h"        // This is our header file!

// RTTL music collection
const char *rtttl_library[]=
{
{"U2Newyears:d=8,o=6,b=125:a5,a5,c,4a5,a5,a5,a5,c,c,e,4c,c,c,e,e,e,g,4e,e,e,a5,e,e,g,e,e,
e,e,e,a5,a5,c,a5,a5,a5,a5,c,c,e,c,c,c,e,e,e,g,e,e,e,2a5"},

{"MissionImp:d=4,o=6,b=150:16d5,16d#5,16d5,16d#5,16d5,16d#5,16d5,16d5,16d#5,16e5,16f5,
16f#5,16g5,8g5,4p,8g5,4p,8a#5,8p,8c6,8p,8g5,4p,8g5,4p,8f5,8p,8p,8g5,4p,4p,8a#5,8p,8c6,
8p,8g5,4p,4p,8f5,8p,8f#5,8p,8a#5,8g5,1d5"},

{"ET:d=2,o=6,b=200:d,a,8g,8f#,8e,8f#,d,1a5,b5,b,8a,8g#,8f#,8g#,e,1c#7,e,d7,8c#7,8b,8a,8g,
f,d.,16d,16c#,16d,16e,f,d,d7,1c#7"},

{"Batman:d=8,o=5,b=160:16a,16g#,16g,16f#,16f,16f#,16g,16g#,4a.,p,d,d,c#,c#,c,c,c#,c#,d,
d,c#,c#,c,c,c#,c#,d,d,c#,c#,c,c,c#,c#,g6,p,4g6"},

{"Axelf:d=8,o=5,b=160:4f#,a.,f#,16f#,a#,f#,e,4f#,c6.,f#,16f#,d6,c#6,a,f#,c#6,f#6,16f#,
e,16e,c#,g#,4f#."},

{"Hogans:d=16,o=6,b=45:f5.,g#5.,c#.,f.,f#,32g#,32f#.,32f.,8d#.,f#,32g#,32f#.,32f.,d#.,
g#5.,c#.,32c,32c#.,32a#5.,8g#5.,f5.,g#5.,c#.,f5.,32f#5.,a#5.,32f#5.,d#.,f#.,32f.,g#.,32f.,
c#.,d#.,8c#."},

{"Jamesbond:d=4,o=5,b=320:c,8d,8d,d,2d,c,c,c,c,8d#,8d#,2d#,d,d,d,c,8d,8d,d,2d,c,c,c,c,
8d#,8d#,d#,2d#,d,c#,c,c6,1b.,g,f,1g."},

{"Pinkpanther:d=16,o=5,b=160:8d#,8e,2p,8f#,8g,2p,8d#,8e,p,8f#,8g,p,8c6,8b,p,8d#,8e,p,8b,
2a#,2p,a,g,e,d,2e"},

{"Countdown:d=4,o=5,b=125:p,8p,16b,16a,b,e,p,8p,16c6,16b,8c6,8b,a,p,8p,16c6,16b,c6,e,p,
8p,16a,16g,8a,8g,8f#,8a,g.,16f#,16g,a.,16g,16a,8b,8a,8g,8f#,e,c6,2b.,16b,16c6,16b,16a,1b"},

{"Adamsfamily:d=4,o=5,b=160:8c,f,8a,f,8c,b4,2g,8f,e,8g,e,8e4,a4,2f,8c,f,8a,f,8c,b4,2g,8f,
e,8c,d,8e,1f,8c,8d,8e,8f,1p,8d,8e,8f#,8g,1p,8d,8e,8f#,8g,p,8d,8e,8f#,8g,p,8c,8d,8e,8f"},

{"TheSimpsons:d=4,o=5,b=160:c.6,e6,f#6,8a6,g.6,e6,c6,8a,8f#,8f#,8f#,2g,8p,8p,8f#,8f#,8f#,
8g,a#.,8c6,8c6,8c6,c6"},

{"Indiana:d=4,o=5,b=250:e,8p,8f,8g,8p,1c6,8p.,d,8p,8e,1f,p.,g,8p,8a,8b,8p,1f6,p,a,8p,8b,
2c6,2d6,2e6,e,8p,8f,8g,8p,1c6,p,d6,8p,8e6,1f.6,g,8p,8g,e.6,8p,d6,8p,8g,e.6,8p,d6,8p,8g,f.6,
8p,e6,8p,8d6,2c6"},

{"GirlFromIpanema:d=4,o=5,b=160:g.,8e,8e,d,g.,8e,8e,8d,g.,e,e,8d,g,8g,8e,e,8e,8d,f,d,d,
8d,8c,e,c,c,8c,a#4,2c"},

{"TakeOnMe:d=4,o=4,b=160:8f#5,8f#5,8f#5,8d5,8p,8b,8p,8e5,8p,8e5,8p,8e5,8g#5,8g#5,8a5,8b5,
8a5,8a5,8a5,8e5,8p,8d5,8p,8f#5,8p,8f#5,8p,8f#5,8e5,8e5,8f#5,8e5,8f#5,8f#5,8f#5,8d5,8p,8b,
8p,8e5,8p,8e5,8p,8e5,8g#5,8g#5,8a5,8b5,8a5,8a5,8a5,8e5,8p,8d5,8p,8f#5,8p,8f#5,8p,8f#5,8e5,
8e5"},

{"She:d=16,o=5,b=63:32b,4c6,c6,c6,8b,8d6,8c6.,8b,a,32b,2c6,p,c6,c6,b,8d6,8c6,8b,a,4c6.,
8c6,b,32c6,4d6,8c6.,b,8a,8g,8f.,e,f,e,32f,2g,8p"},

{"Barbiegirl:d=4,o=5,b=125:8g#,8e,8g#,8c#6,a,p,8f#,8d#,8f#,8b,g#,8f#,8e,p,8e,8c#,f#,c#,p,
8f#,8e,g#,f#"},

{"Entertainer:d=4,o=5,b=140:8d,8d#,8e,c6,8e,c6,8e,2c.6,8c6,8d6,8d#6,8e6,8c6,8d6,e6,8b,d6,
2c6,p,8d,8d#,8e,c6,8e,c6,8e,2c.6,8p,8a,8g,8f#,8a,8c6,e6,8d6,8c6,8a,2d6"},

{"Xfiles:d=4,o=5,b=125:e,b,a,b,d6,2b.,1p,e,b,a,b,e6,2b.,1p,g6,f#6,e6,d6,e6,2b.,1p,g6,f#6,
e6,d6,f#6,2b.,1p,e,b,a,b,d6,2b.,1p,e,b,a,b,e6,2b.,1p,e6,2b."},

{"Autumn:d=8,o=6,b=125:a,a,a,a#,4a,a,a#,a,a,a,a#,4a,a,a#,a,16g,16a,a#,a,g.,16p,a,a,a,a#,
4a,a,a#,a,a,a,a#,4a,a,a#,a,16g,16a,a#,a,g."},

{"Spring:d=16,o=6,b=125:8e,8g#,8g#,8g#,f#,e,4b.,b,a,8g#,8g#,8g#,f#,e,4b.,b,a,8g#,a,b,8a,
8g#,8f#,8d#,4b.,8e,8g#,8g#,8g#,f#,e,4b.,b,a,8g#,8g#,8g#,f#,e,4b.,b,a,8g#,a,b,8a,8g#,8f#,
8d#,4b."},
};
```

```
// The following array stores the frequencies for the musical notes
const unsigned int note[4][12] =
{// C    C#    D    D#    E    F    F#    G    G#    A    A#    B
   262,  277,  294,  311,  330,  349,  370,  392,  415,  440,  466,  494, // 4th octave
   523,  554,  587,  622,  659,  698,  740,  784,  830,  880,  932,  988, // 5th octave
  1047, 1109, 1175, 1244, 1319, 1397, 1480, 1568, 1660, 1760, 1865, 1976, // 6th octave
  2093, 2218, 2349, 2489, 2637, 2794, 2960, 3136, 3320, 3520, 3728, 3951  // 7th octave
};
unsigned int channel_reload_value, duration_timer;
unsigned char sound_playing=0;
unsigned char duration, octave;
unsigned int tempo;

// TPM1 overflow interrupt isr: controls the duration of the sound
void interrupt VectorNumber_Vtpm1ch0 tpm1ch0_isr(void)
{
  TPM1C0SC_CH0F = 0;                      // clear interrupt flag
  TPM1C0V += 4000;                        // next compare in 1 ms
  if (duration_timer) duration_timer--;   // decrement the timer if > 0
  else                                    // if the timer is = 0
  {
    TPM1C1SC = 0;                         // disable TPM1CH1 channel (stops sound)
    sound_playing = 0;                    // clear the flag (no sound is not playing)
  }
}

// TPM1 channel1 compare interrupt isr: control the frequency generation
// the channel pin is automatically toggled on each compare
void interrupt VectorNumber_Vtpm1ch1 tpm1ch1_compare(void)
{
  TPM1C1SC_CH1F = 0;                      // clear interrupt flag
  TPM1C1V += channel_reload_value;        // next compare
}

// Output a sound through the TPM1CH1 pin (PTB5): freq in Hz, dur in ms
void sound(unsigned int freq, unsigned int dur)
{
  while (sound_playing);        // if a sound is playing, wait until if finishes
  // the channel reload value is equal to half the period of the signal
  channel_reload_value = (4000000/freq)/2;
  duration_timer = dur;         // set the total duration time
  // set channel 1 to compare mode (toggling the channel pin)
  TPM1C1V = channel_reload_value; // set the first compare value
  TPM1C1SC = bCHIE | TPM_COMPARE_TOGGLE;
  sound_playing = 1;            // sound is playing (until the flag is cleared)
}

// This function plays a song passed in a string in RTTTL format
void play_song(char *song)
{
  unsigned char temp_duration, temp_octave, current_note, dot_flag;
  unsigned int calc_duration;
  duration = 4;                 // standard duration = 4/4 = 1 beat
  tempo = 63;                   // standard tempo = 63 bpm
  octave = 6;                   // standard octave = 6th
  while (*song != ':') song++;  // find the first ':'
  song++;                       // skip the ':'
  while (*song!=':')            // repeat until find a ':'
  {
    if (*song == 'd')           // if it is the duration setting
    {
      duration = 0;             // set duration to zero (temporarily)
      song++;                   // advance to the next character
      while (*song == '=') song++; // skip the '='
      while (*song == ' ') song++; // skip the spaces
      // if the character is a number, then set the duration
      if (*song>='0' && *song<='9') duration = *song - '0';
      song++;                   // advance to the next character
      // if the character is also a number (duration can be 2 digits long)
      if (*song>='0' && *song<='9')
      { // multiply duration by 10 and add the value of the character
        duration = duration*10 + (*song - '0');
        song++;                 // advance to the next character
      }
      while (*song == ',') song++;  // skip the ','
    }
```

```
        if (*song == 'o')              // if it is the octave setting
        {
          octave = 0;                  // set octave to zero (temporarily)
          song++;                      // advance to the next character
          while (*song == '=') song++;  // skip the '='
          while (*song == ' ') song++;  // skip the spaces
          // if the character is a number, then set the octave
          if (*song>='0' && *song<='9') octave = *song - '0';
          song++;                      // advance to the next character
          while (*song == ',') song++;  // skips the ','
        }
        if (*song == 'b')              // if it is the tempo setting (beats per minute)
        {
          tempo = 0;                   // set tempo to zero (temporarily)
          song++;                      // advance to the next character
          while (*song == '=') song++;  // skips '='
          while (*song == ' ') song++;  // skips spaces
          // now read the tempo setting (can be 3 digits long)
          if (*song>='0' && *song<='9') tempo = *song - '0';
          song++;                      // advance to the next character
          if (*song>='0' && *song<='9')
          {
            tempo = tempo*10 + (*song - '0'); // tempo is two digits
            song++;                    // advance to the next character
            if (*song>='0' && *song<='9')
            {
              tempo = tempo*10 + (*song - '0'); // tempo is three digits
              song++;                  // advance to the next character
            }
          }
          while (*song == ',') song++;  // skip the ','
        }
      while (*song == ',') song++;      // skip the ','
    }
    song++;                            // advance to the next character
    // read the musical notes
    while (*song)                      // repeat until the character is null
    {
      current_note = 255;              // default note = pause
      temp_octave = octave;            // set the octave to the music default
      temp_duration = duration;        // set the duration to the music default
      dot_flag = 0;                    // clear the dot detection flag
      // look for a duration prefix
      if (*song>='0' && *song<='9')
      {
        temp_duration = *song - '0';
        song++;
        if (*song>='0' && *song<='9')
        {
          temp_duration = temp_duration*10 + (*song - '0');
          song++;
        }
      }
      // look for a note
      switch (*song)
      {
        case 'c': current_note = 0; break;     // C (do)
        case 'd': current_note = 2; break;     // D (re)
        case 'e': current_note = 4; break;     // E (mi)
        case 'f': current_note = 5; break;     // F (fa)
        case 'g': current_note = 7; break;     // G (sol)
        case 'a': current_note = 9; break;     // A (la)
        case 'b': current_note = 11; break;    // B (si)
        case 'p': current_note = 255; break;   // pause
      }
      song++;                          // advance to the next character
      // look for a # following the note
      if (*song=='#')
      {
        current_note++;   // increment the note (A->A#, C->C#, D->D#, F->F#, G->G#)
        song++;                        // advance to the next character
      }
```

```
            // look for a '.' (extend the note duration in 50%)
            if (*song=='.')
            {
              dot_flag = 1;              // if a '.' is found, set the flag
              song++;                    // advance to the next character
            }
            // look for an octave postfix
            if (*song>='0' && *song<='9')
            {
              temp_octave = *song - '0';// the temporary octave is set accordingly
              song++;                    // advance to the next character
            }
            if (*song=='.') // a dot can also be found after the octave (???)
            {
              dot_flag = 1;              // if a '.' is found, set the flag
              song++;                    // advance to the next character
            }
            while (*song == ',') song++;    // skip the ','
            // calculate the note duration
            calc_duration = (60000/tempo)/(temp_duration);
            calc_duration *= 4;          // a whole note has four beats
            // check if the dot flag is set, if it is set, extend the duration in 50%
            if (dot_flag) calc_duration = (calc_duration*3)/2;
            // if the current note is not a pause, play the note using the sound function
            if (current_note<255) sound(note[temp_octave-4][current_note],calc_duration);
            else
            { // if the current note = 255 (pause), just wait for the specified amount of time
              duration_timer = calc_duration;
              sound_playing = 1;
            }
            while (sound_playing);       // wait until the current note/pause has finished
          }
        }


        void main(void)
        {
          unsigned char song_sel;
          SOPT1 = bBKGDPE;                // enable the debug pin
          ICSSC = DCO_MID | NVFTRIM;      // configure FTRIM value, select DCO high range
          ICSTRM = NVICSTRM;              // configure TRIM value
          ICSC1 = ICS_FLL | bIREFS;// select FEI mode (ICSOUT = DCOOUT = 1024 * IRCLK)
          ICSC2 = BDIV_1;                 // ICSOUT = DCOOUT / 1
          // BUSCLK = 16MHz
          TPM1SC = TPM_BUSCLK | TPM_DIV4;  // TPMCK = 4MHz
          // set channel 0 to compare mode (interrupt only)
          TPM1C0V = 3999;
          TPM1C0SC = bCHIE | TPM_COMPARE_INT;
          EnableInterrupts;               // enable interrupts (CCR:I=0)
          while(1)
          {
            for (song_sel=0;song_sel<20;song_sel++)
            {
              play_song(rtttl_library[song_sel]); // play the song
              // wait for 2 seconds before start the next song
              duration_timer = 2000;
              sound_playing = 1;
              while (sound_playing);
            }
          }
        }
```

A dotted note multiplies the note duration by 3/2 (1.5 times)

Example 13.8 – RTTTL music player

 This example can be further optimized by using an array with the TPM channel reload values (instead of the note frequencies). Another optimization would be including a specific array for durations (this would save the time spent calculating the duration).

HCS08 Unleashed

13.6. Experiments with Accelerometers

Some demonstration boards (such as the DEMO9S08AW60E, DEMO9S08LC60, DEMO9S08QE128, etc.) include an on-board accelerometer.

An accelerometer is a device that measures acceleration (a rate of change of velocity) and has a large application field: vibration detection, collision detection, inclination measurements, navigation (as part of an inertial navigation system), free fall detection, motion sensing, etc.

Accelerometer chips are based on the MEMS (MicroElectroMechanical Systems) technology[53] and can be found in three different "flavors": single axis, dual axis and triple axis devices. Triple axis accelerometers use three independent accelerometers to measure acceleration on three different axis: X, Y and Z. Such devices can detect movement in any direction.

The following examples are based on the MMA7260QT a three axis MEMS accelerometer with a selectable measuring range and low power consumption. The device is available on a small 16-pin QFN package (6x6mm) and operates from 2.2 up to 3.6V, with a typical supply current of 500μA (3μA in sleep mode).

The MMA7260QT has three accelerometers, internal conditioning circuitry (including switched capacitor single pole filters), range selection circuitry and a power management circuitry (for the sleep mode control).

The device generates voltage signals proportional to the instant acceleration on each axis. The default voltage (at 0g) is nearly 1.65V. Notice that the accelerometer is always under effect of the Earth's gravity field, a nominal 1g (9.81m/s^2). Figure 13.11 shows the three sensitivity axis of the accelerometer.

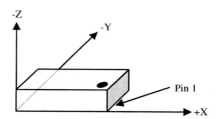

Figure 13.11 – MMA7260QT sensing directions

There are four sensitivity levels available on the MMA7260QT. These levels are selected through the g-select1 and g-select2 input pins, according to the following table:

g-select2	g-select1	g-range*	Sensitivity (mV/g*)
0	0	+/- 1.5g	800
0	1	+/- 2g	600
1	0	+/- 4g	300
1	1	+/- 6g	200

* 1g equals to 9.81m/s^2.

Table 13.7

As shown in figure 13.11 and with a g-range selection of +/- 1.5g, the accelerometer outputs the following voltages (V_{DD}=3.3V): X = 1.65V(0g), Y = 1.65V and Z = 2.45V (+1g).

Figure 13.12 shows the recommended connections for the device input and output pins. The manufacturer recommends the use of low-pass RC filters as shown, to minimize the switching noise of the internal switched filters.

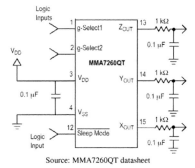

Source: MMA7260QT datasheet

Figure 13.12

13.6.1. A Simple Tilt Indicator

The next example shows how to implement a simple tilt indicator using the DEMO9S08QE128 and the onboard MMA7260QT accelerometer. Figure 13.13 shows the sensitivity axis of the board.

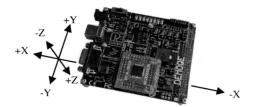

Figure 13.13

The MMA7260QT is connected as shown in figure 13.14. Note that, in order to use the device, the jumper J15 must be placed in the 1-2 position, J13 and J14 must be placed in position 1-2 (sensitivity controlled by pins PTD0 and PTD1 of the MCU). The J16 header must be shorted in order to connect the ZOUT to PTA7, YOUT to PTC7 and XOUT to PTA1.

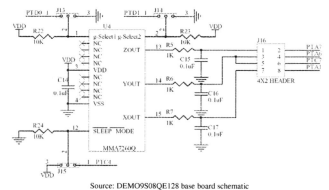

Source: DEMO9S08QE128 base board schematic

Figure 13.14 – MMA7260QT wiring on DEMO9S08QE128 board

The application measures the X axis and turns on the LEDs connected to PTC and PTE, according to the average value read from the accelerometer. After turning the board on, press the PTD3 key to calibrate the indicator (this procedure actually stores the current X axis value into a reference variable).

Once calibrated and if the board is lying still on a horizontal surface, only LED PTC3 should be lit. By tilting the board to the left or to the right, the LEDs change, indicating the relative tilt.

```
/*
   Tilt indicator with the MC9S08QE128 and MMA7260QT (DEMO9S08QE128 board)

   By Fábio Pereira (01/20/08)
   Joinville - SC - Brasil
   www.sctec.com.br

   PTA1 - XOUT
   PTA6 - YOUT
   PTA7 - ZOUT
   PTC4 - MMA7260 sleep mode selection (0 = sleep, 1 = operating)
   PTD0 - MMA7260 g-select1 input
   PTD1 - MMA7260 g-select2 input
   PTC0 to PTC5 - LEDs
   PTE6, PTE7 - LEDs

   ADC:
   - accelerometer measurements (PTA1, PTA6 and PTA7)
*/

#include <hidef.h>          /* for EnableInterrupts macro */
#include "derivative.h"     /* include peripheral declarations */
#include "hcs08.h"          // This is our definition file!

#define ACCEL_ON    PTCD_PTCD4
#define CAL_SW      PTDD_PTDD3

unsigned int x_axis, y_axis, z_axis;
unsigned long int x_sum, y_sum, z_sum;
unsigned int x_cal;
char adc_seq_complete;

// ADC ISR
void interrupt VectorNumber_Vadc adc_isr(void)
{
  static char adc_seq;
  switch (adc_seq)
  {
    case 0:
      x_sum = (x_sum + ADCR) - x_axis;
      ADCSC1 = bAIEN | ADC_CH8; // next channel = 8 (y axis)
      adc_seq++;
      break;
    case 1:
      y_sum = (y_sum + ADCR) - y_axis;
      ADCSC1 = bAIEN | ADC_CH9; // next channel = 9 (z axis)
      adc_seq++;
      break;
    case 2:
      z_sum = (z_sum + ADCR) - z_axis;
      ADCSC1 = bAIEN | ADC_CH1; // next channel = 1 (x axis)
      adc_seq = 0;
      adc_seq_complete = 1;   // signal the end of an ADC sequence
      break;
  }
}

void main(void)
{
  SOPT1 = bBKGDPE;            // configure SOPT1 register, enable pin BKGD for BDM
  ICSSC = NVFTRIM;            // configure FTRIM value
  ICSTRM = NVICSTRM;          // configure TRIM value
  ICSC2 = 0;                  // ICSOUT = DCOOUT / 1
  // enable long sampling, 12-bit mode, ADICLK = 01b, ADCK = BUSCLK/16
  // ADC sampling rate = 25ksps
  ADCCFG = bADLSMP | ADC_12BITS | ADC_BUSCLK_DIV2 | ADIV_8;
  APCTL1 = bADPC1;           // ADP1 in analog mode
  APCTL2 = bADPC8 | bADPC9;  // ADP8 and ADP9 in analog mode
  // ADC channel 1, interrupts enabled
  ADCSC1 = bAIEN | ADC_CH1;
  PTDD = 0;                   // select 1.5g sensitivity
  PTDDD = BIT_0 | BIT_1;
```

```
PTBDD = BIT_6 | BIT_7;      // PTB6 and PTB7 as outputs
PTCDD = 0xFF;               // PTC as outputs
PTEDD = BIT_7 | BIT_6;
x_sum = y_sum = z_sum = 0;
x_axis = y_axis = z_axis = 0;
ACCEL_ON = 1;
x_cal = 0;
EnableInterrupts;          // Enable interruts (CCR:I=0)
while (1)
{
  if (adc_seq_complete)
  {
    x_axis = x_sum/32;     // calculate the x axis
    y_axis = y_sum/32;     // calculate the y axis
    z_axis = z_sum/32;     // calculate the z axis
    adc_seq_complete = 0;
  }
  PTCD = 0xFF;             // turn off all LEDs connected to PTC
  PTED = 0xC0;             // turn off all LEDs connected to PTE
  // if the calibration key is pressed, store the current value of the x axis
  if (!CAL_SW) x_cal = x_axis;
  if (x_axis>(x_cal+500))  PTCD_PTCD0 = 0; else
    if (x_axis>(x_cal+350)) PTCD_PTCD1 = 0; else
      if (x_axis>(x_cal+200)) PTCD_PTCD2 = 0; else
        if (x_axis<(x_cal-500)) PTED_PTED7 = 0; else
          if (x_axis<(x_cal-350)) PTED_PTED6 = 0; else
            if (x_axis<(x_cal-200)) PTCD_PTCD5 = 0; else PTCD_PTCD3 = 0;

}
}
```

Example 13.9 – Tilt indicator example

13.6.2. Music-Shake

Our last example is a modification of the RTTTL player we have just seen in topic 13.5. The application uses the onboard MMA7260QT accelerometer to detect the board movement. When a shake towards the −Y direction is detected (refer to figure 13.13), the current song is finished and the player advances to the next song.

A simple algorithm detects the shake: the Y axis is sampled at every 50ms. If an acceleration higher than 1g is detected, followed by a -1g acceleration (in the next 50ms), the board was shaked towards the −Y axis and the player advances to the next song.

```
/*
    DEMO9S08QE128 - Music-shake - RTTTL player with accelerometer sensing

    By Fábio Pereira (01/20/08)
    Joinville - SC - Brasil
    www.sctec.com.br

    PTA1 - XOUT
    PTA6 - YOUT
    PTA7 - ZOUT
    PTC4 - MMA7260 sleep mode selection (0 = sleep, 1 = operating)
    PTD0 - MMA7260 g-select1 input
    PTD1 - MMA7260 g-select2 input
    PTB5 - piezoelectric buzzer (TPM1CH1 pin)

    ADC:
    - accelerometer measurements (PTA1, PTA6 and PTA7)
    TPM1:
    - channel 0: controls the sound duration and accelerometer sampling
    - channel 1: frequency generation

*/
```

```c
#include <hidef.h>          /* for EnableInterrupts macro */
#include "derivative.h"     /* include peripheral declarations */
#include "hcs08.h"          // This is our header file!

#define ACCEL_ON   PTCD_PTCD4

// RTTL music collection
const char *rtttl_library[]=
{
{"U2Newyears:d=8,o=6,b=125:a5,a5,c,4a5,a5,a5,a5,c,c,e,4c,c,c,e,e,e,g,4e,e,e,a5,e,e,g,e,e,e,
e,e,a5,a5,c,a5,a5,a5,a5,c,c,e,c,c,c,e,e,e,g,e,e,e,2a5"},
{"MissionImp:d=4,o=6,b=150:16d5,16d#5,16d5,16d#5,16d5,16d#5,16d5,16d#5,16e5,16f5,
16f#5,16g5,8g5,4p,8g5,4p,8a#5,8p,8c6,8p,8g5,4p,8g5,4p,8f5,8p,8p,8g5,4p,4p,8a#5,8p,8c6,8p,
8g5,4p,4p,8f5,8p,8f#5,8p,8a#5,8g5,1d5"},
{"ET:d=2,o=6,b=200:d,a,8g,8f#,8e,8f#,d,1a5,b5,b,8a,8g#,8f#,8g#,e,1c#7,e,d7,8c#7,8b,8a,8g,f,
d.,16d,16c#,16d,16e,f,d,d7,1c#7"},
{"Batman:d=8,o=5,b=160:16a,16g#,16g,16f#,16f,16f#,16g,16g#,4a.,p,d,d,c#,c#,c,c,c#,c#,d,d,
c#,c#,c,c,c#,c#,d,d,c#,c#,c,c,c#,c#,g6,p,4g6"},
{"Axelf:d=8,o=5,b=160:4f#,a.,f#,16f#,a#,f#,e,4f#,c6.,f#,16f#,d6,c#6,a,f#,c#6,f#6,16f#,e,
16e,c#,g#,4f#."},
{"Hogans:d=16,o=6,b=45:f5.,g#5.,c#.,f.,f#,32g#,32f#.,32f.,8d#.,f#,32g#,32f#.,32f.,d#.,g#5.,
c#,32c,32c#.,32a#5.,8g#5.,f5.,g#5.,c#.,f5.,32f#5.,a#5.,32f#5.,d#.,f#.,32f.,g#.,32f.,c#.,
d#.,8c#."},
{"Jamesbond:d=4,o=5,b=320:c,8d,8d,d,2d,c,c,c,c,8d#,8d#,2d#,d,d,d,c,8d,8d,d,2d,c,c,c,c,8d#,
8d#,d#,2d#,d,c#,c,c6,1b.,g,f,1g."},
{"Pinkpanther:d=16,o=5,b=160:8d#,8e,2p,8f#,8g,2p,8d#,8e,p,8f#,8g,p,8c6,8b,p,8d#,8e,p,8b,
2a#,2p,a,g,e,d,2e"},
{"Countdown:d=4,o=5,b=125:p,8p,16b,16a,b,e,p,8p,16c6,16b,8c6,8b,a,p,8p,16c6,16b,c6,e,p,8p,
16a,16g,8a,8g,8f#,8a,g.,16f#,16g,a.,16g,16a,8b,8a,8g,8f#,e,c6,2b.,16b,16c6,16b,16a,1b"},
{"Adamsfamily:d=4,o=5,b=160:8c,f,8a,f,8c,b4,2g,8f,e,8g,e,8e4,a4,2f,8c,f,8a,f,8c,b4,2g,8f,e,
8c,d,8e,1f,8c,8d,8e,8f,1p,8d,8e,8f,8g,1p,8d,8e,8f#,8g,p,8d,8e,8f#,8g,p,8c,8d,8e,8f"},
{"TheSimpsons:d=4,o=5,b=160:c.6,e6,f#6,8a6,g.6,e6,c6,8a,8f#,8f#,8f#,2g,8p,8p,8f#,8f#,8f#,
8g,a#.,8c6,8c6,8c6,c6"},
{"Indiana:d=4,o=5,b=250:e,8p,8f,8g,8p,1c6,8p.,d,8p,8e,1f,p..,g,8p,8a,8b,8p,1f6,p,a,8p,8b,
2c6,2d6,2e6,e,8p,8f,8g,8p,1c6,p,d6,8p,8e6,1f.6,g,8p,8g,e.6,8p,d6,8p,8g,e.6,8p,d6,8p,8g,f.6,
8p,e6,8p,8d6,2c6"},
{"GirlFromIpanema:d=4,o=5,b=160:g.,8e,8e,d,g.,8e,e,8e,8d,g.,e,e,8d,g,8g,8e,e,8e,8d,f,d,d,
8d,8c,e,c,c,8c,a#4,2c"},
{"TakeOnMe:d=4,o=4,b=160:8f#5,8f#5,8f#5,8d5,8p,8b,8p,8e5,8p,8e5,8p,8e5,8g#5,8g#5,8a5,8b5,
8a5,8a5,8a5,8e5,8p,8d5,8p,8f#5,8p,8f#5,8p,8f#5,8e5,8e5,8f#5,8e5,8f#5,8f#5,8f#5,8d5,8p,8b,
8p,8e5,8p,8e5,8g#5,8g#5,8a5,8b5,8a5,8a5,8a5,8e5,8p,8d5,8p,8f#5,8p,8f#5,8p,8f#5,8e5,
8e5"},
{"She:d=16,o=5,b=63:32b,4c6,c6,c6,8b,8d6,8c6.,8b,a,32b,2c6,p,c6,c6,b,8d6,8c6,8b,a,4c6.,8c6,
b,32c6,4d6,8c6.,b,8a,8g,8f.,e,f,e,32f,2g,8p"},
{"Barbiegirl:d=4,o=5,b=125:8g#,8e,8g#,8c#6,a,p,8f#,8d#,8f#,8b,g#,8f#,8e,p,8e,8c#,f#,c#,p,
8f#,8e,g#,f#"},
{"Entertainer:d=4,o=5,b=140:8d,8d#,8e,c6,8e,c6,8e,2c.6,8c6,8d6,8d#6,8e6,8c6,8d6,e6,8b,d6,
2c6,p,8d,8d#,8e,c6,8e,c6,8e,2c.6,8p,8a,8g,8f#,8a,8c6,e6,8d6,8c6,8a,2d6"},
{"Xfiles:d=4,o=5,b=125:e,b,a,b,d6,2b.,1p,e,b,a,b,e6,2b.,1p,g6,f#6,
e6,d6,f#6,2b.,1p,e,b,a,b,d6,2b.,1p,e,b,a,b,e6,2b.,1p,e6,2b."},
{"Autumn:d=8,o=6,b=125:a,a,a,a#,4a,a,a#,a,a,a,a#,4a,a,a#,a,16g,16a,a#,a,g.,16p,a,a,a,a#,4a,
a,a#,a,a,a#,a,4a,a,a#,a,16g,16a,a#,a,g."},
{"Spring:d=16,o=6,b=125:8e,8g#,8g#,8g#,f#,e,4b.,b,a,8g#,8g#,8g#,f#,e,4b.,b,a,8g#,a,b,8a,
8g#,8f#,8d#,4b.,8e,8g#,8g#,8g#,f#,e,4b.,b,a,8g#,8g#,8g#,f#,e,4b.,b,a,8g#,a,b,8a,8g#,8f#,
8d#,4b."},
};

// The following array stores the frequencies for the musical notes
const unsigned int note[4][12] =
{//  C     C#    D     D#    E     F     F#    G     G#    A     A#    B
    262,  277,  294,  311,  330,  349,  370,  392,  415,  440,  466,  494,  // 4th octave
    523,  554,  587,  622,  659,  698,  740,  784,  830,  880,  932,  988,  // 5th octave
   1047, 1109, 1175, 1244, 1319, 1397, 1480, 1568, 1660, 1760, 1865, 1976,  // 6th octave
   2093, 2218, 2349, 2489, 2637, 2794, 2960, 3136, 3320, 3520, 3728, 3951   // 7th octave
};

unsigned int channel_reload_value, duration_timer;
unsigned char accel_timer;
char sound_playing=0;
char duration, octave;
unsigned int tempo;
unsigned int y_axis, y_avg;
char next_song_flag = 0;
unsigned int last_y;
```

```c
// TPM1 overflow ISR: controls the duration of the sound
void interrupt VectorNumber_Vtpm1ch0 tpm1ch0_isr(void)
{
  TPM1C0SC_CH0F = 0;                      // clear interrupt flag
  TPM1C0V += 4000;                        // next compare in 1 ms
  if (duration_timer) duration_timer--;  // decrement the timer if > 0
  else                                    // if the timer is = 0
  {
    TPM1C1SC = 0;                         // disable TPM1CH1 channel (stops sound)
    sound_playing = 0;                   // clear the flag (no sound is not playing)
  }
  if (accel_timer) accel_timer--;        // acceleration timer
  else
  {
    last_y = y_axis;                     // sample the y axis
    accel_timer = 50;                    // next sample in 50ms
  }
}


// TPM1 channel1 compare ISR: control the frequency generation
// the channel pin is automatically toggled on each compare
void interrupt VectorNumber_Vtpm1ch1 tpm1ch1_compare(void)
{
  TPM1C1SC_CH1F = 0;                      // clear interrupt flag
  TPM1C1V += channel_reload_value;       // next compare
}


// Output a sound through the TPM1CH1 pin (PTB5): freq in Hz, dur in ms
void sound(unsigned int freq, unsigned int dur)
{
  while (sound_playing);       // if a sound is playing, waits until if finishes
  // the channel reload value is equal to half the period of the signal
  channel_reload_value = (4000000/freq)/2;
  duration_timer = dur;        // set the total duration time
  // set channel 1 to compare mode (toggling the channel pin)
  TPM1C1V = channel_reload_value; // set the first compare value
  TPM1C1SC = bCHIE | TPM_COMPARE_TOGGLE;
  sound_playing = 1;           // sound is playing (until the flag is cleared)
}


// This function plays a song passed in a string in RTTTL format
void play_song(char *song)
{
  unsigned char temp_duration, temp_octave, current_note, dot_flag;
  unsigned int calc_duration;
  duration = 4;               // standard duration = 4/4 = 1 beat
  tempo = 63;                 // standard tempo = 63 bpm
  octave = 6;                 // standard octave = 6th
  while (*song != ':') song++; // find the first ':'
  song++;                     // skip the ':'
  while (*song!=':')          // repeat until find a ':'
  {
    if (*song == 'd')         // if it is the duration setting
    {
      duration = 0;           // set duration to zero (temporarily)
      song++;                 // advance to the next character
      while (*song == '=') song++;  // skip the '='
      while (*song == ' ') song++;  // skip the spaces
      // if the character is a number, set the duration
      if (*song>='0' && *song<='9') duration = *song - '0';
      song++;                 // advance to the next character
      // if the character is also a number (duration can be 2 digits long)
      if (*song>='0' && *song<='9')
      { // multiply duration by 10 and add the value of the character
        duration = duration*10 + (*song - '0');
        song++;               // advance to the next character
      }
      while (*song == ',') song++;  // skip the ','
    }
    if (*song == 'o')         // if it is the octave setting
    {
      octave = 0;             // set octave to zero (temporarily)
```

```
  song++;                     // advance to the next character
  while (*song == '=') song++;  // skip the '='
  while (*song == ' ') song++;  // skip the spaces
  // if the character is a number, then set the octave
  if (*song>='0' && *song<='9') octave = *song - '0';
  song++;                     // advance to the next character
  while (*song == ',') song++;  // skips the ','
}
if (*song == 'b')             // if it is the tempo setting (beats per minute)
{
  tempo = 0;                  // set tempo to zero (temporarily)
  song++;                     // advance to the next character
  while (*song == '=') song++;  // skips '='
  while (*song == ' ') song++;  // skips spaces
  // now read the tempo setting (can be 3 digits long)
  if (*song>='0' && *song<='9') tempo = *song - '0';
  song++;                     // advance to the next character
  if (*song>='0' && *song<='9')
  {
    tempo = tempo*10 + (*song - '0'); // tempo is two digits
    song++;                   // advance to the next character
    if (*song>='0' && *song<='9')
    {
      tempo = tempo*10 + (*song - '0'); // tempo is three digits
      song++;                 // advance to the next character
    }
  }
  while (*song == ',') song++;  // skip the ','
}
while (*song == ',') song++;     // skip the ','
}
song++;                       // advance to the next character
// read the musical notes
while (*song)                 // repeat until the character is null
{
  current_note = 255;         // default note = pause
  temp_octave = octave;       // set the octave to the music default
  temp_duration = duration;   // set the duration to the music default
  dot_flag = 0;               // clear the dot detection flag
  // look for a duration prefix
  if (*song>='0' && *song<='9')
  {
    temp_duration = *song - '0';
    song++;
    if (*song>='0' && *song<='9')
    {
      temp_duration = temp_duration*10 + (*song - '0');
      song++;
    }
  }
  // look for a note
  switch (*song)
  {
    case 'c': current_note = 0; break;     // C (do)
    case 'd': current_note = 2; break;     // D (re)
    case 'e': current_note = 4; break;     // E (mi)
    case 'f': current_note = 5; break;     // F (fa)
    case 'g': current_note = 7; break;     // G (sol)
    case 'a': current_note = 9; break;     // A (la)
    case 'b': current_note = 11; break;    // B (si)
    case 'p': current_note = 255; break;   // pause
  }
  song++;                     // advance to the next character
  // look for a # following the note
  if (*song=='#')
  {
    current_note++;   // increment the note (A->A#, C->C#, D->D#, F->F#, G->G#)
    song++;                   // advance to the next character
  }
  // look for a '.' (extend the note duration in 50%)
  if (*song=='.')
  {
    dot_flag = 1;             // if a '.' is found, set the flag
    song++;                   // advance to the next character
  }
```

```
        // look for an octave postfix
        if (*song>='0' && *song<='9')
        {
            temp_octave = *song - '0';// the temporary octave is set accordingly
            song++;                    // advance to the next character
        }
        if (*song=='.')               // a dot can also be found after the octave (???)
        {
            dot_flag = 1;              // if a '.' is found, set the flag
            song++;                    // advance to the next character
        }
        while (*song == ',') song++;   // skip the ','
        // calculate the note duration
        calc_duration = (60000/tempo)/(temp_duration);
        calc_duration *= 4;            // a whole note has four beats
        // check if the dot flag is set, if it is set, extend the duration in 50%
        if (dot_flag) calc_duration = (calc_duration*3)/2;
        // if the current note is not a pause, play the note using the sound function
        if (current_note<255) sound(note[temp_octave-4][current_note],calc_duration);
        else
        { // if the current note = 255 (pause), just wait for the specified amount of time
            duration_timer = calc_duration;
            sound_playing = 1;
        }
        while (sound_playing);         // wait until the current note/pause has finished
        if (next_song_flag) break;     // if a shake was detected, skip to the next song
    }
}

// ADC ISR
void interrupt VectorNumber_Vadc adc_isr(void)
{
    y_avg = (y_avg + ADCR) - y_axis;   // average the accelerometer reading
    y_axis = y_avg >> 3;               // current Y axis value is y_avg/8
    // if an acceleration towards +Y is detected, followed by an acceleration towards -Y
    // skip to the next song
    if (y_axis<800 && last_y>3000) next_song_flag = 1;
}

void main(void)
{
    unsigned char song_sel;
    SOPT1 = bBKGDPE;                   // enable the debug pin
    ICSSC = DCO_MID | NVFTRIM;         // configure FTRIM value, select DCO high range
    ICSTRM = NVICSTRM;                 // configure TRIM value
    ICSC1 = ICS_FLL | bIREFS;          // select FEI mode (ICSOUT = DCOOUT = 1024 * IRCLK)
    ICSC2 = BDIV_1;                    // ICSOUT = DCOOUT / 1
    // BUSCLK = 16MHz
    TPM1SC = TPM_BUSCLK | TPM_DIV4;    // TPMCK = 4MHz
    // set channel 0 to compare mode (interrupt only), 1ms interrupts
    TPM1C0V = 3999;
    TPM1C0SC = bCHIE | TPM_COMPARE_INT;
    // enable long sampling, 12-bit mode, ADICLK = 01b, ADCK = BUSCLK/16
    // ADC sampling rate = 25ksps
    ADCCFG = bADLSMP | ADC_12BITS | ADC_BUSCLK_DIV2 | ADIV_8;
    APCTL2 = bADPC8;                   // ADP8 in analog mode
    // ADC channel 8, interrupts enabled, continuous conversion
    ADCSC1 = bAIEN | ADC_CH8 | bADCO;
    PTDD = 0;                          // select 1.5g sensitivity
    PTDDD = BIT_0 | BIT_1;             // PTD0 and PTD1 as outputs
    PTBDS = BIT_5;                     // turn on drive strengh on PTB5 (buzzer output)
    PTCDD = BIT_4;                     // PTC4 as an output
    y_avg = 0;
    y_axis = 0;
    accel_timer = 50;                  // force sampling
    ACCEL_ON = 1;                      // turn on the accelerometer (sleep mode off)
    EnableInterrupts;                  // enable interrupts (CCR:I=0)
    while(1)
    {
        for (song_sel=0;song_sel<20;song_sel++)
        {
            duration_timer = 250;
            sound_playing = 1;
            while (sound_playing);     // wait for 250ms before starting the next song
            play_song(rtttl_library[song_sel]); // play the song
```

```
            // wait for 2 seconds before start the next song
            duration_timer = 2000;              // wait for 2000ms before continuing
            sound_playing = 1;
            // skip to the next song if the duration timer times out or the song is skipped
            while (sound_playing && !next_song_flag);
            next_song_flag = 0;
        }
    }
}
```

Example 13.10 – Music-shake

A

HCS08 Instruction Set

The following abbreviations were used in the tables below:

opr8 – 8-bit operand;
opr16 – 16-bit operand;
ii – 8-bit immediate constant;
jjkk – 16-bit immediate constant;
dd – 8-bit immediate address;
hhll – 16-bit extended immediate address;
ff – 8-bit signed immediate constant;
eeff – 16-bit signed immediate constant;
n – Immediate constant between 0 and 7;
rel – 8-bit signed immediate address (relative addressing).

Format	Description	Operation	V	H	I	N	Z	C	Addressing Mode	Opcode	Operand	BUSCLK Cycles
ADC #opr8									IMM	A9	ii	2
ADC opr8									DIR	B9	dd	3
ADC opr16									EXT	C9	hhll	4
ADC ,X	Addition with carry	A = (A) + (M) + C	?	?	-	?	?	?	IX	F9		3
ADC opr8,X									IX1	E9	ff	3
ADC opr16,X									IX2	D9	eeff	4
ADC opr8,SP									SP1	9EE9	ff	4
ADC opr16,SP									SP2	9ED9	eeff	5
ADD #opr8									IMM	AB	ii	2
ADD opr8									DIR	BB	dd	3
ADD opr16									EXT	CB	hhll	4
ADD ,X	Addition without carry	A = (A) + (M)	?	?	-	?	?	?	IX	FB		3
ADD opr8,X									IX1	EB	ff	3
ADD opr16,X									IX2	DB	eeff	4
ADD opr8,SP									SP1	9EEB	ff	4
ADD opr16,SP									SP2	9EDB	eeff	5
AIS #opr8	Add a signed immediate constant to the SP	SP = (SP) + (16<<M)	-	-	-	-	-	-	IMM	A7	ii	2
AIX #opr8	Add a signed immediate constant to the H:X pair	H:X = (H:X) + (16<<M)	-	-	-	-	-	-	IMM	AF	ii	2
AND #opr8									IMM	A4	ii	2
AND opr8									DIR	B4	dd	3
AND opr16									EXT	C4	hhll	4
AND ,X	Logical AND	A = (A) & (M)	0	-	-	?	?	-	IX	F4		3
AND opr8,X									IX1	E4	ff	3
AND opr16,X									IX2	D4	eeff	4
AND opr8,SP									SP1	9EE4	ff	4
AND opr16,SP									SP2	9ED4	eeff	5
ASL opr8									DIR	38	dd	5
ASLA									INH	48		1
ASLX	Arithmetic shift to the left (same as LSL)		?	-	-	?	?	?	INH	58		1
ASL ,X									IX	78		4
ASL opr8,X									IX1	68	ff	5
ASL opr8,SP									SP1	9E68	ff	6
ASR opr8									DIR	37	dd	5
ASRA									INH	47		1
ASRX	Arithmetic shift to the right		?	-	-	?	?	?	INH	57		1
ASR ,X									IX	77		4
ASR opr8,X									IX1	67	ff	5
ASR opr8,SP									SP1	9E67	ff	6
BCC rel	Branch if C=0	PC = (PC) + 2 + rel ? C = 0	-	-	-	-	-	-	REL	24	rr	3

Format	Description	Operation	V	H	I	N	Z	C	Addressing Mode	Opcode	Operand	BUSCLK Cycles
BCLR n, opr8	Clear the bit n in the specified direct memory address	Mn = 0	-	-	-	-	-	-	DIR (b0)	11	dd	5
									DIR (b1)	13	dd	5
									DIR (b2)	15	dd	5
									DIR (b3)	17	dd	5
									DIR (b4)	19	dd	5
									DIR (b5)	1B	dd	5
									DIR (b6)	1D	dd	5
									DIR (b7)	1F	dd	5
BCS rel	Branch if C=1 (same as BLO)	PC = (PC) + 2 + rel ? C = 1	-	-	-	-	-	-	REL	25	rr	3
BEQ rel	Branch if Equal	PC = (PC) + 2 + rel ? Z = 1	-	-	-	-	-	-	REL	27	rr	3
BGE rel	Branch if Greater or Equal to (signed operands)	PC = (PC)+2+rel ? (N ⊕ V) = 0	-	-	-	-	-	-	REL	90	rr	3
BGND	Enter the active background debug mode (if ENBDM=1)	Wait for debug commands	-	-	-	-	-	-	INH	82		>=5
BGT rel	Branch if Greater Than (signed operands)	PC=(PC)+2+rel ? (Z)|(N⊕V)=0	-	-	-	-	-	-	REL	92	rr	3
BHCC rel	Branch if HC = 0	PC=(PC)+2+rel ? (H) = 0	-	-	-	-	-	-	REL	28	rr	3
BHCS rel	Branch if HC = 1	PC=(PC)+2+rel ? (H) = 1	-	-	-	-	-	-	REL	29	rr	3
BHI rel	Branch if higher (unsigned operands)	PC=(PC)+2+rel ? (C) | (Z) = 0	-	-	-	-	-	-	REL	22	rr	3
BHS rel	Branch if higher or same (same as BCC)	PC=(PC)+2+rel ? (C) = 0	-	-	-	-	-	-	REL	24	rr	3
BIH rel	Branch if IRQ pin = 1	PC=(PC)+2+rel ? IRQ = 1	-	-	-	-	-	-	REL	2F	rr	3
BIL rel	Branch if IRQ pin = 0	PC=(PC)+2+rel ? IRQ = 0	-	-	-	-	-	-	REL	2E	rr	3
BIT #opr8									IMM	A5	ii	2
BIT opr8									DIR	B5	dd	3
BIT opr16									EXT	C5	hhll	4
BIT ,X									IX	F5		3
BIT opr8,X	Bit test	(A) & (M)	0	-	-	?	?	-	IX1	E5	ff	3
BIT opr16,X									IX2	D5	eeff	4
BIT opr8,SP									SP1	9EE5	ff	4
BIT opr16,SP									SP2	9ED5	eeff	5
BLE opr8	Branch if Lower than or Equal to (signed operands)	PC=(PC)+2+rel ? (Z)|(N⊕V)=1	-	-	-	N	Z	-	REL	93	rr	3
BLO rel	Branch if Lower (same as BCS) (unsigned operands)	PC=(PC)+2+rel ? (C) = 1	-	-	-	-	-	-	REL	25	rr	3
BLS rel	Branch if Less than or Equal to (unsigned operands)	PC=(PC)+2+rel ? (C)|(Z) = 1	-	-	-	-	-	-	REL	23	rr	3
BLT rel	Branch if Less Than (signed operands)	PC=(PC)+2+rel ? (N⊕V)=1	-	-	-	-	-	-	REL	91	rr	3
BMC rel	Branch if I = 0	PC=(PC)+2+rel ? (I) = 0	-	-	-	-	-	-	REL	2C	rr	3
BMI rel	Branch if Minus	PC=(PC)+2+rel ? (N) = 1	-	-	-	-	-	-	REL	2B	rr	3
BMS rel	Branch if I = 1	PC=(PC)+2+rel ? (I) = 1	-	-	-	-	-	-	REL	2D	rr	3
BNE rel	Branch if Not Equal	PC=(PC)+2+rel ? (Z) = 0	-	-	-	-	-	-	REL	26	rr	3
BPL rel	Branch if Positive	PC=(PC)+2+rel ? (N) = 0	-	-	-	-	-	-	REL	2A	rr	3
BRA rel	Branch Always	PC=(PC) + 2 + rel	-	-	-	-	-	-	REL	20	rr	3
BRCLR n,opr8,rel	Branch if the bit n in the specified memory position is clear	PC=(PC)+3+rel ? Mn = 0	-	-	-	-	-	?	DIR (b0)	01	ddrr	5
									DIR (b1)	03	ddrr	5
									DIR (b2)	05	ddrr	5
									DIR (b3)	07	ddrr	5
									DIR (b4)	09	ddrr	5
									DIR (b5)	0B	ddrr	5
									DIR (b6)	0D	ddrr	5
									DIR (b7)	0F	ddrr	5
BRN rel	Branch Never	PC=(PC) + 2	-	-	-	-	-	-	REL	21	rr	3
BRSET n,opr8,rel	Branch if the bit n in the specified memory position is set	PC=(PC)+3+rel ? Mn = 1	-	-	-	-	-	1	DIR (b0)	00	ddrr	5
									DIR (b1)	02	ddrr	5
									DIR (b2)	04	ddrr	5
									DIR (b3)	06	ddrr	5
									DIR (b4)	08	ddrr	5
									DIR (b5)	0A	ddrr	5
									DIR (b6)	0C	ddrr	5
									DIR (b7)	0E	ddrr	5
BSET n,opr8	Set the bit n in the specified direct memory address	Mn = 1	-	-	-	-	-	-	DIR (b0)	10	dd	5
									DIR (b1)	12	dd	5
									DIR (b2)	14	dd	5
									DIR (b3)	16	dd	5
									DIR (b4)	18	dd	5
									DIR (b5)	1A	dd	5
									DIR (b6)	1C	dd	5
									DIR (b7)	1E	dd	5

Format	Description	Operation	V	H	I	N	Z	C	Addressing Mode	Opcode	Operand	BUSCLK Cycles
BSR rel	Branch to subroutine	PC=PC+2 push (PCL) SP=SP-1 push (PCH) SP=SP-1 PC=PC + rel	-	-	-	-	-	-	REL	AD	rr	5
CALL page, opr16[1]	Call subroutine	PC = PC + 4 push (PCL) SP=SP-1 push (PCH) SP=SP-1 push (PPAGE) SP=SP-1 PPAGE=page PC = opr16	-	-	-	-	-	-	EXT	AC	iihhll	8
CBEQ opr8, rel CBEQA #opr8,rel CBEQX #opr8,rel CBEQ X+,rel CBEQ opr8,X+,rel CBEQ opr8,SP,rel	Compara and Branch if Equal	PC=(PC)+3+rel ? (A) – (M) = 0 PC=(PC)+3+rel ? (A) – (M) = 0 PC=(PC)+3+rel ? (X) – (M) = 0 PC=(PC)+2+rel ? (A) – (M) = 0 PC=(PC)+3+rel ? (A) – (M) = 0 PC=(PC)+4+rel ? (A) – (M) = 0	-	-	-	-	-	-	DIR IMM IMM IX+ IX1+ SP1	31 41 51 71 61 9E61	ddrr iirr iirr ffrr ffrr	5 4 4 5 5 6
CLC	Clear C	C = 0	-	-	-	-	-	0	INH	98		1
CLI	Clear I (enable interrupts)	I = 0	-	-	0	-	-	-	INH	9A		1
CLR opr8 CLRA CLRX CLRH CLR ,X CLR opr8,X CLR opr8,SP	Clear	M = 0 A = 0 X = 0 H = 0 M = 0 M = 0 M = 0	?	-	-	?	?	?	IMM INH INH INH IX IX1 SP1	3F 4F 5F 8C 7F 6F 9E6F	dd ff ff	5 1 1 1 4 5 6
CMP #opr8 CMP opr8 CMP opr16 CMP ,X CMP opr8,X CMP opr16,X CMP opr8,SP CMP opr16,SP	Compare A with memory	(A) – (M)	?	-	-	?	?	?	IMM DIR EXT IX IX1 IX2 SP1 SP2	A1 B1 C1 F1 E1 D1 9EE1 9ED1	ii dd hhll ff eeff ff eeff	2 3 4 3 3 4 4 5
COM opr8 COMA COMX COM ,X COM opr8,X COM opr8,SP	1's complement	M = ($\overline{M}$) = $FF – (M) M = ($\overline{A}$) = $FF – (M) M = ($\overline{X}$) = $FF – (M) M = ($\overline{M}$) = $FF – (M) M = ($\overline{M}$) = $FF – (M) M = ($\overline{M}$) = $FF – (M)	0	-	-	?	?	?	DIR INH INH IX IX1 SP1	33 43 53 73 63 9E63	dd ff ff	5 1 1 4 5 6
CPHX #opr16 CPHX opr16 CPHX opr8 CPHX opr8,SP	Compare H:X with memory	H:X – ((M):(M + 1))	?	-	-	?	?	?	IMM EXT DIR SP1	3E 65 75 9EF3	jjkk hhll dd ff	3 6 5 6
CPX #opr8 CPX opr8 CPX opr16 CPX ,X CPX opr8,X CPX opr16,X CPX opr8,SP CPX opr16,SP	Compare X with memory	(X) – (M)	?	-	-	?	?	?	IMM DIR EXT IX IX1 IX2 SP1 SP2	A3 B3 C3 F3 E3 D3 9EE3 9ED3	ii dd hhll ff eeff ff eeff	2 3 4 3 3 4 4 5
DAA	Decimal Accumulator Adjust	(A)$_{10}$	U	-	-	?	?	?	INH	72		1
DBNZ opr8,rel DBNZA rel DBNZX rel DBNZ X,rel DBNZ opr8,X,rel DBNZ opr8,SP,rel	Decrement and Branch if Not Zero	A=(A)–1 or M=(M)-1 or X=(X)-1 PC=(PC)+3+rel ? (result)≠0 PC=(PC)+2+rel ? (result)≠0 PC=(PC)+2+rel ? (result)≠0 PC=(PC)+2+rel ? (result)≠0 PC=(PC)+3+rel ? (result)≠0 PC=(PC)+4+rel ? (result)≠0	-	-	-	-	-	-	DIR INH INH IX IX1 SP1	3B 4B 5B 7B 6B 9E6B	ddrr rr rr ffrr ffrr	7 4 4 6 7 8
DEC #opr8 DECA DECX DEC ,X DEC opr8,X DEC opr8,SP	Decrement	M = (M) –1 A = (A) – 1 X = (X) – 1 M = (M) – 1 M = (M) – 1 M = (M) – 1	?	-	-	?	?	-	DIR INH INH IX IX1 SP1	3A 4A 5A 7A 6A 9E6A	dd ff ff	5 1 1 4 5 6
DIV	Divide	A = (H:A)/(X) H = remainder	-	-	-	-	?	?	INH	52		6
EOR #opr8 EOR opr8 EOR opr16 EOR ,X EOR opr8,X EOR opr16,X EOR opr8,SP EOR opr16,SP	Exclusive-OR operation between A and memory	A = (A⊕M)	0	-	-	?	?	-	IMM DIR EXT IX IX1 IX2 SP1 SP2	A8 B8 C8 F8 E8 D8 9EE8 9ED8	ii dd hhll ff eeff ff eeff	2 3 4 3 3 4 4 5

Format	Description	Operation	V	H	I	N	Z	C	Addressing Mode	Opcode	Operand	BUSCLK Cycles
INC opr8		M = (M) + 1	?	-	-	?	?	-	DIR	3C	dd	5
INCA		A = (A) + 1							INH	4C		1
INCX		X = (X) + 1							INH	5C		1
INC ,X	Increment	M = (M) + 1							IX	7C		4
INC opr8,X		M = (M) + 1							IX1	6C	ff	5
INC opr8,SP		M = (M) + 1							SP1	9E6C	ff	6
JMP opr8			-	-	-	-	-	-	DIR	BC	dd	3
JMP opr16									EXT	CC	hhll	4
JMP ,X	Jump	PC = destination address							IX	FC		3
JMP opr8,X									IX1	EC	ff	3
JMP opr16,X									IX2	DC	eeff	4
JSR opr8		PC = (PC) + n (n=1,2 ou 3)	-	-	-	-	-	-	DIR	BD	dd	5
JSR opr16		push (PCL); SP = (SP) – 1							EXT	CD	hhll	6
JSR ,X	Jump to subroutine	push (PCH); SP = (SP) – 1							IX	FD		5
JSR opr8,X		PC = subroutine address							IX1	ED	ff	5
JSR opr16,X									IX2	DD	eeff	6
LDA #opr8			0	-	-	?	?	-	IMM	A6	ii	2
LDA opr8									DIR	B6	dd	3
LDA opr16									EXT	C6	hhll	4
LDA ,X									IX	F6		3
LDA opr8,X	Load A with memory	A = (M)							IX1	E6	ff	3
LDA opr16,X									IX2	D6	eeff	4
LDA opr8,SP									SP1	9EE6	ff	4
LDA opr16,SP									SP2	9ED6	eeff	5
LDHX #opr16			0	-	-	?	?	-	IMM	45	jjkk	3
LDHX opr8									DIR	55	dd	4
LDHX opr16									EXT	32	hhll	5
LDHX ,X	Load H:X with memory	H:X = (M:M + 1)							IX	9EAE		5
LDHX opr8,X									IX1	9ECE	ff	5
LDHX opr16,X									IX2	9EBE	eeff	6
LDHX opr8,SP									SP1	9EFE	ff	5
LDX #opr8			0	-	-	?	?	-	IMM	AE	ii	2
LDX opr8									DIR	BE	dd	3
LDX opr16									EXT	CE	hhll	4
LDX ,X									IX	FE		3
LDX opr8,X	Load X with memory	X = (M)							IX1	EE	ff	3
LDX opr16,X									IX2	DE	eeff	4
LDX opr8,SP									SP1	9EEE	ff	4
LDX opr16,SP									SP2	9EDE	eeff	5
LSL opr8			?	-	-	?	?	?	DIR	38	dd	5
LSLA									INH	48		1
LSLX	Logical Shift to the								INH	58		1
LSL ,X	Left (same as ASL)								IX	78		4
LSL opr8,X									IX1	68	ff	5
LSL opr8,SP									SP1	9E68	ff	6
LSR opr8			?	-	-	0	?	?	DIR	34	dd	5
LSRA									INH	44		1
LSRX	Logical Shift to the								INH	54		1
LSR ,X	Right								IX	74		4
LSR opr8,X									IX1	64	ff	5
LSR opr8,SP									SP1	9E64	ff	6
MOV opr8,op8		(M)destination = (M)source	0	-	-	?	?	-	DIR/DIR	4E	dddd	5
MOV opr8,X+	Move	H:X = (H:X) + 1 (IX+D, DIX+)							DIR/IX+	5E	dd	5
MOV #opr8,opr8									IMM/DIR	6E	iidd	4
MOV X+,opr8									IX+/DIR	7E	dd	5
MUL	Unsigned Multiply	X:A = (X) x (A)	-	0	-	-	-	0	INH	42		5
NEG opr8		M = -(M) = 0 – (M)	?	-	-	?	?	?	DIR	30	dd	5
NEGA		A = -(A) = 0 – (M)							INH	40		1
NEGX	2's Complement	X = -(X) = 0 – (X)							INH	50		1
NEG ,X		M = -(M) = 0 – (M)							IX	70		4
NEG opr8,X		M = -(M) = 0 – (M)							IX1	60	ff	5
NEG opr8,SP									SP1	9E60	ff	6
NOP	No Operation	none	-	-	-	-	-	-	INH	9D		1
NSA	Swap nibbles in the accumulator	A = (A[3:0]:A[7:4])	-	-	-	-	-	-	INH	62		1
ORA #opr8			0	-	-	?	?	-	IMM	AA	ii	2
ORA opr8									DIR	BA	dd	3
ORA opr16									EXT	CA	hhll	4
ORA ,X	Logical OR operation								IX	FA		3
ORA opr8,X	between A and	A = (A)	(M)							IX1	EA	ff	3
ORA opr16,X	memory								IX2	DA	eeff	4
ORA opr8,SP									SP1	9EEA	ff	4
ORA opr16,SP									SP2	9EDA	eeff	5
PSHA	Push A onto the stack	Push (A); SP = (SP) + 1	-	-	-	-	-	-	INH	87		2
PSHH	Push H onto the stack	Push (H); SP = (SP) + 1	-	-	-	-	-	-	INH	8B		2
PSHX	Push X onto the stack	Push (X); SP = (SP) + 1	-	-	-	-	-	-	INH	89		2
PULA	Pull A from the stack	SP = (SP) + 1); pull (A)	-	-	-	-	-	-	INH	86		3
PULH	Pull H from the stack	SP = (SP) + 1); pull (H)	-	-	-	-	-	-	INH	8A		3
PULX	Pull X from the stack	SP = (SP) + 1); pull (X)	-	-	-	-	-	-	INH	88		3

Format	Description	Operation	V	H	I	N	Z	C	Addressing Mode	Opcode	Operand	BUSCLK Cycles
ROL opr8 ROLA ROLX ROL ,X ROL opr8,X ROL opr8,SP	Rotate left through carry	(C ← b7...b0 ←)	?	-	-	?	?	?	DIR INH INH IX IX1 SP1	39 49 59 79 69 9E69	dd ff ff	5 1 1 4 5 6
ROR opr8 RORA RORX ROR ,X ROR opr8,X ROR opr8,SP	Rotate right through carry	(→ b7...b0 → C)	?	-	-	?	?	?	DIR INH INH IX IX1 SP1	36 46 56 76 66 9E66	dd ff ff	5 1 1 4 5 6
RSP	Reset stack pointer (SP)	SP = $FF	-	-	-	-	-	-	INH	9C		1
RTC[1]	Return from CALL	SP = SP + 1 pull (PPAGE) SP = SP + 1 pull (PCH) SP = SP + 1 pull (PCL)	-	-	-	-	-	-	INH	8D		7
RTI	Return from Interrupt	SP = SP + 1; pull CCR SP = SP + 1; pull A SP = SP + 1; pull X SP = SP + 1; pull PCH SP = SP + 1; pull PCL							INH	80		9
RTS	Return from Subroutine	SP = SP + 1; pull PCH SP = SP + 1; pull PCL	-	-	-	-	-	-	INH	81		6
SBC #opr8 SBC opr8 SBC opr16 SBC ,X SBC opr8,X SBC opr16,X SBC opr8,SP SBC opr16,SP	Subtraction with borrow	A = (A) − (M) − C	?	-	-	?	?	?	IMM DIR EXT IX IX1 IX2 SP1 SP2	A2 B2 C2 F2 E2 D2 9EE2 9ED2	ii dd hhll ff eeff ff eeff	2 3 4 3 3 4 4 5
SEC	Set C	C = 1	-	-	-	-	-	1	INH	99		1
SEI	Set I (disable interrupts)	I = 1	-	-	1	-	-	-	INH	9B		1
STA opr8 STA opr16 STA ,X STA opr8,X STA opr16,X STA opr8,SP STA opr16,SP	Store A into memory	M = (A)	0	-	-	?	?	-	DIR EXT IX IX1 IX2 SP1 SP2	B7 C7 F7 E7 D7 9EE7 9ED7	dd hhll ff eeff ff eeff	3 4 3 3 4 4 5
STHX opr8 STHX opr16 STHX opr8,SP	Store H:X into memory	(M:M +1 = (H:X)	0	-	-	?	?	-	DIR EXT SP1	35 96 9EFF	dd hhll ff	4 5 5
STOP	Enable interrupts and enter low power mode	I = 0	-	-	0	-	-	-	INH	8E		>=2
STX opr8 STX opr16 STX ,X STX opr8,X STX opr16,X STX opr8,SP STX opr16,SP	Store X into memory	M = (X)	0	-	-	?	?	-	DIR EXT IX IX1 IX2 SP1 SP2	BF CF FF EF DF 9EEF 9EDF	dd hhll ff eeff ff eeff	3 4 2 3 4 4 5
SUB #opr8 SUB opr8 SUB opr16 SUB ,X SUB opr8,X SUB opr16,X SUB opr8,SP SUB opr16,SP	Subtraction	A = (A) − (M)	?	-	-	?	?	?	IMM DIR EXT IX IX1 IX2 SP1 SP2	A0 B0 C0 F0 E0 D0 9EE0 9ED0	ii dd hhll ff eeff ff eeff	2 3 4 3 3 4 4 5
SWI	Software interrupt	PC = (PC) + 1; push (PCL) SP = (SP) − 1; push (PCH) SP = (SP) − 1; push (X) SP = (SP) − 1; push (A) SP = (SP) − 1; push CCR SP = (SP) − 1; I = 1 PCH = Interrupt vector MSB PCL = Interrupt vector LSB	-	-	1	-	-	-	INH	83		11
TAP	Transfer A to CCR	CCR = (A)	?	?	?	?	?	?	INH	84		1
TAX	Transfer A to X	X = (A)	-	-	-	-	-	-	INH	97		1
TPA	Transfer CCR to A	A = (CCR)	-	-	-	-	-	-	INH	85		1

Format	Description	Operation	CCR Affected Flags						Addressing Mode	Opcode	Operand	BUSCLK Cycles
			V	H	I	N	Z	C				
TST opr8									DIR	3D	dd	4
TSTA									INH	4D		1
TSTX	Test if negative or zero	(A) – 0 or (X) – 0 or (M) - 0	0	-	-	?	?	-	INH	5D		1
TST ,X									IX	7D		3
TST opr8,X									IX1	6D	ff	4
TST opr8,SP									SP1	9E6D	ff	5
TSX	Transfer SP to H:X	H:X = (SP) + 1	-	-	-	-	-	-	INH	95		2
TXA	Transfer X to A	A = (X)	-	-	-	-	-	-	INH	9F		1
TXS	Transfer H:X to SP	(SP) = (H:X) - 1	-	-	-	-	-	-	INH	94		2
WAIT	Enable interrupts and enter low power mode	I = 0; stops the CPU	-	-	0	-	-	-	INH	8F		>=2

[1] these instructions are available only on Flexis devices.

B

ASCII Table

The table below shows the American Standard Code for Information Interchange (ASCII) encoding for text and control characters. Using the table is quite simple: the ASCII encoding for "A" is 41 (hex) or 64+1=65 (decimal). The encoding for DEL is 7F (hex) or 112+15=127 (decimal).

Dec		0	1	2	3	4	5	6	7	8	9	10	11	12	13	14	15
	Hex	0	1	2	3	4	5	6	7	8	9	A	B	C	D	E	F
0	0	NUL	SOH	STX	ETX	EOT	ENQ	ACK	BEL	BS	HT	LF	VT	FF	CR	SO	SI
16	1	DLE	DC1	DC2	DC3	DC4	NAK	SYN	ETB	CAN	EM	SUB	ESC	FS	GS	RS	US
32	2	SP	!	"	#	$	%	&	'	(	)	*	+	,	-	.	/
48	3	0	1	2	3	4	5	6	7	8	9	:	;	<	=	>	?
64	4	@	A	B	C	D	E	F	G	H	I	J	K	L	M	N	O
80	5	P	Q	R	S	T	U	V	W	X	Y	Z	[	\	]	^	_
96	6	`	a	b	c	d	e	f	g	h	i	j	k	l	m	n	o
112	7	p	q	r	s	t	u	v	w	x	y	z	{			}	~	DEL

INDEX

Symbols and Numbers

W

X

References

1. PREDKO, M. **Handbook of Microcontrollers**. USA: McGraw-Hill, 1999.
2. PEREIRA, F. **Microcontroladores HC908Q: Teoria e Prática**. São Paulo: Érica, 2004.
3. GANSSLE, J. **The Firmware Handbook: The Definitive Guide to Embedded Firmware Design and Applications**. USA: Elsevier, 2004.
4. PEREIRA, F. **Microcontroladores HCS08: Teoria e Prática**. São Paulo: Érica, 2005.

Manuals:

5. METROWERKS. **HC08/HCS08 Compiler REV 1.2**. USA: Metrowerks, 2003.
6. FREESCALE. **MC9S08GB60A Data sheet Rev 1.00**. USA: Freescale, 2005.
7. _____. **MC9S08RG60 Data sheet v1.11**. USA: Freescale, 2005.
8. _____. **MC9S08AW60 Data sheet Rev 2**. USA: Freescale, 2006.
9. _____. **MC9S08DZ60 Data sheet Rev 1 Draft E**. USA: Freescale, 2006.
10. _____. **MC9S08EL32 Data sheet Rev 0**. USA: Freescale, 2006.
11. _____. **MC9S08EN32 Data sheet Rev 1 Draft E**. USA: Freescale, 2006.
12. _____. **MC9S08QD4 Data sheet Rev 1**. USA: Freescale, 2006.
13. _____. **MC9S08SG8 Data sheet Rev 0**. USA: Freescale, 2006.
14. _____. **HCS08 Family Reference Manual HCS08RMv1/D Rev. 2**. USA: Freescale, 2007.
15. _____. **MC9S08AC16 Data sheet Rev 0**. USA: Freescale, 2007.
16. _____. **MC9S08JM60 Data sheet Rev 1**. USA: Freescale, 2007.
17. _____. **MC9S08LC60 Data sheet Rev 2**. USA: Freescale, 2007.
18. _____. **MC9S08QE128 Data sheet Rev 3**. USA: Freescale, 2007.
19. _____. **MC9S08QE128 Reference Manual Rev. 2**. USA: Freescale, 2007.
20. _____. **MC9S08SH8 Data sheet Rev 1**. USA: Freescale, 2007.
21. _____. **MC9S08QG8 Data sheet Rev. 4**. USA: Freescale, 2008.

Other Documents:

22. SIBIGTROTH, J. M. M68HC05 Understanding Small Microcontrollers. USA: Freescale, 1998.
23. ROBB, S. **AN2093:** Creating Efficient C Code for the MC68HC08. USA: Freescale, 2000.
24. LUCAS, B.; PAPE, S. **AN2494:** Configuring the System and Peripheral Clocks in the MC9S08GB/GT. USA: Freescale, 2003.
25. KIKUCHI, K.; SUCHYTA, J. **AN2497:** HCS08 Background Debug Mode versus HC08 Monitor Mode. USA: Freescale, 2003.
26. FREESCALE. **BRHCS08/D:** HCS08 Brochure. USA: Freescale, 2003.
27. GARCIA, D.; PAPE, S. **AN2493:** MC9S08GB/GT Low Power Modes. USA: Freescale, 2004.

28. MONTAÑEZ, E. **AN2596:** Using the HCS08 Family On-Chip In-Circuit Emulator (ICE). USA: Freescale, 2004.

29. PICKERING, S. **AN2616:** Getting Started with HCS08 and Codewarrior Using C. USA: Freescale, 2004.

30. PAPE, S. **AN3041:** Internal Clock Source (ICS) Module on the HCS08s in Depth. USA: Freescale, 2005.

31. PERALEZ, R.; GARCIA, D. **AN3031:** Temperature Sensor for the HCS08 Microcontroller Family. USA: Freescale, 2006.

32. FREESCALE. **HCS08QRUG:** HCS08 Peripheral Module Quick Reference. USA: Freescale, 2006.

33. FREESCALE. **AN3552**: Analog Comparator Tips and Tricks for the MC9S08QG MCU. USA: Freescale, 2007.

Internet Sites:

34. www.freescale.com – Semiconductor manufacturer (HC08, HCS08, Coldfire and many other chips).

35. www.sctec.com.br – Design house specialized on Freescale microcontrollers.

36. http://en.wikipedia.org/wiki/C_%28programming_language%29 – Wikipedia page about the C language

37. http://en.wikipedia.org/wiki/Computer - Computer page at Wikipedia

38. http://en.wikipedia.org/wiki/Von_Neumann_architecture - Von Neumann architecture

39. http://en.wikipedia.org/wiki/Harvard_architecture - Harvard architecture

40. http://cslibrary.stanford.edu/101/EssentialC.pdf - A good review of the C language

41. http://en.wikipedia.org/wiki/Nyquist%E2%80%93Shannon_sampling_theorem – Nyquist-Shannon theorem

42. http://www.nxp.com/acrobat_download/literature/9398/39340011.pdf - I^2C protocol specifications

43. hp.vector.co.jp/authors/VA002416/teraterm.html – Home of the TeraTerm terminal emulation software

44. http://www.lynxmotion.com/Product.aspx?productID=179&CategoryID=15 – ServoMotion software

45. http://www.brookshiresoftware.com/ - Visual Show Automation software (demo version available for download)

46. http://www.rentron.com/Mini-ssc.htm - Visual Basic tutorial showing how to write a Windows application to control MINI-SSC controllers (such as the one presented in this book)

47. http://www.csoft.co.uk/ringtones/rtttl_rtx.htm - RTTTL and .RTX specification

48. http://www.beyondlogic.org/pic/ringtones.htm - RTTTL player and information

49. http://overtonez.co.uk/ - Website for RTTTL ringtones download

50. http://home.iae.nl/users/pouweha/lcd/lcd.shtml - Character LCD module tutorial

51. http://pinouts.ru/ - Pinouts of hundreds of electronic devices and connectors

52. http://en.wikipedia.org/wiki/Electronic_speed_control - ESC page at Wikipedia

53. http://en.wikipedia.org/wiki/Microelectromechanical_systems - MEMS page at wikipedia

Registered Trademarks

Codewarrior, HC05, HC08, HCS08, HCS12 and Coldfire are trademarks of Freescale Semiconductor Inc.

Delphi and Code Gear are registered trademarks of Borland Software Corporation.

MSP430 is a registered trademark of Texas Instruments Inc.

PEMICRO is a registered trademark of P&E Microcomputer Systems.

PIC is a registered trademark of Microchip Technology Inc.

Windows and Visual Basic are registered trademarks of Microsoft Corporation.

Z-80 and Z8-Encore are registered trademarks of Zilog Inc.

All other product or service names are property of their respective owners.